Waste Rubber Recycling Routes

Waste Rubber Recycling Routes

Editors

Lucia Asaro
Nourredine Aït Hocine

CWP
Central West Publishing

This edition has been published by Central West Publishing, Australia

For more information about the books published by Central West Publishing, please visit https://centralwestpublishing.com

Disclaimer
Every effort has been made by the publisher, editors and authors while preparing this book, however, no warranties are made regarding the accuracy and completeness of the content. The publisher, editors and authors disclaim without any limitation all warranties as well as any implied warranties about sales, along with fitness of the content for a particular purpose. Citation of any website and other information sources does not mean any endorsement from the publisher, editors and authors. For ascertaining the suitability of the contents contained herein for a particular lab or commercial use, consultation with the subject expert is needed. In addition, while using the information and methods contained herein, the practitioners and researchers need to be mindful for their own safety, along with the safety of others, including the professional parties and premises for whom they have professional responsibility. To the fullest extent of law, the publisher, editors and authors are not liable in all circumstances (special, incidental, and consequential) for any injury and/or damage to persons and property, along with any potential loss of profit and other commercial damages due to the use of any methods, products, guidelines, procedures contained in the material herein.

NATIONAL LIBRARY OF AUSTRALIA

A catalogue record for this book is available from the National Library of Australia

ISBN (print): 978-1-922617-62-0

Preface

The growing consumption of elastomers, particularly in the automotive, offshore, and aeronautic industries, creates a large amount of waste rubber. Because of the strong three-dimensional network of rubbers produced by the vulcanization process, and their particular composition, the recycling of such materials represents a scientific and industrial challenge.

The book is devoted to the rubber recycling through the devulcanization process, using different methods including chemical, microwave, ultrasound, thermo-mechanical, etc. All chapters are well structured and easy to read, making the book accessible to a large audience of different sectors i.e. scientific researchers, teachers, engineering students, engineers, and industry staff. The used approaches are justified and the results are discussed and argued.

About the Editors

Prof. Nourredine AÏT HOCINE has received his PhD degree in Mechanical Engineering from the University of Lille, France, in 1996. He was lecturer at the University of Brest – France, for 10 years. He is full Professor in mechanics since 2009, at INSA Centre Val de Loire, Blois, France. His scientific research field includes recycling waste polymers and rubbers, experimental characterization of physical and mechanical properties of polymer-based micro- nano-composites, finite element modeling of polymer mechanical behavior, and fracture of rubbers. He has published more than 100 papers in reputed journals and conferences. He is currently EBM and reviewer for several international journal.

Dr. Lucia ASARO has received her PhD degree in Materials Science from the National University of Mar del Plata, Argentina in 2016. She did a postdoc in the field of rubber recycling between 2017 and 2018 at INSA Centre Val de Loire, Blois, France. Currently, she is researcher and assistant professor at INTEMA-National University of Mar del Plata, Argentina. She is working on the recycling of waste rubbers, and in the development of micro and nano composites for the aerospace industry. She has published more than 30 scientific articles in international journals and conferences.

Table of Contents

1

Enhancing Rubber Recycling through Improved Waste Tyre Granulation Technologies

Alex J. Bowles* and **Geoffrey D. Fowler**
Department of Civil and Environmental Engineering, Skempton Building, South Kensington Campus, Imperial College, London, UK, SW7 2AZ.
**alex.bowles15@imperial.ac.uk*

1.1 Introduction

1.1.1 Sustainability of tyres

Tyres are a critical component of automotive transportation. Around 1.53 billion were manufactured and discarded worldwide in 2016 [1]. It is recognised that the entire lifecycle of tyres must be improved to enable their sustainable use in the future. The environmental burden of tyre production is high, with 55 kg of hydrocarbon feedstock being required per kg of tyre [2]. The calorific and material recovery potential of tyres is thus higher than for many other wastes. Over 14 megatonnes (Mt) of natural rubber (NR) [3] and 5.1 Mt of artificial styrene-butadiene rubber (SBR) were produced in 2019 [4], suggesting that recycling has the potential to contribute significant quantities of material in the rubber market. Currently in the EU and USA, <10% of tyre rubber is recycled to high-quality rubber materials [5], a figure which should be increased in the near-future.

The complex structure and geometry of tyres present many challenges for disposal. Tyres do not stack well, are difficult to transport, store economically and pose a fire risk. Historically, tyre disposal by landfilling was widely employed, similar to other wastes which are difficult to recycle. However, this method has proven unsuitable for tyre disposal because the tyre void space creates landfill structural instability and accumulates landfill gas, thereby presenting explosion risks [6]. Tyre void spaces also fill with rainwater and in warmer climates can incubate insect disease vectors such as mosquitos [7].

The main processing pathways for an end of life (EoL) tyre are material and/or energy recovery. Tyre material recovery by established and emerging technologies will be the focus of this chapter, with granulation being a core step for most of these subsequent processes. The core aim of granulation is to reduce the particle size of an EoL tyre to a level where it is classified as Ground Tyre Rubber (GTR) or Crumb Rubber (CR), with a particle diameter (D) that is dependent upon end-use but generally below 5 cm. GTR has a wide range of applications which are summarised in Figure 1.1 and will be discussed in detail in Section 1.1.3, with a focus on its production through commercial and research activities which aim to increase the yield and product quality of GTR for recycling applications.

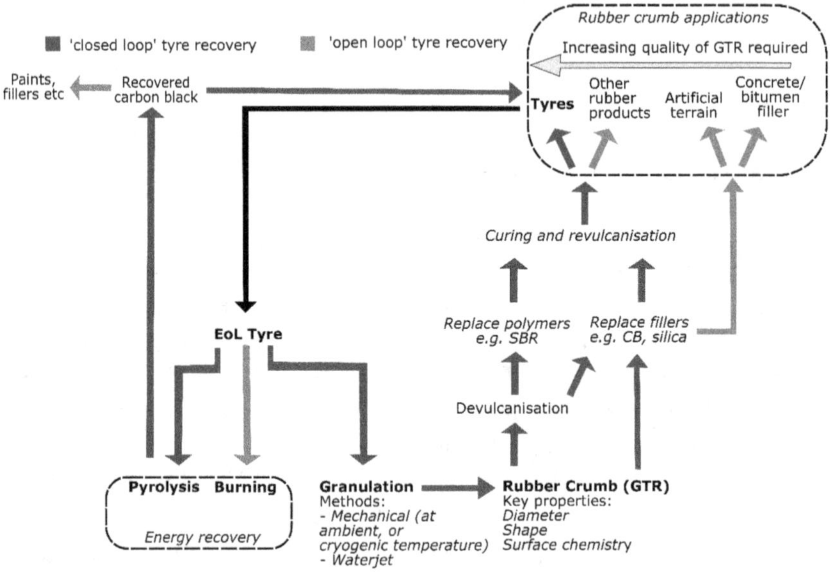

Figure 1.1 *Overview of a sustainable lifecycle for tyre rubber based on preferable disposal technologies. An emphasis has been placed on ground tyre rubber (GTR). Adapted from* [8], *SBR = Styrene Butadiene Rubber. CB = Carbon Black. Reproduced under license 4878250012437.*

Granulation is required to make a product from tyres which is processable, flowable and with a high surface area. The intended end-use of the recovered material determines the selected method of granulation and further development is needed in order to realise the full potential of this rubber recycling methodology. Granulation can be achieved in 3 ways: mechanically at ambient or cryogenic temperatures, or by water jetting (Figure 1.2). Chipping, defined as the preliminary stage(s) of tyre size reduction, is a pre-requisite for some granulation, reuse, and energy recovery processes and generally produces particles with a diameter of around 5 cm. Other products of the chipping/granulation processes include metal wiring from the belts and beads, fibre from the textile cord layers in the tyre and GTR (Figure 1.2).

Rubber typically accounts for >75% of the tyres mass. Current granulation systems yield extremely limited quantities of high value GTR which can be directly recycled in quality products, limiting the commercial opportunities. Generally, the highest value and most useful rubber products require GTR with the lowest particle sizes. Multiple processing steps can be used to reduce particle size but at the expense of more complex plant equipment and processing energy costs. Thus, tyre recyclers must fulfil the specification for GTR applications, thereby meeting market need, without excessive processing. Small-sized recycled rubber particles are commonly added to rubber blends as a filler material but can compromise the performance characteristics of the recycled product (see Section 1.1.3) and may be substitutable with other unrecycled feedstock materials. Feedstock for GTR is not limited to EoL tyres. Other rubber products, such as conveyor belts, seals, gaskets and rubber weather strips, can also be granulated to produce GTR [9], but the quantities of crumb produced are negligible when compared to tyre-derived GTR, and so will not be discussed further.

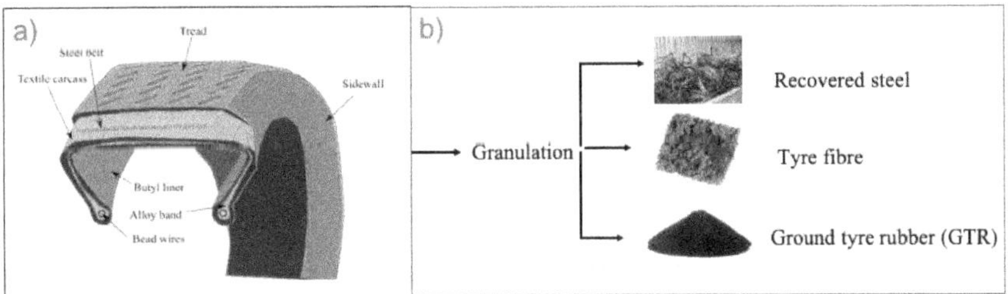

c)

Granulation conditions	GTR product	Abbreviation
Mechanical grinding at ambient temperatures	Mechanically Ground Crumb	MGC
Mechanical grinding at cryogenic temperatures	Cryogenically Ground Crumb	CGC
Waterjet impact	Waterjet Crumb	WJC

Figure 1.2 *a) Cross section of an automotive tyre with illustrated components. b) Products of the waste tyre granulation process. c) The different granulation processes, products and abbreviations for products used in this chapter.*

1.1.2 Tyres: A challenging waste

Progress in tyre recycling is dependent on recognition of the challenges associated with utilising a waste stream as a feedstock material. Raw materials for manufacturing processes are almost invariably required to be of high quality and purity, with defined and reliable physical characteristics dependent on specific end-application. In contrast, waste streams are often very heterogenous, comprising a wide range of materials with varying physical properties. An example of such heterogeneity is given by the difference in the material composition of car and truck tyres which is illustrated in Figure 1.3 [1].

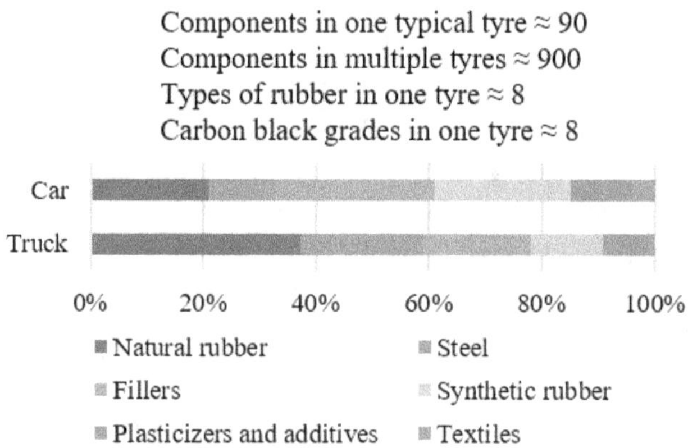

Figure 1.3 *Heterogeneity of tyres. Adapted from data in* [1].

Tyre feedstock also varies according to tyre model, age and purpose and includes brand new sub-standard products, damaged and part-worn tyres and extremely worn ones with a much lower rubber content. All are classed as EoL tyres and are likely to exist in the same waste stream. Some tyres may contain contaminants such as stones or oil. Moreover, tyres are constructed with different rubber formulations in each part of the tyre with the tread often containing more natural rubber (NR) than artificial rubber (SBR). There is also variation in the composition of carbon black and silica fillers within the carcass, sidewall, and inner liner (Figure 1.1). These are all factors which tyre recyclers cannot control but must be considered in subsequent recovery stages and product recycling strategies.

There are processes which tyre recyclers can implement to address these challenges, the most important of which are outlined in Figure 1.4. Firstly, feedstock material can be managed to limit batch heterogeneity by selection of a specific type of tyre, e.g. passenger-car or truck. Feedstock variation can be limited further by processing just one part of the tyre, but this requires pre-processing and reduces the quantity of material available for recycling. Either the entire EoL tyre or the tread alone can be granulated. More material is made available by granulating entire tyres (a car tyre is typically 8-12 kg), but by selecting just the tread, heterogeneity of the GTR can be limited by exclusion of rubber from the sidewall, carcass and inner liner [9]. The tread usually contains the highest quality compounds, rendering it more desirable for reapplication in recycled rubber products. Recovery techniques (Figure 1.4c) and final application (Figure 1.4d) are the focus of ongoing research. In the remainder of this chapter, we will describe how the granulation mechanism (Figure 1.4b) significantly influences the overall recovery opportunities for GTR and may be the pivotal component of the overall recycling strategy.

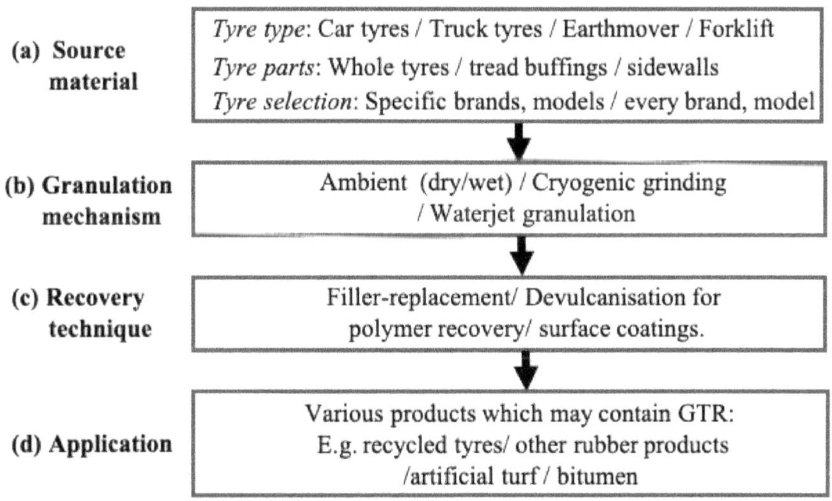

Figure 1.4 *Flow diagram of a typical waste tyre disposal process.*

1.1.3 Applications of GTR

Current commercial opportunities for GTR, categorised according to tyre recycling level, are summarised in Table 1.1 [5], with an approximate indication of the diameter of GTR used in each application. Artificial turf is currently the most important of the diverse range of commercial applications for GTR [10].

Table 1.1 *Commercial opportunities for GTR, categorised according to tyre recycling level, adapted from* [5].

Recycling level	Description	Treatment	Application examples	Approx. GTR size (mm)
Level 0	Disposal; minimum recovery	Collection, disposal	Dumping	Whole tyres
			Landfilling	Whole tyres
Level 1	Mechanical processing to degrade tyre structure	Bead removal	Construction bales	100+
		Sidewall removal	Artificial reefs	100+
		Cutting	Landfill engineering	100+
		Compression	Fluvial reinforcement	100+
			Civil engineering stabilisation	100+
			Acoustic dampening	100+
Level 2a	Size reduction to separate material elements	Shredding	Landfill engineering	10-100+
		Chipping	Drainage	10-100+
			Insulation	10-100+
			Lightweight fill	10-100+
Level 2b	Further size reduction	Ambient grinding	Sports/play surfaces	1-50
		Vibrating separation belts	Paving blocks	1-50
			Roofing/carpet underlays	1-50

			Road surfaces	1-50
			Moulded/extruded parts	1-50
Level 3	Multi-treatment procedures to further process material	Devulcanisation	Carbon products	0.1-1
		Cryogenic grinding	Gaskets, hoses, rings, seals	0.1-1
		Waterjet grinding	Asphalt additives	0.1-1
		Surface modification	Moulded/extruded parts	0.1-1
Level 4	Post-treatment processes to upgrade the material	Thermoplastic elastomers	Coatings	0.005-0.5
		Carbon products	Sealants	0.005-0.5
		Recycled carbon black	Inks	0.005-0.5
		Cryogenic grinding	Activated carbons	0.005-0.5
		Waterjet grinding	Tyres	0.005-0.5

The current markets for GTR artificial turf products are almost saturated and profit margins are tight [1]. This is due to inherent limited demand which constrains production far more than the availability of the globally abundant EoL tyre source material. Secondly, there is growing concern around artificial play surfaces regarding the exposure to contaminants, such as polyaromatic hydrocarbons, which has limited the wider development of this application [11]. Thirdly, rubber-derived microplastic release is significant, with a recent study estimating that artificial turf contributes between 12-50% of all rubber-derived microplastics in the environment [12].

As a result of these concerns, these particles are being targeted by recent upcoming European legislation from the European Chemical Health Agency, which has proposed legislation against the intentional addition of microplastics to rubber products such as artificial turf, and may extend to prohibition of the use of GTR in turf material as early as 2021 [13].

The use of tyres as an supplementary fuel in cement kilns remains a competitive pathway to GTR production. However, this pathway is in decline mainly as a result of recent fluctuations in the gate fee paid for EoL tyres (i.e. the amount paid by the consumer for disposal) which affect business modelling in cement manufacture [14]. Opportunities for GTR application as a filler include: concrete manufacture [15], bitumen [16], railway sleepers [17], drainage in green roofs [18], and reapplication in tyre manufacture [8].

There are two processes commonly used to recycle GTR, as illustrated in Figure 1.5. One of these processes involves the granulate being suspended in a matrix (Figure 1.5a) which is usually a virgin binder mix and includes, but is not limited to: rubber polymers, bitumen and cement. The GTR in suspension chemically bonds to binder, forming a part-recycled material. This mechanism of recycling has been commercialised for low-value products such as artificial turf, carpet underlays and concrete blocks [19]. The advantages of this process include its relative simplicity and low costs. The GTR material can add preferential properties to the matrix, such as cold-weather cracking resistance in asphalt roads. The main disadvantages relate to limitations associated with suspending a recovered solid particle in a virgin matrix, as frequently the recycled filler-matrix bonding is not as strong as that between the virgin filler and matrix. This can result in detrimental impacts upon the properties and durability of products. This is especially problematic for where performance demands in premium applications are significant [8].

Devulcanisation technologies are being developed to 'reverse engineer' GTR, thereby reforming its original polymers (Figure 1.5b). This entails selective breakage of the C-S and S-S sulphide bonds between polymer chains, whilst keeping the C-C bonds intact. This eliminates the need for a matrix thereby producing a homogenous product based upon the original ingredients, which could essentially be fully recycled. This technology is still in the research phase and possible solutions to selectively break C-S and S-S bonds using a variety of techniques are being developed and are discussed elsewhere [20]. Successful application of this technology will convert GTR to a devulcanisate, which can then be used directly to manufacture new polymer products, including tyres. The form of the GTR which is used as a feedstock in the devulcanisation process will be a critical determinant of the success of the technology and the granulation process effects this form.

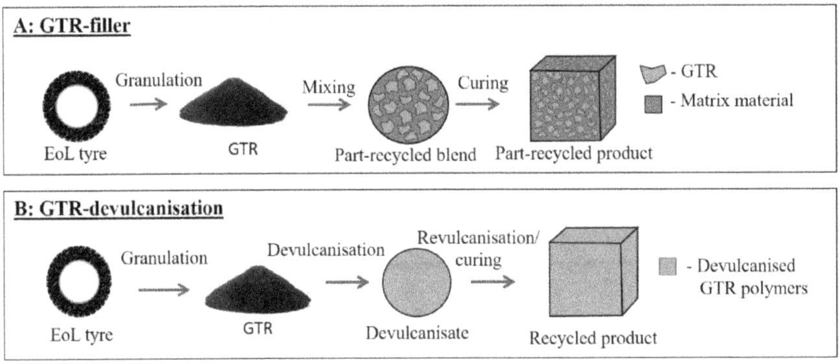

Figure 1.5 *Primary recycling routes for rubber from tyres: (a) GTR-filler, (b) GTR-devulcanisation.*

1.2 Granulation technologies

1.2.1 Ambient mechanical granulation

1.2.1.1 Granulation system

Ambient grinding is typically achieved using a combination of knives, crackers and mills which grind whole tyres down to a GTR product with a particle diameter of ≥ 0.4 mm. It is a fully commercialised, multi-step, continuous process which normally takes place at ambient temperature. The feedstock consists of whole or sliced tyres or those which have undergone prior removal of certain components such as the bead, sidewall or steel wire. The ambient mechanical grinding process is usually sequential, with the shredders (cutting systems) and mills (grinding systems) arranged in series. Typically, this involves the use of 3-6 particle size reduction units for most GTR applications [21], with one of these machines having the general layout as illustrated in Figure 1.6. Whole tyres or shred are granulated and screened with separation of fibre and steel particles before the product is fed to the next machine. Particles exceeding a specified size are fed back to the grinder system via the reclaim feed. *Eco Green Equipment LLC*, a leading manufacturer in this sector, produces a system which initially cuts whole tyres into ~50 cm shreds, and then sequentially reduces GTR particle size to ~16 mm, ~4 mm, and either ~0.85 mm or 0.42 mm in the final stage [22]. Sieves and screens are incorporated in these size reduction processes to ensure consistent particle size and quality of the product.

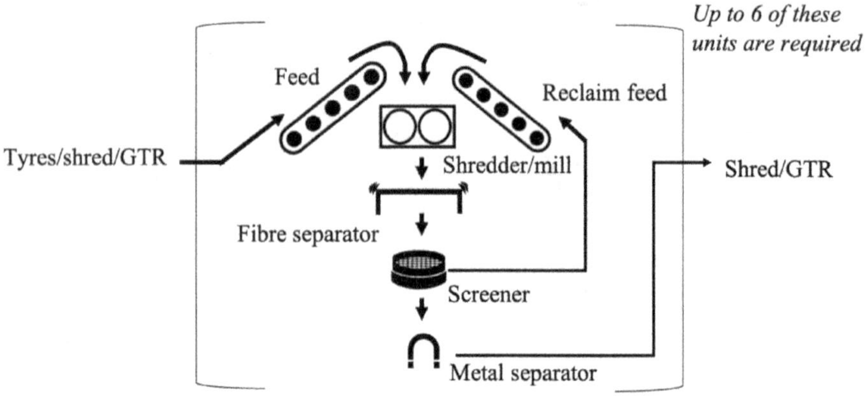

Figure 1.6 *Ambient tyre mechanical size-reducing unit. The conversion of whole tyres to crumb with a diameter of <1 mm typically requires between 3 and 6 of these units in series within a granulation plant [23].*

Alternative features of the grinding system include secondary granulators, screw presses, cracker mills, high speed rotary mills and extruders [24]. Production of GTR with a diameter of < 0.5 mm is difficult due to heating from friction. This temperature rise in ambient systems can exceed 130°C which can shorten polymer chains through thermal degradation and cause the re-arrangement of sulphide linkages between the chains, causing undesirable chemical degradation, thereby, compromising physical properties of the rubber [25]. In addition, the elastomeric properties of GTR at ambient temperature can allow the rubber to be compressed between blades instead of being fractured by them, particularly as the blades become blunt. Grinding at too small particle sizes can also be challenging as the surface charge of the small particles may cause them to aggregate [26].

Shredders use knife systems and *Eco Green Equipment LLC* have patented a triangular knife design which can be rotated when one knife-apex becomes blunted thereby improving operational longevity [22]. Such rotating knife systems produce rough shred, which may be adequate for some civil engineering applications, pyrolysis systems, and tyre derived fuel. However, for most recycling applications, a lower particle size is required, and this is achieved by one of two processes (illustrated in Figure 1.7):

1) Coaxial rollers (cracker mills), which can incorporate dry or wet feedstock (through dry or wet grinding, respectively).
2) A single or twin screw extruder, i.e. Solid State Shear Extrusion (SSSE) [1, 27].

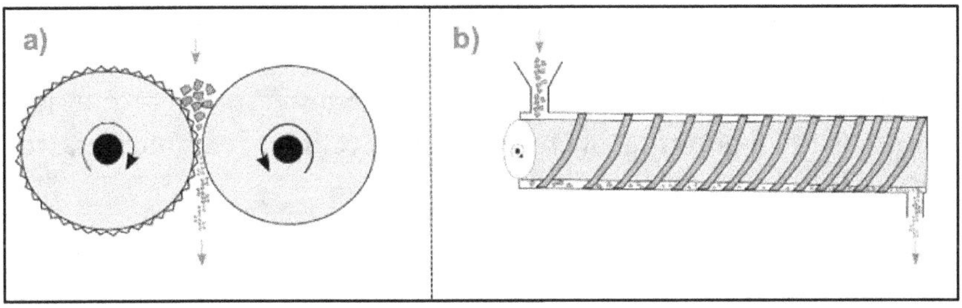

Figure 1.7 *Ambient temperature production mechanisms for mechanically ground crumb (MGC) of fine grade (D <3 mm): (a) Co-axial cracker mill, (b) Solid State Shear Extrusion (SSSE) single screw system. Dual screws can also be used.*

In the cracker mill system (Figure 1.7a), two adjacent rollers counter rotate at speeds of up to 50 RPM. One of the rollers is corrugated causing the rubber to crack and granulate. Tear surfaces can be observed from granulate produced in such systems. The process is relatively simple and commercial quantities of mechanically ground crumb (MGC) can be produced easily [8]. However, it can be difficult to granulate tyres beyond a certain diameter using this method, due to the high strength and elasticity of the rubber [28]. Another major disadvantage of this grinding method is that the heat released from splitting the rubber particles is primarily absorbed by the rubber granulate, which can increase the temperature to 130°C, causing thermal degradation of the rubber surface [29].

Very small particle sizes can be produced by wet grinding, where wet crumb particles of 0.65-2 mm are fed between two grinding stones, resulting in the production of GTR particles which can be as fine as 0.087 mm [30, 31]. An advantage of this process is the production of GTR with a less vulcanized surface chemistry due to water absorbing the heat of friction rather than the rubber. This method allows finer particles of a much higher surface area : volume ratio to be produced which results in improved processability, particularly extrusion, and more consistent bonding properties in new blends [32]. Moreover, the application of a liquid medium facilitates separation of tyre fibre from the rubber crumb due to the density difference between the two products, resulting in a more homogenous product. In spite of these advantages, the process has not yet been widely commercialised as the GTR product requires drying.

Solid-state shear extrusion processes (Figure 1.7b) currently in the research phase of development, are an alternative to established ambient grinding processes [27]. However, particle agglomeration during shear extrusion can reduce efficiency. This limitation can be improved by the addition of a second extruder screw [1]. SSEE has recently shown potential in closed loop recycling of technical elastomers [33].

1.2.1.2 Mechanically ground crumb properties

MGC-GTR, produced from ambient grinding, is generally associated with limitations to its reuse in high performance recycled materials. These limitations include:

1) A moderate surface area:
   This is due to large particle sizes with a low surface roughness, which limit the effective bonding surface area and bonding affinity at the filler-matrix interface.
2) Product impurity:
   MGC commonly contains contaminants such as tyre fibre and metal fragments. Even small amounts of impurities reduce the effectiveness of the bonding and product quality, particularly in higher performance rubber products which require high strength, elasticity and abrasion-resistance [9].
3) Excessive vulcanisation:
   This can occur if the temperature of the GTR exceeds approximately 70°C during the grinding process, resulting the rubber entering the "overcure phase" with the formation of extra sulfidic links between polymer chains. This causes the crumb to become brittle with the loss of its elastomeric properties, reducing performance in blends.

The main impact of these limitations is to compromise both the bonding of GTR in new filler-matrix blends and to hinder the development of emerging devulcanisation techniques. An increased surface area and a more reactive, less degraded GTR interface would increase the efficiency of the developing devulcanisation techniques as a result of smaller diffusion pathways to the centre of each rubber particle [1]. This would minimise non-uniform exposure to the selected devulcanisation method whereby the surface of each particle is much more devulcanised than the centre. These core limitations of MGC limit the wider application of recycled GTR, particularly as a feedstock in higher value products. Improvements to GTR properties are essential to facilitate the wider uptake and use of recycled rubber, which justifies investigation into alternative EoL tyre granulation mechanisms in order to assess whether properties of this material can be improved.

1.2.2 Cryogenic mechanical granulation

1.2.2.1 Granulation system

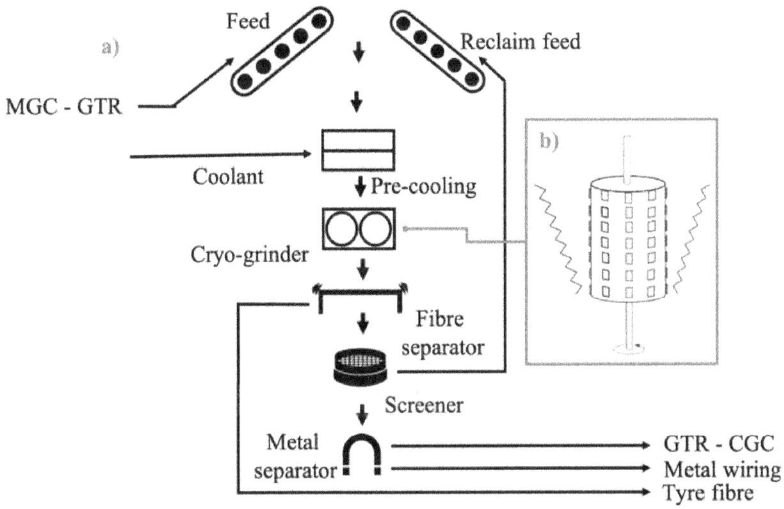

Figure 1.8 *a) Cryogenic crumb production. b) Illustration of a typical cryo-grinder.*

Cryogenic grinding involves cooling rubber particles (usually larger sized MGC) to below the glass transition temperature of rubber (between -30°C and -80°C depending on the rubber formulation) before pulverisation in a hammer mill or grinder system [34]. The process is continuous and very similar to an ambient mechanical grinding system although has not been as widely deployed commercially. An illustrated schematic is shown in Figure 1.8. The cooling is normally achieved through the use of liquid nitrogen or commercial refrigerants [5]. The cooling system freezes the crumb, before it is ground against a serrated edge using a bladed device spinning at up to 1800 RPM [35]. Very small particles can be produced from this method, down to 50 μm diameter and below. A broad particle size distribution (PSD) is generated from cryogenic systems in contrast to ambient methods. Nevertheless, in cryogenic systems, particles can be size-separated by screening and re-processed until they are below the target diameter thereby achieving a constant and narrow PSD according to end-product specification. Cryogenic systems have recently been combined with methods to increase the feedstock moisture content prior to grinding in order to enhance the brittleness and effectiveness of the grinding mechanism [36]. Centrifuges have also been included to facilitate the rapid separation of superfine crumb from the remainder of the products [37].

Although cryogenic technology has been applied to rubber crumb since the 1970s [38], methods capable of granulating particles to a very low particle size have only been developed recently [35]. The acquisition of *Lehigh Technologies Inc.*, a producer of cryogenic 'micronized rubber powder (MRP)' by the *Michelin Group* in 2017 heralded the emerging commercial importance of a cryogenic grinding. In 2020, four cryogenic crumb products were offered, including additives for tyres and asphalt [39].

Hoyer's group recently investigated the efficiency of cryogenic systems [1]. From their analysis, 1.75 kg liquid nitrogen was required to cryogenically grind 1 kg rubber, with an energy demand to produce the liquid nitrogen of 1 KWh. Therefore, cooling 1 tonne of rubber would cost ~$100 (assuming $0.10 / KWh). However, this figure excludes the costs of operating the grinding machinery required to reduce a whole tyre to the maximum in-feed size of the CGC system, to provide the force required for shattering the frozen rubber and for particle size separation.

Manufacturers have shown continued interest in cryogenic crumb production technology. This is mainly due to production advantages, such as the capacity to re-process ground crumb until consistent particle size can be achieved, high production rates, and compatibility with existing MGC granulation technologies. Micronised Rubber Powder (MRP) is available in the particle size range from <50 μm to <850 μm and *Lehigh Technologies Inc* have a current capacity of 64,000 tonnes of MRP-CGC per year [39].

1.2.2.2 Cryogenically ground crumb properties

There are three important differences between cryogenic ground crumb (CGC) and MGC:

1) Particle size:
   MGC is typically very difficult to produce in commercially viable quantities with particle diameters < 0.4 mm due to the elasticity of rubber at ambient temperature, whereas CGC with diameters of < 0.05 mm can be produced at commercial levels.
2) Surface area:
   The shattering of frozen CGC generally results in smoother particle surfaces with more angular edges than the rougher tear surfaces observed in MGC, which are discernible in the scanning electron microscopy (SEM) images shown in Figure 1.9.
3) Surface chemistry:
   CGC is associated with reduced degradation with respect to MGC. This is attributable to the temperature in a cryogenic system being sufficiently low to avoid heat degradation and lower oxygen availability thereby minimising degradation by oxidisation.

The cryogenic grinding process promotes brittle fracturing and thus a product with small, processable particle sizes and smooth surfaces with a low surface area. This has been demonstrated by gas adsorption investigations [40] and SEM image comparisons of crumb produced by different methods (Figure 1.9). The relatively low surface area of CGC was reported over 30 years ago [41], which has discouraged the application of cryogenic crumb to new blends, including asphalt, due to demonstrated limitations in bonding strength between the granulate and the new matrix [23, 41, 42]. Another major drawback of cryogenic grinding is the operating expenditure required for the coolant. For example, liquid nitrogen requires specialised storage, transport and production facilities. Often secondary processing cycles are required to reduce the particles to the required size which can emerge damp, requiring an additional drying stage [24]. Lastly, there is a significant risk of asphyxiation for plant employees, due to the large quantities of liquid cryogenic gas which are required and its constant evaporation during process operation.

Figure 1.9 *SEM images of rubber crumb. a) CGC* [23], *b) CGC* [24], *c) CGC of D < 0.074 mm* [42], *d) MGC of D = 0.4 mm* [42]. *Reproduced under license CC BY-NC-SA 4.0.*

Novel part-recycled rubber formulations containing 45-85 weight % (wt%) CGC have been produced [43], but further comparative studies are required before a conclusive assessment of the material can be made. In recent progress, Hassim compared the performance of part-recycled rubber blends comprising a matrix of either natural rubber (NR), butyl rubber (BR) or a 70/30 NR/BR blend. To these matrices, either CGC of ~ 0.074 mm diameter or MGC of 0.4 mm diameter, was incorporated at 8-21 wt% [42]. The analysis of physical parameters such as tear strength and abrasion resistance demonstrated that CGC generally performed worse as a filler. Transmission Electron Microscopy (TEM) network visualisation, a technique used to observe crosslinking in rubber formations [44], allowed investigation of the interface between MRP particles and the virgin rubber matrix. The appearance of undesirable 'void' spaces at the bonding interface, particularly for matrices containing NR, indicated a lower surface activity of CGC for rebonding than MGC.

Conversely, for the butyl rubber (BR) matrix, better physical performance was associated with inclusion of MGC in spite of the larger particle size than the CGC [42]. These results imply that for blends containing natural rubber derived from car tyres and particularly those from buses, trucks and earthmoving equipment, the weak surface activity and smoothness of the CGC particles cause undesirable physical properties in new blends. These findings suggest that for natural rubber blends, surface roughness is a more important parameter, allowing the particle 'grip' in the matrix, giving enhanced strength and abrasion resistance properties. Other groups have also shown that rubber containing MGC has better tensile fatigue resistance than that containing CGC [1]. This finding was attributed to the rough, disrupted structure of the MGC crumb which allows a smooth transition between the relatively stiff crumb particle and the softer matrix.

Brunauer–Emmett–Teller (BET) theory is an important analytical technique for specific surface area determination [45] which has been applied to GTR. This technique has shown that for similar particle size distributions, CGC has a much (at least 25%) lower surface area than that of MGC [1, 46]. However, other researchers published an article which concluded that CGC had a higher surface area for a given particle size [47]. The authors claimed that MGC had a higher fibre content than CGC which gave the impression of a greater surface area. Nevertheless, more recently it has been demonstrated that for all ambient MGC particle sizes, BET surface area was 1.5 to 2 times greater than that of CGC (See Figure 1.9b vs Figure 1.9c) [46]. Only at very low particle sizes (around <0.09 mm) were the CGC surface area measurements similar to or greater than MGC. The authors neither measured fibre content in the CGC or the MGC.

The main challenge in cryogenic grinding system development is to ensure that the energy cost associated with granulating to a small particle diameter to compensate for the smooth particle surface area is economically justifiable. Cryogenic systems are well suited to continuous processes. Initially the tyres can be shred into "chips" of 0.5-10 cm. Moreover, grinding systems are well-understood, trusted and are compatible with current commercial ambient grinding systems. In contrast, industry know-how of novel technologies, such as waterjets, has much further to develop. Despite the apparent advantages of CGC systems, more comparative assessments of MGC and CGC in new rubber blends are required to assess whether the energy cost of cooling the rubber is sufficient compensation for the superior bonding properties exhibited by the smaller CGC particles which are produced. Research findings suggest that the smooth shattered surface area of CGC is an inherent limitation in its bonding capability in new blends for very high-performance applications, such as incorporation into recycled tyres. Further research is required to delineate the role of CGC in sustainable rubber recycling in the future.

1.2.3 Waterjet granulation

1.2.3.1 Granulation system

Waterjet (WJ) granulation involves disassembling an EoL tyre through the impact of an ultra-high-pressure water (UHPW) jet to produce water jet crumb (WJC) GTR. The principle components of a waterjet granulation system are a UHPW pump-set capable of generating water pressures of 1500-3000 bar, a nozzle to focus the water at the target tyre, and a suitable carrier system for the EoL tyre. Either the nozzle or the carrier is required to move to enable disassembly of the tyre by allowing the jet to be aimed at the required area for granulation. Tyres can be entirely disassembled using the method outlined in Figure 1.10, although most current systems only remove the high-quality tread material.

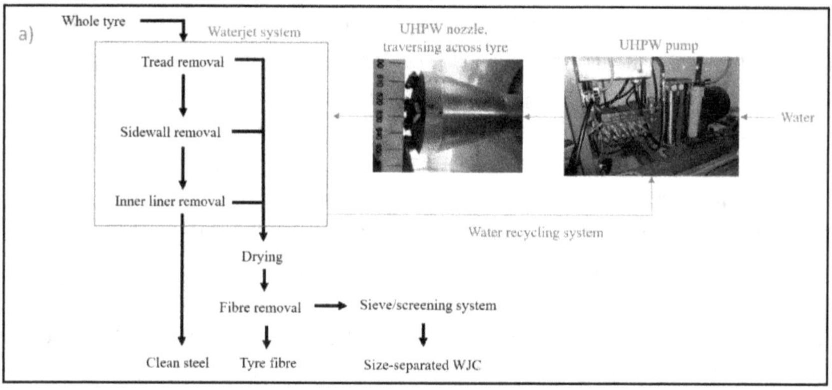

Figure 1.10 *Illustration of the waterjet tyre granulation process.*

WJ granulation has received increasing attention in the last decade but commercial systems are still under development and have been mainly limited to batch systems operating on larger tyres such as those from buses, trucks and earthmoving equipment. *Aquajet Zrt.*, based in Hungary, produces WJC from truck tyres with a particle diameter < 0.4 mm [48]. *Big Tyre Recycling Corporation* (BTRC), based in Belgium, has removed rubber economically in a single-step operation from large earth moving equipment tyres [49]. Similarly,

Rubberjet Valley SRL, based in Italy, has removed tread buffings from large earth moving tyres, in the mining industry [50].

Larger tyres are inherently well suited to currently available batch-waterjet systems due to their relatively thick tread, which has a higher rubber content per tyre thereby reducing the interruption of operation from loading and unloading each tyre, in contrast to processes involving smaller tyres such as car tyres. The tread component of tyres usually contains the highest quality rubber, and the WJ process can separate this material from other components of the tyre such as the sidewall and carcass. Finally, the mines where the largest tyres are used are in remote locations which makes recycling more challenging and offers an opportunity for novel solutions such as WJ granulation. There are also groups investigating WJ granulation on car tyres [8, 28]. Post granulation stages, such as drying, are very similar to other systems. For example, the fibre removal stage at *Rubberjet SRL* involves a vibrating screen accompanied by air flows to produce a fluidised bed which is similar to those incorporated in some MGC systems [50]. Waterjet granulation systems have several advantages over other tyre granulation methods:

1) Clean separation:
   Metal wires, GTR and other tyre components can be separated from a tyre in one cycle using waterjets [51]. This is particularly the case for most systems which focus on the removal of just the tread, as the steel belt acts as a barrier between the high value tread and the lower quality rubber which lies underneath. Thus, it is possible to recover a higher purity grade of material than non-selective cutting methods such as bulk ambient or cryogenic mechanical grinding.
2) Single disassembly step:
   A very low crumb particle size (below 0.1 mm diameter) can be readily produced in a single processing step. This both reduces the number of separate stages of size reduction and the overall footprint of the plant. Furthermore, it facilitates the development of mobile plants to recycle tyres in remote environments, such as mining or military operations, particularly as a UHPW pump system can either operate on electricity or via direct drive from a diesel engine.
3) High quality GTR:
   Several recent research studies have shown that WJC is associated with a relatively low particle size, a very high active surface

area, and a devulcanised undegraded surface chemistry [46, 52, 53]. This implies that WJC is the highest quality crumb available when incorporated in new blends for a given particle diameter [8, 46, 54]. This finding is crucial to increasing the potential for significant quantities of recycled rubber to enter the market and will be discussed further in Section 1.2.3.2.

There are also some disadvantages associated with waterjet systems:

1) Water pressure requirement:
   In spite of UHPW pumps becoming cheaper and more accessible, the extreme pressures required are associated with an energy demand which contributes significantly to the operating costs. UHPW also makes the system noisy, and vapour must be controlled by means of an enclosure around the tyre to protect the electronic components of the robotic traversing system. Nevertheless, there are neither blades to replace nor a cryogenic coolant, so these costs are avoided in a WJ system.
2) Addition of a drying stage:
   Initially, the WJC product is wet and must be dried to avoid particle clumping during the screening separation stage. This can be achieved using conventional, low cost drying equipment operating at a maximum temperature of 40°C to avoid thermal degradation of the WJC rubber polymers. The UHPW generates heat which can be recovered to reduce the energy demand of the drying process, thereby minimising energy costs.
3) A wide particle size distribution:
   Unlike CGC or MGC, WJC cannot be fed back into the system through a reclaim feed, as the rubber must remain attached to the tyre in order to be fragmented by the waterjet. As a result, the particle size distribution (PSD) for WJC is wider than for either MGC or CGC. However, recent studies have shown that particle diameter is not the most important parameter for WJC, and that the PSD is partially controllable by appropriate selection of waterjet operational parameters (see Section 1.2.3.3) [8, 55]. Nevertheless, the largest particles produced are typically < 5 mm and at least 50% are < 0.4 mm [46].

1.2.3.2 Waterjet crumb properties

Waterjet crumb (WJC) is a relatively new material, the first batches having been produced around 2005 (See Figure 1.11). Early research mainly focused on the separation mechanism, whereas recent investigations have focussed on the superior properties of WJC in comparison to MGC or CGC in new material blends [8, 46, 51].

Figure 1.11 *a) WJC viewed at the micro scale b) MGC viewed at the micro-scale. c) SEM of MGC at the nanoscale, d) SEM of WJC at the nanoscale. Images source: a, b) from* [8], *Reproduced under license 4878250012437. c, d) Novel unpublished images taken by the authors.*

It has been highlighted that the performance of GTR as a filler is dependent primarily on the strength of bonding at the rubber filler-matrix interface [42]. Various studies have shown the increased surface area of WJC relative to other crumb types (MGC and CGC). Consequently, the main advantage of WJC is that the rubber-filler interface is larger per unit weight. Evidence for the increased WJC surface area relative to CGC or MGC includes: BET analyses showing greater surface area values at all measured particle sizes, [46], lower volumetric densities for a given particle diameter [56], and 2D shape quantification results for surface roughness via image analysis [8]. BET analysis showed that at particles of 0.1 mm diameter, WJC had double the surface area (~0.2 m^2/g) of CGC, whilst MGC had a value of 0.17 m^2/g [46]. At the largest measured particle size (D = 0.55 mm), this difference was greater, with 0.095 m^2/g for WJC, 0.02 m^2/g for CGC and 0.045 m^2/g for MGC. Surface area and roughness have significant implications for bonding characteristics in new rubber blends. Further evidence of a high surface area to volume ratio is demonstrated in the preliminary research which has compared WJC with MGC as a filler in new material matrices.

The main evidence attesting to the superior bonding properties of WJC derive from studies of their potential as a filler material when compared with MGC. In 2018, Loderer et al. compared MGC, CGC and WJC as a filler material in new bitumen blends [54]. They found that in addition to outperforming virgin 70/100 bitumen and SBS-modified bitumen, WJC-modified bitumen proved to be stronger at -10°C when compared to CGC and MGC-modified bitumen. Interestingly, comparatively large WJC particles (<1 mm as opposed to 0.35 mm) increased the ductility of modified bitumen, thereby improving the extreme-weather performance by reducing brittleness of the road surface at cold temperatures and thus susceptibility to cracking. This benefit was attributed to the larger particles causing a greater crack path deviation from a straight line, thereby increasing strength at cold temperatures (< 5°C) by up to 15% when compared to unfilled bitumen. This contrasts with studies which investigated MGC-modified bitumen in which smaller particle sizes performed better in bitumen blends. This was attributed to the relatively smooth surfaces of the MGC and CGC which caused cracks to propagate along the bitumen with much less resistance, leading to a low fracture toughness.

Another research group have evaluated the performance of WJC in bitumen blends relative to MGC [57]. They also showed that bitumen blends incorporating WJC, as opposed to MGC, were more resistant to temperature fluctuations and exhibited high storage stability characteristics. This was attributable to the WJC blend having a greater penetration index which is indicative of a higher degree of softening than the MGC blend. Therefore, WJC-modified bitumen was found to have a higher temperature resistance when compared to virgin and MGC-modified bitumen.

There is growing evidence of the superior bonding properties of WJC in part-recycled tyre rubber formulations. Recent work by Bowles et al. entailed substitution of virgin materials of a standard SBR/NR tyre rubber formulation at 10-30 wt.% with either WJC or MGC [8]. It was shown that for a given particle size, WJC caused the formation of a part-recycled tyre rubber that was 6-26% stronger, more flexible and more abrasion resistant than the same formulation with MGC substituted as a filler (see Figure 1.12a). These findings demonstrate the potential for significant quantities of WJC-GTR to be incorporated into high-performance recycled rubber products such as tyres, a product which requires novel solutions to become compatible with the developing circular economy approach to resources and materials use. Moreover, WJC particles of a much larger diameter were also tested, and similarly outperformed MGC particles which were less than half the particle size diameter of the WJC samples (see Figure 1.12b).

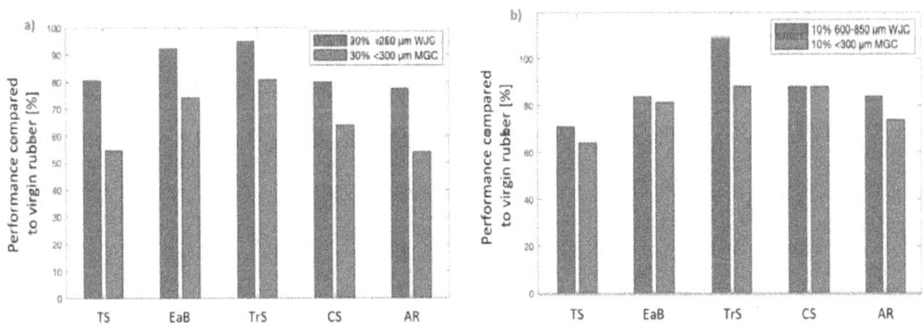

Figure 1.12 *Physical strength parameters of NR/SBR blends containing MGC and WJC at different concentrations and particle diameters. TS = tensile strength, EaB = elongation at break, TrS = Tear strength, CS = compression strength, AR = abrasion resistance. Adapted from* [8]. *Reproduced under license 4878250012437.*

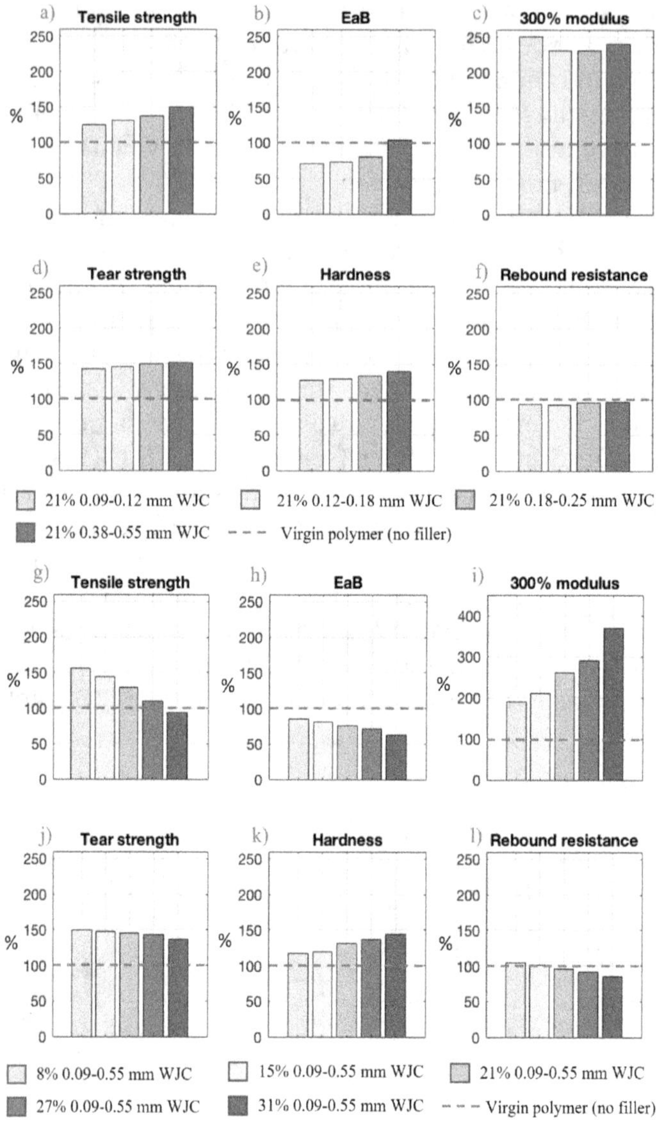

***Figure 1.13** Physical performance parameters of the NR/GTR waterjet crumb blends at differing particle sizes (a-f) and different proportions of incorporation (g-l). Adapted from data in* [46].

The incorporation of WJC into new rubber mixes was investigated by Zefeng et al. (2018). WJC of different particles sizes (0.09-0.55 mm diameter, see Figure 1.13a-f for details) and different blend proportions (8-31 weight %, see Figure 1.13g-l for details) were added to a standard NR tyre rubber mix, with subsequent determination of tensile strength, elongation at break, 300% modulus, tear strength, hardness and rebound resistance. Tensile strength at 21% WJC incorporation increased linearly with decreasing particle diameter and was at least 20% higher at all WJC particle sizes relative to virgin rubber (See Figure 1.13a). With increasing WJC incorporation, various physical parameters such as elongation at break, tensile strength, tear strength and rebound resistance all decreased linearly, whereas 300% modulus and hardness both increased linearly. These results suggest that the incorporation of WJC as a filler material at increasing proportions in part-recycled NR compounds results in a weaker, less flexible and harder rubber product. Nevertheless, the performance characteristics of the rubber blends in this study actually improved with WJC incorporation for many characteristics (e.g. tensile strength and rebound resistance) when compared to the virgin unrecycled polymer, particularly at lower WJC incorporation values. Results also suggest that WJC bonds more preferentially with natural rubber [46] than artificial rubber (styrene butadiene) [8].

Figure 1.13, sections g-l show that the characteristics of adding WJC to blends can be further enhanced if smaller particle sizes are used. For almost all of the physical characteristics of the blend containing the smallest WJC particle size (0.09-0.12 mm diameter) the modified blend outperformed the virgin rubber, indicating that part-recycled rubbers containing very small diameter WJC actually *improve* rubber formulations and performance.

The increased bonding performance of WJC in new rubber blends may cause a step-change in the range and quality of products into which recycled GTR can be incorporated. Recent studies suggest the increased performance of WJC, relative to MGC and CGC, could lead to substantially greater GTR incorporation into new, high performance rubber and other polymer blends [8, 46, 52, 58]. In the future, WJC may contribute to 'closed-loop' tyre recycling by its use in combination with recovered carbon black from pyrolysis and recycled tyre wire [59, 60]. These pathways may allow significant quantities of recycled materials to substitute hydrocarbon derived polymers and re-enter the market in new tyres and other high-performance rubber products. This is crucial for circular economy principles as tyre manufacture consumed 45.2 billion tonnes of raw materials in 2017, most of which were unsustainable hydrocarbons derived from fossil fuels [61].

1.2.3.3 Modelling and optimising the waterjet system

In the previous section, evidence was presented showing that WJC has the potential to be a disruptive and innovative material which may contribute to significant quantities of GTR entering the market as recycled or part-recycled high-performance rubber products. However, the mechanism by which WJC is produced is complex and warrants further discussion, as the production of commercially viable quantities should be the focus of current research and development. The first stage is to understand the physical mechanics of the waterjet process.

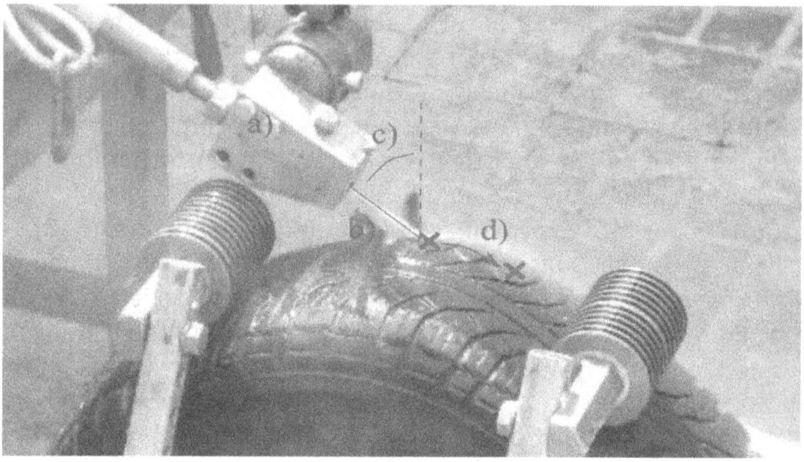

***Figure 1.14** Photograph of waterjet system* [58], *with illustrated annotations of the critical parameters which affect WJC production: a) flow rate, pressure of UHPW. b) standoff distance. c) jet angle. d) traverse pattern and traverse speed of jet. Reproduced with permission under license 4878310543276.*

The controllable parameters associated with EoL tyre granulation using a waterjet are outlined in Figure 1.14 and include the diameter of the nozzle, pressure of the waterjet, distance from the tyre surface, traverse speed and pattern of the jet over the tyre surface, and the angle between the tyre surface and the waterjet [28]. The system efficiency can be optimised by careful adjustments of these parameters. The limiting factor for waterjet systems is the size of the water pump, as industrial ultra-high pressure water (UHPW) pumpsets have a maximum flow rate for a given pressure [62]. These need to be carefully optimised, as there can be a complex inverse relationship between stripping rate (mass WJC/ unit time) and the average particle size produced. For instance, an increase in the standoff distance, i.e. the gap between the jet nozzle and the tyre, can reduce the average particle size by allowing more air entrainment into the jet, creating higher quality WJC, as shown by Figure 1.13g-l, but at the cost of a reduced stripping rate and a lower efficiency due to energy-loss of the waterjet by the air. Groups in the UK, Czech Republic and China are working to optimise the parameters outlined in Figure 1.14 with the aim of generating the highest quality product at an economically acceptable rate.

The energy demand of the WJ system is determined by the volumetric flow rate and the pressure of UHPW utilised, a factor which also affects the stripping rate and the average WJC particle size. A WJ granulation system should be set up to produce WJC below a target particle size whilst maximising the stripping rate with minimal use of water. System efficiency can be quantified as the ratio of the mass of WJC to the mass of UHPW. In a recent study, 50-70 kg of WJC was produced per hour at an energy demand of 5-8 kWh [54], giving an energy cost of roughly $80-150/tonne of WJC (assuming an energy price of $0.10/KWh). This cost is generally acceptable and, as explained in Section 1.2.2.1, is comparable to the cost of the coolant for a CGC-granulation system. Furthermore, unlike mechanical grinding systems, WJ granulation is still in its early phase and therefore there is significant opportunity for optimisation in the future. For example, the conversion to a fully continuous system, which has to the best of our knowledge not yet been achieved for WJ granulation, would be likely to result in further increases in the efficiency.

In a WJC production system, energy losses primarily occur through two processes: deflection of high-velocity water after impact with the tyre and air resistance between the nozzle exit and the tyre surface. Operating parameters must be adjusted to minimise both of these losses. In 2018, it was shown that the nozzle diameter of the waterjet must be carefully considered. For example, a nozzle with a diameter of 1 mm requires 2.75 times more water flow rate than a diameter of 0.4 mm [28]. The selection of a lower diameter nozzle, with multiple nozzles, was found to be more effective than a single nozzle of large diameter at stripping rubber from the tyre carcass.

The degradation patterns of rubber strips subjected to an ultra-high pressure waterjet have been investigated [55]. A near-linear correlation between cutting depth and pump pressure was demonstrated and a traverse speed across the rubber surface of < 2 mm/s resulted in the greatest depth of penetration. However, higher stripping rates may be attainable by selecting traverse speeds greater than this value. This study also showed that a lower standoff distance resulted in higher cutting rates, although in practice this distance may limited by potential irregularities in the tyre surface which may impinge on the nozzle when traversing. Therefore a minimum standoff distance of 5-15 mm is usually applied to protect the system from damage [62]. Moreover, increasing the standoff distance may induce disruption of the waterjet by air entrainment causing dispersion into droplets, which may cause more irregular particles to be produced [8].

Improvements of the efficiency and product quality of WJ granulation requires detailed consideration of the fluid mechanics of the extraordinary conditions under which the system operates. There has been significant progress towards this goal since 2014. Ultra-high pressure waterjets have been utilised to cut a range of materials with different properties for more than three decades [57, 63]. However, cutting of tyre rubber is uniquely challenging because of its elasticity and rebound resistance relative to other materials. The interaction between the waterjet and the tyre rubber induces failure of the rubber which is determined by both the jet conditions and the physical characteristics of the rubber.

The waterjet granulation system can be reduced to three core components: a high-pressure water reservoir, an extremely fast-moving jet in atmospheric conditions, and a static tyre surface. The assumption of complete energy transfer between the core components without any losses allows for a simplified analysis of the physical system. In 2014, five possible mechanisms of fracture were proposed: cavitation, impact of the jet, dynamic pressure within the jet, fatigue failure, and effects associated with water jets [55]. Two of these effects were investigated in detail by the authors: water hammer pressure, and hydro-dynamical pressure (stagnation pressure). More recently, our group proposed and numerically derived three main mechanisms, which comprise six effects by which waterjets degrade and granulate the surface of a tyre [8]. The proposed underlying effects are described below.

The laminar stagnation point flow scenario is a model based on Bernoulli's equation suggesting that all of the water pressure (2000-3000 bar depending on pump used) impinges at a single point on the tyre surface, i.e. $P_s = P_w$ where P_s = stagnation pressure and P_w is water pressure in the pump. This creates a significant tangential pressure across the surface of the tyre which causes stresses to the rubber beyond its elastic deformation point, causing fracture (illustrated in Figure 1.15a). This assumes a perfectly efficient jet in a closed system, which would not suffer any disruption from the air entrainment between the nozzle and the tyre.

In the second model, there is significant air entrainment which causes droplet break-up within the jet. This would generate a fully disrupted jet which contains droplets travelling at supersonic velocities resulting in a shockwave impacting the tyre surface prior to droplet contact. Following droplet contact various destructive fluid effects occur such as the water hammer effect, stagnation pressure deformation, and tangential shear which applies stress to the tyre surface.

The third model is based on cavitation effects which are caused by an extreme pressure differential between the waterjet and the air causing vapour pockets to open as the jet expands radially. These microbubbles can collapse with enormous velocity and force, and if they do so within the vicinity of the tyre surface, micro-jets and shockwaves may form causing "pitting" effects on the tyre surface. This mechanism is known to cause damage to ship propellers, although it can be used beneficially in underwater mining techniques [64]. These last two models are similar to a mechanism of degradation proposed by Hu, namely, fatigue failure which degrades the tyre structure as a result of repetitive loading.

The determination of the relative contribution of these fluid effects remains the principle challenge for research groups. This could be achieved through a variety of methods, including high-speed imaging and shape-factor analysis of WJ and WJC generated under different operating conditions (standoff distance, traverse speed, degree of air entrainment, etc.). This would allow for optimisation of the WJ system. This understanding, along with the increasing availability of more efficient UHPW pumps, may allow economic production of WJC and enable it to become a commercially viable recycled sustainable material, replacing traditional fossil-fuel derived virgin polymers in new high-performance rubber mixtures.

1.3 Conclusions

Tyres are an indispensable component of modern society, yet they are complex and energy intensive to produce. Due to their structural and chemical complexity, novel solutions are required to maximise material recovery and minimise the environmental impacts associated with tyre production and disposal. The critical stage of EoL tyre

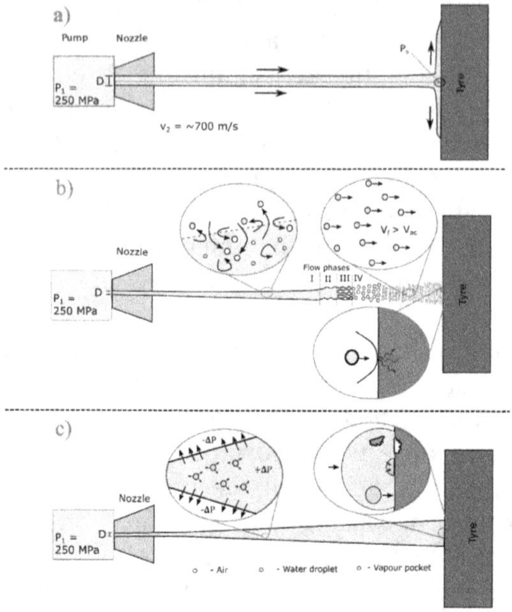

Figure 1.15 *a) Laminar stagnation point flow waterjet model (P_s = stagnation pressure point), b) Disrupted waterjet with impact cratering model (V_f = Velocity of waterjet flow, V_{ac} = acoustic velocity), c) Continuous waterjet with cavitation model. Adapted from* [8]. *Reproduced under license 4878250012437.*

recovery is the separation of the major tyre components (steel, rubber and fibre) via granulation. The efficient conversion of a whole tyre into suitably sized particles, which exhibit appropriate physiochemical properties to ensure maximum reuse potential, is essential. We have outlined existing and developing technologies for the production of GTR. It is anticipated that one or several of these methods will eventually help contribute to the widespread production of high-quality rubber granulate with appropriate physical and chemical characteristics for tyre recycling. These granulates include:

- MGC: The most established technology is probably reaching the limit of its development due to the inherent concerns relating to how the physical and chemical GTR products are changed during the successive processing stages. Primary limitations with current commercially available MGC-GTR include: i) an impure product (with fibre/metal contamination), ii) a moderate surface area: volume ratio, and iii) a scorched, chemically inactive and overvulcanised surface chemistry which has unpreferable bonding performance in recycled material applications. These factors greatly influence and limit the quantity and quality of products available containing recycled rubber. Moreover, the energy and space-intensive requirements of the operating equipment suggest the process may have reached the limit of its development and could be superseded by other systems.

- CGC: A proven technology due to: i) the compatibility with the pre-developed continuous mechanical grinding system, ii) the very low particle size which can be achieved with a narrow particle distribution, and iii) the devulcanised particle surfaces produced. However, the main disadvantage is the very smooth particle surface texture. This means particles must be reduced to very low (<0.1 mm) diameters to have a similar surface area to MGC and smaller still for comparable WJC performance. Size reduction requires repeated processing cycles and thus CGC production is energy intensive.
- WJC: The most recently developing technology which is still at a pre-commercial stage is associated with preferential granulate properties in relation to MGC and CGC: i) the surface roughness of these particles means that not only do similarly sized WJC particles outperform MGC in recycled tyre rubber blends but also larger sized WJC particles may do so, ii) the size reduction process is performed as a single step, and iii) the crumb shows consistent and strong devulcanised surface properties without significant metal contamination. These advantageous characteristics mean that high quality WJC-recycled blends have been produced. However, technical knowledge of the processing method is currently limited. Moreover, a high performance ultra-high-pressure water pump and an extra drying stage is required, and a range of particle sizes are produced. Nevertheless, process control can facilitate the selection of particles within a specified size range. Further process optimisation is required to maximise the potential applications and realise the full value of the WJC process and its products.

The characteristics of GTR are very important in rubber recycling and can be controlled, to some extent, by the method of granulation applied to an EoL tyre. Commercially available GTR is currently unsuitable for incorporation into most high-performance rubber products greatly hindering 'closed-loop' tyre recycling. Current commercialised technologies include low-value products such as rubber-modified road surfaces and artificial turf, with a much lower value than the tyres from which the GTR was originally derived from.

Further research into alternative granulation methods is essential in order to improve quality of GTR available for a range of manufacturers to incorporate into their recycled products. Future work, specifically comparing CGC, improved MGC, and WJC in a range of new recycled material applications is required to accurately determine how to improve rubber recycling from tyres. This is a prerequisite for more sustainable rubber usage in the future.

References

1. Hoyer, S., Kroll, L. & Sykutera, D. Technology comparison for the production of fine rubber powder from end of life tyres. Procedia Manuf. 43, 193–200 (2020).
2. Amari, T., Themelis, N. J. & Wernick, I. K. Resource recovery from used rubber tires. Resour. Policy 25, 179–188 (1999).
3. Garside, M. Natural rubber global production 2018. Statista.com (2019).
4. Ceresana. Styrene Butadiene Rubber (SBR) Market Report, 2025. (2019).
5. Shulman, V. Tire recycling. in Waste 489–515 (Elsevier, 2019).
6. Aderemi, A. & Otitoloju, A. An assessment of landfill fires and their potential health effects-a case study of a municipal solid waste landfill in Lagos, Nigeria. Int. J. Environ. Prot. 2, 22–26 (2012).
7. Rajesh, K., Dhanasekaran, D. & Tyagi, B. K. Survey of container breeding mosquito larvae (Dengue vector) in Tiruchirappalli district, Tamil Nadu, India. J. Entomol. Zool. Stud. 1, 88–91 (2013).
8. Bowles, A. J., Fowler, G. D., O'Sullivan, C. & Parker, K. Sustainable rubber recycling from waste tyres by waterjet: A novel mechanistic and practical analysis. Sustain. Mater. Technol. 25, e00173–e00173 (2020).
9. Forrest, M. Recycling and re-use of waste rubber. (Smithers Rapra, 2014).
10. Cheng, H., Hu, Y. & Reinhard, M. Environmental and health

impacts of artificial turf: a review. Environ. Sci. Technol. 48, 2114–2129 (2014).

11. Perkins, A. N. et al. Evaluation of potential carcinogenicity of organic chemicals in synthetic turf crumb rubber. Environ. Res. 169, 163–172 (2019).
12. Kole, P. J., Löhr, A. J., Van Belleghem, F. G. A. J. & Ragas, A. M. J. Wear and tear of tyres: a stealthy source of microplastics in the environment. Int. J. Environ. Res. Public Health 14, 1265 (2017).
13. Clausen, L. P. W., Hansen, O. F. H., Oturai, N. B., Syberg, K. & Hansen, S. F. Stakeholder analysis with regard to a recent European restriction proposal on microplastics. PLoS One 15, e0235062–e0235062 (2020).
14. TRA. Tyre Recovery Association (private correspondance). https://tyrerecovery.org.uk/ (2020).
15. Bravo, M. & De Brito, J. Concrete made with used tyre aggregate: durability-related performance. J. Clean. Prod. 25, 42–50 (2012).
16. Zoorob, S. E. & Al-Bahar, S. Preliminary Rheological Characterization of Tyre Derived Crumb Rubber Blended with Kuwaiti Bitumen. in MATEC Web of Conferences vol. 203 5001 (EDP Sciences, 2018).
17. Saroja, V. K., Kumarb, A. & Singhc, V. P. Design and Development of Railway Sleepers using Recycled Waste Tyre Rubber. J. Basic Appl. Eng. Res. 6, 309–312 (2020).
18. Vila, A., Pérez, G., Solé, C., Fernández, A. I. & Cabeza, L. F. Use of rubber crumbs as drainage layer in experimental green roofs. Build. Environ. 48, 101–106 (2012).
19. Murfitts-Industries. Murfitts Industries. https://murfittsindustries.com/home (2020).
20. Asaro, L., Gratton, M., Seghar, S. & Hocine, N. A. Recycling of rubber wastes by devulcanization. Resour. Conserv. Recycl. 133, 250–262 (2018).
21. Trojak, L. Shredding Equipment Focus--New Values. Recycling Today https://www.recyclingtoday.com/article/shredding-equipment-focus--new-values/ (2006).
22. Green, E. Eco Green Company Brochure. https://ecogreenequipment.com/wp-content/uploads/2020/05/2016_ECO-Brochure2_1.pdf (2016).
23. Presti, D. Lo. Recycled tyre rubber modified bitumens for road asphalt mixtures: A literature review. Constr. Build. Mater. 49,

863–881 (2013).

24. Pehlken, A. & Essadiqi, E. Scrap tire recycling in Canada. CANMET Mater. Technol. Lab. 8, (2005).
25. Bockstal, L., Berchem, T., Schmetz, Q. & Richel, A. Devulcanisation and reclaiming of tires and rubber by physical and chemical processes: A review. J. Clean. Prod. 117574 (2019).
26. Forrest, M. Overview of the world rubber recycling market. Recycling and Re-Use of Waste Rubber (Smither Rapra Technology, 2014).
27. Chaubey, T. & Arastoopour, H. Studying the pulverization mechanism of rubber with a modified design of the solid-state shear extrusion process. J. Appl. Polym. Sci. 119, 1075–1083 (2011).
28. Holka & Jarzyna. Selected structural issues of the waterjet method in industrial recycling of tires. Eng. Mech. 2018 305–308 (2018) doi:10.21495/91-8-305.
29. Karger-Kocsis, J., Mészáros, L. & Bárány, T. Ground tyre rubber (GTR) in thermoplastics, thermosets, and rubbers. J. Mater. Sci. 48, 1–38 (2013).
30. Adhikari, B., De, D. & Maiti, S. Reclamation and recycling of waste rubber. Prog. Polym. Sci. 25, 909–948 (2000).
31. Klingensmith, B. Recycling, production and use of reprocessed rubbers. Rubber World;(United States) 203, (1991).
32. Abraham, E., Cherian, B. M., Elbi, P. A., Pothen, L. A. & Thomas, S. Recent advances in the recycling of rubber waste. Recent Dev. Polym. Recycl. 47, 100 (2011).
33. Kroll, L. & Hoyer, S. Zero-waste production: Technology for the in-house recycling of technical elastomers. Procedia Manuf. 33, 335–342 (2019).
34. Lievana, E. J. Recycling of ground tyre rubber and polyolefin wastes by producing thermoplastic elastomers. (2005).
35. Cialone, A. M., Waznys, P. J. & Meckert, G. W. European Patent number EP1685163B1 - Process and apparatus for manufacturing crumb and powder rubber. (2008).
36. Wan, Z. Chinese Patent number CN206913515U - A kind of low temperature breaking equipment recycled for junked tire. (2017).
37. Shaotong, L. Chinese Patent number CN107213970A - Superfine powder process units. (2018).
38. Kay, E. L. & Laman, J. R. A review of scrap tire disposal processes. SAE Tech. Pap. 790301, 16 (1979).

39. Lehigh. Lehigh Technologies product overview. https://lehightechnologies.com/products_services/overview (2020).
40. Wang, Z., Kang, Y. & Wang, Z. Pulverization of end-of-life tires by ultra-high pressure water jet process. J. Polym. Eng. 37, 211–225 (2017).
41. Roberts, F. L., Kandhal, P. S., Brown, E. R. & Dunning, R. L. Investigation and evaluation of ground tire rubber in hot mix asphalt. NCAT Rep. 83–89 (1989).
42. Hassim, D. H. A. I., Abraham, F., Summerscales, J. & Brown, P. The effect of interface morphology in waste tyre rubber powder filled elastomeric matrices on the tear and abrasion resistance. Express Polym. Lett. 13, 248 (2019).
43. Kulkaski, D. & Simmering, T. D. Patent number US9840613B1 - Elastomeric composition having high impact strength. (2017).
44. Cook, S., Cudby, P. E. F., Davies, R. T. & Morris, M. D. The microstructure of natural rubber latex films. Rubber Chem. Technol. 70, 549–559 (1997).
45. Brunauer, S., Emmett, P. H. & Teller, E. Adsorption of gases in multimolecular layers. J. Am. Chem. Soc. 60, 309–319 (1938).
46. Zefeng, W., Yong, K., Zhao, W. & Yi, C. Recycling waste tire rubber by water jet pulverization: powder characteristics and reinforcing performance in natural rubber composites. J. Polym. Eng. 38, 51–62 (2018).
47. Ayyer, R., Rosenmayer, T., Schreiber, W. & Colton, J. Effects of micronized rubber powders on structure and properties of polypropylene composites. Waste and Biomass Valorization 4, 65–71 (2013).
48. Aquajet. Aquajet- The power of water - Products. http://aquajet.hu/ (2020).
49. BTRC. Big Tyre Recycling Corporation. http://www.btrc.be/ (2020).
50. Rubberjet. Technology | Rubberjet. https://www.rubberjet-group.com/hpwj-technology (2020).
51. Verri, R. Patent number US20110168818A1 - Plant for recycling tires. (2011).
52. Wang, Z., Jiang, Y. & Pan, C. Mechanochemical devulcanization of waste tire rubber in high pressure water jet pulverization. Prog. Rubber, Plast. Recycl. Technol. 14777606211019428–14777606211019428 (2021).

53. Wang, Z. & Zeng, D. Preparation of devulcanized ground tire rubber with supercritical carbon dioxide jet pulverization. Mater. Lett. 282, 128878 (2021).
54. Loderer, C., Partl, M. N. & Poulikakos, L. D. Effect of crumb rubber production technology on performance of modified bitumen. Constr. Build. Mater. 191, 1159–1171 (2018).
55. Hu, Y. et al. Mechanism and experimental investigation of ultra high pressure water jet on rubber cutting. Int. J. Precis. Eng. Manuf. 15, 1973–1978 (2014).
56. Memon, N. A. Characterisation of conventional and chemically dispersed crumbrubber modified bitumen and mixtures. (2011).
57. Fiore, N., Caro, S., D'Andrea, A. & Scarsella, M. Evaluation of bitumen modification with crumb rubber obtained through a high pressure water jet (HPWJ) process. Constr. Build. Mater. 151, 682–691 (2017).
58. Holka, H. & Jarzyna, T. Recycling of car tires by means of Waterjet technologies. in AIP Conference Proceedings vol. 1822 20008 (AIP Publishing, 2017).
59. Norris, C. J., Hale, M. & Bennett, M. Pyrolytic carbon: factors controlling in-rubber performance. Plast. Rubber Compos. 43, 245–256 (2014).
60. Moulin, L., Da Silva, S., Bounaceur, A., Herblot, M. & Soudais, Y. Assessment of Recovered Carbon Black Obtained by Waste Tires Steam Water Thermolysis: An Industrial Application. Waste and Biomass Valorization 8, 2757–2770 (2017).
61. Bostock, J. Global Tire Industry Volume to Reach 2.7 Billion Units by 2022 | Smithers. https://www.smithers.com/resources/2017/dec/global-industry-tire-volume-to-reach-2-7-billion (2020).
62. Wang, Z., Kang, Y. & Cheng, Y. Multiresponse optimization of process parameters in water jet pulverization via response surface methodology. Int. J. Precis. Eng. Manuf. 18, 1855–1871 (2017).
63. Hughes-Pumps. Waterjet cutting applications. https://www.hughes-pumps.co.uk/water-jetting-pumps/ultra-high-pressure-plunger-pump/ (2015).
64. Peng, K., Tian, S., Li, G., Huang, Z. & Zhang, Z. Cavitation in water jet under high ambient pressure conditions. Exp. Therm. Fluid Sci. 89, 9–18 (2017).

2

Pyrolysis of Waste Tire Rubber towards High Value-Added Products

Zhen Zhang[a], **Lan Ma**[a], **Shuang Liu**[a], **Krzysztof Formela**[b] and **Shifeng Wang**[a]

[a]*Department of Polymer Science and Engineering, Shanghai Key Lab. of Electrical Insulation and Thermal Aging, Shanghai Jiao Tong University, No.800, Dong Chuan Rd. Shanghai, 200240, China.*
[b]*Department of Polymer Technology, Faculty of Chemistry, Gdańsk University of Technology, Gabriela Narutowicza 11/12, 80–233 Gdańsk, Poland.*

2.1 Introduction

Due to the covalently cross-linked network, it is particularly difficult in recycling of waste tire rubbers at the end of the service life, resulting in a large accumulation of waste rubbers [1-4]. Landfills of waste rubber occupy a great amount of land resources and it is easy there to breed mosquitoes to spread diseases, threatening both the environment and human health [5, 6]. To solve the problem, some researchers have tried different treatment methods, such as refurbishment, grinding and energy recovery [7, 8]. The energy recovery methods include incineration and pyrolysis processes. The direct incineration of waste tires can recycle part of the energy and replace non-renewable resources such as coal and petroleum [9]. However, it will release dioxins, polycyclic aromatic hydrocarbons (PAHs) and volatile toxic pollutants, which are harmful to the environment and human health. Compared with direct incineration, pyrolysis of waste rubber can transform these waste rubbers into high calorific gases, oils, chemicals and other valuable materials [3, 10, 11]. Moreover, pyrolysis is an efficient and eco-friendly process by reducing the waste tire environmental impact. Therefore, pyrolysis of waste tires has received more and more attention due to its unique advantages in the last decades.

The pyrolysis of waste rubber began in the 1970s. Nowadays, it develops into vacuum pyrolysis, catalytic pyrolysis, biomass pyrolysis, plasma pyrolysis and light pyrolysis [12-14]. Many types of reactors have also been developed in recent years, mainly including fluidized bed, fixed bed, rotary kiln, spouted bed and microwave reactors [15]. At the chemical reaction level, pyrolysis process can be regarded as the reverse process of polymerization, in which the polymer chains are continuously broken down into smaller molecules at high temperature. The small molecules ultimately evaporate and release from the reaction vessel, which can be directly used or condensed into liquid as fuel. The fuel has advantages in the high calorific value ranging from 35 to 45 MJ kg^{-1}, which is more than that of direct incineration of waste tires [16]. Moreover, the pyrolysis process can also separate carbon black (CB) of the tire from the volatile matters, which can be mainly added into new rubber products as additives to improve the material performance. The yield and quality of the pyrolytic CB strongly depends on the pyrolysis parameters, such as pyrolysis temperature, time, and rubber particle size.

Pyrolysis plays a key role in the treatment of waste tires, reducing environmental pollution and producing high value-added products. This chapter is divided into five parts. In the second part, it summarizes the main advances of pyrolysis technologies such as fluidized bed, fixed bed, rotary kiln, microwave reactor and other designs. The third part shows the types of reactors and the key parameters of the pyrolysis technologies. The fourth part discusses the obtained products from the pyrolysis of waste tire rubber in terms of their respective parameters (temperature, heating rate, gas flow rate and residual time, etc.). The fifth part analyzes the composition, characteristics and the potential applications of the pyrolytic materials. In addition, light pyrolysis process of waste rubber and the application of obtained products are described in detail. Finally, the advantages and disadvantages of the different pyrolysis methods are summarized and the future development is discussed.

2.2 Waste tire rubber pyrolysis technologies

The pyrolysis of waste rubber contains a lot of complicated reaction, such as primary degradation and secondary reactions of polymer. The pyrolysis behavior of waste rubber is different from other polymeric materials due to the different structures and compositions. The degree of primary and secondary degradations and the properties of the obtained product are related to pyrolysis parameters, such as temperature, heating rate, residence time and tire rubber particle size [12, 17, 18].

The key factor of rubber pyrolysis is the type of reactor, which has significant impact on the pyrolysis parameters. In fact, some pyrolysis conditions, such as temperature and heating rate, strongly depend on the reactor design. The common reactors are grouped into fixed bed, fluidized bed, rotary kiln, microwave reactor and other reactors. The main features of the pyrolysis reactors mentioned above are discussed in the following sections.

2.2.1 Fixed bed reactors

Fixed bed reactors (FBR) have been extensively applied in waste tire rubber pyrolysis, particularly in lab-scale [19, 20]. The main advantages of FBR lie in: (i) simple design and operation; (ii) less mechanical wear on the pyrolysis catalyst; (iii) residual time can be rigorously controlled and the temperature can be properly regulated, which play important role in achieving high selectivity and conversion rate of the resulting products [21, 22]. In this type of reactors, tire samples are often adjusted to a relatively slow heating speed (5 ~ 40 °C min^{-1}) and the gases can be cleaned by inert flow.

Laresgoiti et al. [23] carried out the waste tire pyrolysis experiments in the stainless steel FBR (3.5 dm^3) in the atmosphere of nitrogen. The pyrolysis temperature was separately set at 300, 400, 500, 600 and 700 °C while the heating rate was set as 15 °C min-1 and then the tests were carried out during 30 min. They found that the production of pyrolysis gas and liquid hardly changed at temperatures above 500 °C. Islam et al. [24] investigated the effect of the temperature, tire particles size and the gas residual time on the pyrolytic products in a continuous fixed bed reactor. The best liquid yield temperature of the fixed-bed fire-tube heating reactor system was set at 475 °C and feed size of the reactor was 4 cm^3 and gases residual time was set as 5 s. Cunliffe and Williams et al. [25] studied how the final temperature between 450 °C and 600 °C influenced on the products in a fixed batch reactor with a capacity of 3 kg to. The results showed that the oils involve quite a lot of polycyclic aromatic hydrocarbons (PAHs). Some of these gases proved to be either carcinogenic or mutagenic. The content of PAHs was increasing from 1.5 to 3.4 wt.% of oil when the reactor temperature varied from 450 °C to 600 °C. Mkhize et al. [26] studied the influence of the temperature and heating speed on the pyrolysis in a fixed bed reactor. The influence of temperature was obvious on the pyrolysis oil yield of waste tire, while the influence of both factors was obvious on the chemical structure of the pyrolysis oil.

However, the FBR has disadvantages in poor heat transfer rate and complicated handling parameters, particularly temperature controlling. The commercial process of the FBR is a significant challenge due to the poor heat transfer rates when the reactor become larger.

2.2.2 Fluidized bed reactors

The fluidized bed reactor was firstly put forward by Kaminsky from the University of Hamburg [27]. The advantages of this reactor lay in fast pyrolysis conditions, short volatile retention time and high heat transfer rates. These conditions ensures that the pyrolysis process is controlled and reproducible [28]. Moreover, this type of reactors may perform the continuous operation, which is significant to the scale expansion.

The main advances of the Kaminsky' research group is to obtain more pyrolysis monomers (ethylene, propylene, butadiene, isoprene, limonene and BTX (benzene, toluene, xylene, etc.)) and high-quality carbon black from tire rubber powder. Kaminsky et al. [27, 29] used fluidized bed reactors in the laboratory to conduct experiments at 1 Kg h^{-1}, 30 Kg h^{-1} and 200 Kg h^{-1} scales. In 2009, they found that the pyrolysis oil yield of 1 Kg h^{-1} was 30.2 wt.% at 740 °C. At 200 Kg h^{-1} pilot scale, the pyrolysis oil production of the entire tire was 26.8 wt.% at 700 °C. The oil production rate of steel-free tires was 30.9 wt.%, which is equivalent to that of laboratory reactors. In addition, the entire tire was indirectly heated in a fluidized bed to a typical temperature of 500 ~ 780 °C in a large-scale, in which the pyrolysis gas combustion provides energy. The researchers got rid of ash accumulated on the fluidized bed using a screw extruder and the gas enters through a cyclone to get rid of some matter for storage and pyrolysis. At last, the oil was condensed into two fractions, heavy oil and light oil. Other contribution of this research group is the application of catalysts to pyrolyze waste rubber in situ. Dai et al. [30, 31] investigated the continuous pyrolysis of waste rubber in a circulating fluidized bed reactor with a capacity of 5 Kg h^{-1}. This research mainly focused on the effect of particle size (0.32 mm and 0.8 mm), temperature (360 - 810 °C) and gas residual time (1 - 5 s) on the final products.

The advantage of the fluidized bed is the large flow-solid phase boundary, which is conducive to heterogeneous reactions. The reactor improves the utilization rate of the catalyst and can realize continuous pyrolysis. However, the disadvantages are that there is no difference of temperature and concentration in the axial direction of the bed and the gas flow may go through the bed at fast speed, resulting in the poor contact with the solid. Thus, the conversion rate of the fixed bed is often higher than the fluidized bed. In addition, violent collisions occur between the catalyst particles, increasing the depletion of the catalyst and the challenge of ash removal. The sand as the heat-carrier cause serious wear on the equipment, which limits its production scale to a certain extent. These collisions also cause abrasion to the pipes and containers.

Compared with the tubular furnace in petroleum pyrolysis, the fluidized bed consumes more energy. In addition, the design and operation of the reactors are complex and require more investment than other technologies, which hinders the further application of the technology.

2.2.3 Rotary kilns

The rotary kiln reactors are divided into external heating type (indirect heating) and internal heating type (direct contact heating) [32]. The external heating pyrolysis process has a large pyrolysis oil yield, high calorific value and better quality of carbon black. In addition, less polluting emissions can be realized compared with internal thermal pyrolysis. Therefore, external heating pyrolysis processes is widely used in the industry.

Recently, Antoniou and Zabaniotou et al. [33] designed a rotary kiln reactor capable of 100 Kg h-1. The continuous pyrolysis was conducted at 550 °C for 100 h. The yields of different products achieved 14.5 ± 2 wt.%, 37.5 ± 1 wt.% and 48.0 ± 2 wt.%, respectively. Ayanoglu and Yumrutas et al. [34, 35] investigated the pyrolysis process of waste tires in a rotary kiln reactor with a scale of 4 tons/batch at 400 °C. The waste tire rubber was pyrolyzed in a rotary kiln reactor, in order to obtain more gas, light liquid, heavy liquid, wax products, and less carbon black. The yields of these products were 25%, 12%, 8%, 35% and 20% of the total weight, respectively. The final analysis, such as density, viscosity, heating value, FTIR feature peak of the obtained heavy oil and light oil were similar to the commercial fuels. However, the sulfur content was very high, and the temperature of flash point was low, which caused a serious problem. Syamsiro et al. [36] investigated the effect of speed on the pyrolysis of waste tires in the laboratory. The rotational rate and lifter greatly affected the product distribution and oil properties. The reactors without lifter showed that the highest liquid product was achieved at the rotational speed of 15 rpm. However, the liquid materials obtained in this reactor was relatively low in these conditions. The application of lifter decreased the yield of liquid and solid materials when the gaseous product increased. The properties of the oil were still approval for application in the engine. Li et al. [37] from Tsinghua University and Zhejiang University studied the continuous pyrolysis of waste rubber by a rotary kiln on a pilot scale. The diameter of the rotary kiln is 0.3 m and the total length is 3.0 m. By adjusting and changing the rotation rate of the rotary kiln, they controlled the residence time of the pyrolysis reaction. The kiln was under negative pressure (-20 ~ 10 Pa) and the pyrolysis reaction temperature was between 450 and 650 °C. The production of pyrolysis products is mainly analyzed in the physical and chemical properties of the oil and the surface area of the pyrolytic CB. Luo and Feng et al. [38] developed a continuous rotary furnace (200 g min^{-1}) pyrolysis process using blast furnace slag as the heat carrier and catalyst. This technology upgraded industrial by-products into heat carriers and catalysts, which was a sustainable process.

Compared with fluidized bed and fixed bed pyrolysis processes, the rotary kiln reactor design and structure are simple. Continuous rotation of the drum can elevate the blending of materials and improve the mass transfer and heat-transfer efficiency. Moreover, the residence time is easy to control by changing the pace of rotary kiln and continuous feeding is conducive to the progress of the pyrolysis reaction. This type of reactor is suitable for most of solid waste material due to the wide adaptability to the shape and size of the waste material. The rotation of pyrolysis enables the materials to achieve better mixing, which makes uniform pyrolysis char. However, the main disadvantages lie in poor thermal-transfer efficiency and more raw materials and energy-consuming.

2.2.4 Microwave reactors

Microwave is an electromagnetic wave with a wavelength between 0.001 and 1 m and a frequency between 0.3 and 300 GHz [39-42]. Microwaves with frequency of 2.45 (±0.05) GHz are widely used nowadays. Rubber does not have the ability to absorb microwaves and convert them into heat, and thus it is necessary to use some microwave absorbing substances together with the rubber. The tire rubbers composed of carbon black as high as 30% and a certain amount of metal wire, are good microwave absorbing materials. It is reported that carbon black can reached 1200 °C for only 120 s, when the input energy is 600 W. Therefore, the microwave pyrolysis of waste rubbers has unique superiority [43].

Song et al. [44] investigated the microwave pyrolysis in 30 g scale. Their research concentrated on the effect of operating parameters (for instance specific power) on the chemical component of cracked oil and gas materials. The influence of the steel wires and the promotion of emissions on microwave technology of waste tire rubber was also analyzed. The major compositions of the pyrolytic gases were H2, CH4, and C2H4. The liquid materials were obtained as terms of the blends of organic chemicals with large parts of aromatics and limonene. The solid residues with high proportion of carbon black and sulfur were almost obtained at 20 min, and the rubbers in the center of the reactor were pyrolyzed earlier than those in another place. At last, about 43% of the solid residues, 45% of the liquid oils and 12% of the pyrolytic gases were achieved. Undri et al. [45] proposed a new composite parameter P/M^2 when studying microwave cracking liquid products. P stands for microwave power and M stands for mass per unit area. The study found that the viscosity, density and calorific value of the liquid pyrolysis product dropped sharply when the value of P/M^2 lowered. The quality of the liquid product increased after condensation to 473 K. They also explored the influence of microwave pyrolysis of the pyrolytic CB [46]. When the value of P/M^2 is lower than 55.1×10-3 W/g^2, hydrogen content of the carbon black obtained from microwave pyrolysis is higher than 1 %. The Brunauer-Emmett-Teller (BET) of specific surface area of the pyrolytic CB after microwave pyrolysis is higher and the steel mesh in the tire form the metal sulfide ZnS with the sulfur in the rubber during the pyrolysis. Frediani et al. [45] studied the decomposition of waste tires by microwave with a frequency of 2.45 GHz and a maximum input power of 6 kW. They paid close importance to the effect of pyrolysis conditions on the properties of pyrolysis oil and found that the best pyrolysis conditions can be reached when the input energy is 3 kW and the mass of pyrolysis product is 0.2 kg.

The interesting things about microwave pyrolysis are that it enhances the heat transfer to the tire particles and allows reactors under the same temperature, relating to the uniform heat distribution [47-49]. Compared with thermal pyrolysis, microwave pyrolysis has the following advantages: (i) fast heating speed; (ii) energy-saving and high efficiency; (iii) low temperature. The degree of microwave pyrolysis at 350 °C is equivalent to that under ordinary thermal pyrolysis conditions at 500 °C.

2.2.5 Other reactor designs

Other pyrolysis reactor designs for tire pyrolysis have also been studied including spouted bed, vacuum pyrolysis, dropper, steel wire mesh and other rapid heating reactors [50-53].

The application of conical spouted bed (CSBR) for waste tire pyrolysis is proposed by Martin [54]. Its advantage lies in the conical geometric structure of CSBR, which has superior hydrodynamic characteristics. Different carrier-gas flow rates can keep stable in the CSBR reactor. Though rubber is a poor thermal conductor, the violent circular motion between the solid particles in the CSBR improves the heat and mass transfer rates in different phases. Therefore, heat-energy utilization rate of the CSBR is high and the occurrence of secondary pyrolysis reactions can be reduced at the same time, and thus the pyrolysis oil yield is higher. Olazar et al. [55] developed a conical spouted bed reactor in order to investigate the pyrolysis of tire rubbers and found that the pyrolysis of solid particles produced an 62 wt.% oil yield, char production rate of 35 wt.%, and 3 wt.% gas production. At the same time, they studied the application of catalysts to test the influence on the components of oil and gas. They also studied spouted bed reactor under vacuum pyrolysis parameters [56]. The vacuum conditions have an effect on increasing the yield of diesel fraction and isoprene [57]. The main influence of vacuum reactor is the increasing yield of the liquid product. A highest yield of isoprene has reached higher than 7 wt.% under vacuum. Conesa et al. [58] studied the pyrolysis of waste rubber granules (4 mm) in a dropper pyrolysis reactor. The tire particles were fed into a reactor at 450, 750 or 1000 °C in 5 g batches with a height of 65.4 cm. It was found that the increase of temperature significantly reduced the yield of the pyrolytic oil from 37.8 wt.% at 450 °C to 10.9 wt.% at 750 °C. Zabaniotou et al. [15] placed about 200 mg of ground tire rubber (0.18 ~ 1.0 mm) on a steel wire mesh, and the current passed through the mesh quickly increased the temperature of the mesh (70 - 90 °C/s) to a temperature between 390 ~ 890 °C. It was found that the pyrolytic oil yield was lower (5 wt.%) and the gas production varied from 22 wt.% at 450 °C to 73 wt.% at 860 °C. Recently, Mkhize et al. [59] investigated the effect of residual time and heating rate on product yield and its components in different reactors, such as fluidized bed reactor, fixed bed and a conical spouted bed reactor.

There are very few studies on the catalytic pyrolysis of tires in fluidized bed reactors. Williams and Brindle et al. [60] have investigated the pyrolysis of tire catalysts in continuous fed bench-scale fluidized bed reaction vessel. The oil yield decreased while an increase of the gas yield occurred in the presence of the catalyst. The chemicals in the oils, such as benzene, toluene and xylenes, significantly increased when using both catalysts., Although surface acidities and pore sizes of the two catalysts were different, similar proportions of benzene, toluene, xylene and limonene were produced. Li et al. [131] studied the catalytic influence of biochar catalyst on the pyrolysis of waste rubbers. They found great influence of the biochar catalyst on the changing process of gas products, liquid products and properties of the solid residual, while the total yields of these materials were not kept the same as before. The biochar produced from the gasification of poplar wood is a catalyst candidate for the pyrolysis of tires. The continuous development of new pyrolysis reactors overcomes the problems of low heat utilization, slow heating rate and reduction of secondary reactions existing in large-scale reactors, but they are still in the laboratory stage and need to be further explored for industrial applications.

2.3 Factors influencing tire pyrolysis

Tires are mainly contained rubber (60~65 wt.%), carbon black (25~35 wt.%) and zinc oxide, sulfur and other organic components. After thermal pyrolysis, rubber and other organic components are cracked into oil and combustible gas. The residual solid substances after pyrolysis includes the pyrolytic CB, inorganic ash and steel wire [61].

Pyrolysis temperature and pressure are two important parameters affecting pyrolysis products [62, 63]. Generally, waste tires start to decompose at 290 °C and ends at about 570 °C [64]. The generated pyrolysis gas will be quickly removed by vacuum pump, which avoids the subsequent reactions of pyrolysis gas. The higher pyrolysis temperature can hardly effect affect the yield of pyrolysis products [65]. However, elevated temperature enhances secondary reactions under atmospheric pressure. The rubber hydrocarbons will crack into small molecular gaseous hydrocarbons, which will be deposit on the surface of the pyrolytic CB and will cover the active points and holes of the CB and leads to the reduced value of the pyrolytic CB. With the increasing pyrolysis temperature, the yield of oil decrease while the yield of gas and carbon black increase. The surface chemical properties of the pyrolytic CB obtained under vacuum or reduced pressure are closer to the commercial carbon black used as filler in the rubber [15]. In addition, the carrier gas and residence time will also have an important impact on pyrolysis and product composition [26].

Therefore, this part of the manuscript will introduce the following important factors affecting pyrolysis: (1) the type of composition and particle size of tires; (2) temperature; (3) stress; (4) carrier gas and the residence time.

2.3.1 Effect of tire type composition and particle size

Different kinds of tires have different compositions and characteristics of pyrolytic products due to their different compositions and processing methods. the tires can be generally classified into truck tires, car tires and motorcycle tires according to the different contents of natural rubber and synthetic rubber [69].

Ucar et al. [70, 71] measured different rubber contents in waste truck and car tires. The result showed that truck tires contained 51 (wt.) % natural rubber (NR), 39 (wt.)% styrene-butadiene rubber (SBR) and 10 (wt.) % polybutadiene rubber (BR). However, car tires consisted of 35(wt.) % NR and 65 (wt.) % BR. Kyari et al. [72] found that components and concentration in pyrolysis oil and gas have a big difference in the different tires. Islam et al. [24] made industrial and elemental analysis on motorcycle tires and automobile tires. They found that there are several differences in elemental composition and component contents between them. Motorcycle tires have less volatile components and much ash content, which has a direct impact on the yield properties and yield of pyrolysis oil, gas and carbon.

The main types of rubber contained in tires are SBR, NR and BR [73-75]. Seidelt et al. [76] studied the thermal decomposition of different types of rubbers by using a coupled TG-MS analysis. They determined the correlation of the composition with thermal degradation temperature of rubber. The main degradation products of NR are xylene and isoprene dimer, while the main degradation products of SBR are ethylbenzene, styrene and cumene.

2.3.2 Effect of temperature

Temperature is the main factor influencing pyrolysis reactions [77-79]. Generally, increasing the temperature will promote the pyrolysis reaction. Kim et al. [80] reported that the increase of temperature benefit the further pyrolysis of liquid components when waste rubber was cracked in the fixed bed. When the pyrolysis temperature reached 500 °C, the liquid and gas phase yields were 38.3 wt.% and 22.6 wt.%, respectively. When the cracking temperature rises to 800 °C, the pyrolysis oil components decrease to 29.8 wt.%, while the pyrolysis gas components increased to 30.1 wt.%[81]. Li et al. [82] found that the liquid yield was decreased while the gas yield was increased through rotary kiln pyrolysis process when temperature increased from 550 to 650 °C. Pyrolysis temperature is beneficial to the composition of oil [83].

Rodriguez et al. [84, 85] reported that the yield of oil and gas was nearly unchanged when the temperature exceeded 500 °C in autoclave. As temperature increases, waste rubber can be completely cracked at 550 °C. The increase of temperature enhances secondary cracking of pyrolysis vapors, which is favorable to the gas formation rather than liquid hydrocarbons.

The most appropriate temperature range for increasing the yield of cracked oil is 450-550 °C in thermal pyrolysis [86]. At lower temperature, the conversion/removal of compounds for three main types of rubber is not completely because they remain in the char (mainly consisted of carbon black and inorganic substances) and has the properties of non-uniform adhesive. However, higher temperature enhances the secondary reactions of pyrolysis steam, which is more conducive to the formation of gas hydrocarbons than liquid ones [87]. Moreover, the residence time of primary pyrolysis products in the reactor has a great influence on the reaction degree.

The main rubber hydrocarbons (NR, SBR, BR) remain in the pyrolytic carbon when decomposed at lower temperature [88]. Therefore, low temperature makes the surface of the pyrolytic carbon black to have an uneven sticky morphology. As the char content varies with pyrolysis temperature, the results show that 450~550 °C is the most proper temperature range for research. The main tire compounds cannot be separated and they remain in the pyrolytic carbon black at poor pyrolysis temperature, resulting in low purity [89]. Furthermore, higher temperature hardly has an influence on the yield of pyrolytic char.

2.3.3 Influence of pressure

Although temperature is the main determinant of waste tires pyrolysis process, the final pyrolysis products are also affected by other factors, including heating rate, pressure and carrier gas flow rate. These facts play an important role in controlling the secondary reactions [90].

Bueken et al [91] studied on the pyrolysis of waste plastic and the results that the increase of pyrolysis pressure enhances dehydrogenation and char formation. The pyrolytic oil thus becomes more viscous. The residence time of volatile become shorter under vacuum pressure when compared with that under atmospheric pressure, leading fewer secondary reactions to occur. They found that carbonaceous deposits generated on the pyrolytic carbon black decreased. Thus, the improved surface of the pyrolytic carbon black and the increased number of surface-active sites can be obtained, which is closer to the characteristics of commercial carbon black.

Lopez et al. [50] found that vacuum pressure enhances the diffusion of volatile components on the surface between solid particles. The diffusion minimizes the generation of carbonaceous deposits and reduces the blockage of pore structure on the surface of the pyrolytic CB. Roy et al. [92] stated that the reduction of pyrolysis temperature is another major advantage of pyrolysis under vacuum. In addition, pyrolysis under vacuum condition reduces the use of condensing devices and slows down the flow rate of inert carrier gas, which makes the pyrolysis process more feasible [93].

2.3.4 Influence of carrier gas flow and residence time

The inert carrier gas is generally used to control secondary reactions such as repolymerization and char formation in the pyrolysis process [94]. The flow velocity of the carrier gas is closely related to the residence time of volatile matter released during pyrolysis [27]. Generally, the high-velocity carrier gas rapidly moves the volatiles out of the high temperature reaction zone, which limits secondary reactions. Cunliffe and Williams et al. [95] found that the yield of liquid components increased with the flow velocity of carrier gas during pyrolysis. Meanwhile, they stated that the pyrolysis steam had been removed from the reaction zone by N2 before carbonization. Islam et al. [24] found that the residence time of pyrolysis steam and the yield of pyrolysis gas increased while the output of pyrolysis oil and pyrolysis carbon decreased with the decrease of velocity of carrier gas flow.

Different carrier gas types, such as steam and pyrolysis gas, greatly influenced the composition and properties of the pyrolysis products [96]. Ogasawara et al. [97] studied water vapor and helium as carrier gases when the temperature was 900 °C. They found that the pyrolytic carbon blacks were only 9 wt.% but had a quite high surface area (1260 m2/g). The pyrolysis gas mainly consisted of H2 and CO while the pyrolysis oil mainly consisted of aliphatic hydrocarbons and alkylbenzene. However, thermal pyrolysis under H2 atmosphere greatly influenced on the composition of the liquid products. The free radicals produced in the pyrolysis process become more stable due to the addition of H2, which prevents the secondary reactions. The saturated components in the pyrolysis oil increased while the aromatic and polar compounds decreased. Moreover, the yield of oil increased during the process. Therefore, the use of non-inert carrier gas in the pyrolysis process leads to a series of homogeneous and heterogeneous reactions. Therefore, the composition and properties of the products were changed.

Roy et al. [92] have been engaged in the research on vacuum pyrolysis of waste tires for many years. In 1987, they established a pilot test of a small-scale treatment plant and pilot plant with a capacity up to 200 kg/h. The waste tire was cracked into large pieces and sent to the heating furnace. The temperature successively decreases from top to bottom with the top temperature controlled at about 510 °C. The combustion heat of gas and light pyrolysis oil produced in the process of pyrolysis supplies the heat required by cracking furnace. The mixture of pyrolysis gas and oil is separated from heavy and light oil by secondary cooling. The pyrolytic solid products fall into water to realize the direct separation of semi-coke, steel wire and curtain layer.

It is reported that pyrolysis under vacuum was characterized by short residence time of pyrolysis gas. The reason was that vacuum conditions can quickly discharge gas from the high temperature region of the reactor, thus avoiding secondary reactions. They also reported that vacuum conditions accelerated the diffusion of the products generated in the tire, because the pressure inside and outside of the particle was different. The time of secondary reactions was reduced due to the rapid diffusion of volatile gases.

The pyrolysis of waste rubber is a very complex process. The composition and properties of pyrolysis products are also greatly impacted by the particle size, temperature rise rate and cracking time [98]. The parameters of the pyrolysis process are also restricted by the characteristics of the reactor. Tire type determines the composition and properties of pyrolysis products. The increase of temperature will further promote the pyrolysis reactions. Therefore, the liquid component decreases and the gas component increases with rising pyrolysis temperature. With the increasing pressure, liquid product is more viscous and char tendency is improved [32]. Pyrolysis under vacuum reduces the residence time, therefore the secondary reactions is limited. In addition, the residence time of the vapors is closely related to the flow rate of gas. Secondary reactions will be reduced with the shorten of residence time [99].

In conclusion, pyrolysis process, pyrolysis parameters and the properties of raw materials affect the composition and properties of the pyrolysis products.

2.4 The products of tire pyrolysis and their applications

Waste tire pyrolysis is usually carried out in an inert atmosphere. The main products are gaseous products, pyrolytic oil, and pyrolytic carbon black. A detailed introduction of different pyrolytic products above are demonstrated in the following part.

2.4.1 Pyrolytic gas

The pyrolytic gas is mainly a mixture of alkanes and alkenes, and its main components are methane, ethane, ethylene, propane, acetylene, butane, butane, 1,3-butadiene, pentane, hydrogen, carbon monoxide, carbon dioxide and sulfur hydrogen etc. [87].

The composition and output of pyrolytic gas are greatly affected by raw materials, pyrolysis process and temperature [100, 101]. When the pyrolysis temperature increases, hydrocarbons with larger molecular weight undergo further pyrolysis and dehydrogenation reactions generate small molecular gases, such as methane and hydrogen etc. At present, most pyrolysis processes produce less pyrolytic gas but with a higher calorific value. The pyrolytic gases are often considered as fuel to provide energy for the treat of waste tires. Rodriguez et al. [102] used an autoclave to pyrolyze waste tires, and the resulting pyrolytic gas has a high calorific value (68~84 MJ/m3). Galvagno et al. [103] found that when the pyrolysis temperature is 550 °C, the pyrolytic gas has a heat value of 22 MJ/kg, which is 29

MJ/kg under 680 °C. Kaminsky et al. [92] used pyrolysis gas to provide energy for the pyrolysis process and built a semi-self-sufficient energy supply system for the process. Roy et al. obtained less than 10 wt.% pyrolytic gas by vacuum moving bed due to the short resident time of pyrolytic steam. The steam residence time in rotary kiln was longer, and the production of pyrolysis gas is 18 wt.% as a result. The gas contains valuable olefins such as ethylene and propylene. Thus, it plays a crucial role in the production of basic petrochemical raw materials. Therefore, it is possible to increase the content of pyrolytic gas and the content of ethylene and propylene in it by increasing the temperature and adding catalysts (HZSM-5, etc.) [104].

2.4.2 Pyrolytic oil

Pyrolytic oil is a kind of brown liquid mixture which has a relatively complex composition including paraffins, olefins and aromatic hydrocarbons [82]. Its calorific value is higher than direct combustion of waste tires, which is about 43 MJ/kg. Therefore, pyrolytic oil can be applied as alternatives for fuel [105, 106].

Pyrolytic oil is composed of different hydrocarbons, such as aliphatic, aromatic and heteroatom compounds [107]. Pyrolytic oil can be decomposed into a wide range of chemical distillates consisting of aliphatic, aromatic, heteroatom and polar distillates. Dai et al. reported the composition of pyrolytic oil obtained during tire pyrolysis in a circulating fluidized bed, in which alkanes accounted for 26.77 wt.%, aromatics accounted for 42.09 wt.%, non-hydrocarbons accounted for 26.64 wt.%, and asphalt accounted for 4.05 wt.% [40]. In addition, pyrolytic oil also contains high proportions of specific valuable chemicals. The advantage of the liquid fuel is its high calorific value, which is higher than that of the original tire.

Pyrolytic oil has high calorific value and can be used as fuel. However, aromatic compounds could be found in the pyrolytic oil because of a certain amount of styrene-butadiene rubber in waste tires. At the same time, the composition and properties of the pyrolytic oil are affected by the pyrolysis conditions, mainly by temperature. When temperature increases, the content of aromatic components in the pyrolytic oil will increase due to the secondary reaction of cyclization or aromatization occurring in the primary products of pyrolysis. According to Rodriguez's [85] analysis of the pyrolytic oil composition, it was found that the H/C of pyrolytic oil is 1.4 and aromatic compounds are the main component. This is because of the reaction of aliphatic and aromatic free radicals generated during the pyrolysis process, and also because the aliphatic chain segment underwent a cyclization reaction.

Pyrolysis is performed at the elevated temperature, which consumes high energy and has a certain impact on its industrialization. On the basis of detailed analysis of the pyrolysis products, they should be further processed and refined to obtain higher value products. In addition, with reference to petroleum cracking and coal gasification, the pyrolysis process conditions should be further optimized.

Comparing pyrolytic oil with the commercial diesel, the high sulfur content and low flash point of the pyrolytic oil also limit the application as fuel, in addition to the higher content of olefins and aromatics. Generally, the sulfur content of commercial diesel is less than 0.05 wt.%, while the sulfur content of pyrolytic oil is around 1.0 wt.%. The flash point of pyrolytic oil is generally lower than 30 °C. For example, the flash point of the pyrolysis oil obtained in vacuum pyrolysis is 22-28 °C, and the flash point of oil obtained in the rotary kiln is 13.5-30 °C. The reason for lower flash point of pyrolytic oil is the more complex composition and the content of low-boiling hydrocarbons in it. The lower the flash point is, the lower its safety is. As a results, the low flash point brought pyrolytic oil storage difficulty. Therefore, the pyrolytic oil must be further refined (desulfurization, vacuum distillation, etc.) to improve its fuel characteristics. The treated pyrolytic oil can be used together with commercial diesel, or it can be further processed into light fuel oil or petrochemical products through catalytic pyrolysis.

Different pyrolytic oil optimization processes have been proposed in the literature, and hydrotreating and fluid catalytic cracking (FCC) are considered to be feasible and stable routes. Pedro et al. hydrotreated the pyrolytic oil using NiMo as catalyst and found that the sulfur content was reduced from 11600 mg/L to 1000~1500 mg/L after being hydrotreated, which was reduced by 92 % [108].

2.4.3 Pyrolytic char

As a reinforcing filler of rubbers, carbon black takes up 25-30 % of the rubber weight. Therefore, the recycling of carbon black has great commercial value, which will bring considerable benefits to the pyrolysis industry [10, 109-111]. Pyrolytic carbon black is widely used in two aspects: one way is often used as a reinforcing filler or pigment in rubber products, in which it requires that the pyrolytic carbon black have suitable particle size and surface chemical properties to ensure the bonding strength between carbon black and rubber polymers. The other way is used as activated carbon, which is activated and modified by carbon dioxide or water vapor at a higher temperature from pyrolytic carbon black.

So far, the solid product obtained from pyrolysis is mainly the pyrolytic carbon black [112, 113]. Darmstadt et al. [114] compared the composition and structure of commercial CB with the pyrolytic CB produced at a temperature of 500 to 700 °C and pressure of 0.2 to 20 kPa. The main structure of these carbon blacks was almost same, which was characterized by 13C-NMR (carbon-13 nuclear magnetic resonance spectroscopy) and XRD (X-Ray Diffractometer). Meanwhile, the component analysis and elemental analysis of the pyrolytic carbon black proved that carbonaceous deposits content is relatively high on its surface (9-14 wt. %, mainly including Zn, Ca, Si), while commercial carbon black usually contains ash content of less than 0.7 wt.% [115]. This detail distinguishes the performance of the pyrolytic carbon black and the commercial carbon black. The structure and performance of the pyrolytic CB are affected by the composition of raw materials and pyrolysis conditions. However, the carbonaceous deposits cover a certain number of active sites on the surface, which greatly affects the performance of the pyrolytic carbon black as rubber reinforcing filler.

Roy et al. [92] found that the yield of the pyrolytic carbon black lies on the pressure of the reactor by comparing the yield of the pyrolytic CB produced under different pressure. The pyrolytic carbon black obtained under vacuum was 36.5 wt.%, which was 4 % lower than that under normal atmospheric pressure [92]. This may be attributed to the short residual time of the primary pyrolytic materials in the system under vacuum. Pyrolysis stream was space drawn out of the reactor under vacuum conditions, which reduced the probability of carbonaceous deposits generation. Some hydrocarbons were adsorbed on the surface during the pyrolysis process and polymerization, or condensation reactions occurred, which eventually resulted in the generation of larger molecular carbonaceous deposits on the surface. Therefore, a certain number of carbonaceous deposits inevitably covered the surface of carbon black under pressure. At the same time, they also analyzed surface chemical properties of the pyrolytic carbon black. On the ESCA (electron spectroscopy for chemical analysis) spectrum, the C1S element corresponds to six peaks:

graphitized peak (C0), aliphatic and small aromatic ring compounds peak (C1), three different oxidation valence peaks (C2, C3, C4) and the cytoplasmic cluster peak (C5). The C1 peak appeared on the ESCA spectrum of the pyrolytic CB, while the commercial CB mainly has C0 and C4 peaks, which further showed that harmful deposits exist on the surface of pyrolytic CB during the pyrolysis process.

Lopez et al. [50] used CSBR to pyrolyse waste tires (the carbon black N220 used as fillers in their research) and found that pyrolytic carbon black has a higher sulfur content (about 3 wt. %), which limited its reuse as rubber reinforcing fillers. CSBR is a type of fast pyrolysis reactor with short resident time as the steam can quickly diffuse and reduce deposition of carbonaceous deposits. The BET of pyrolytic carbon black is 46.5 m^2/g at a low pyrolysis temperature of 425 °C, while the specific surface area significantly increased to 116.3 m^2/g at a higher temperature of 600 °C. This may be due to the fewer hydrocarbons that were adsorbed on the surface of the carbon black when the temperature increased. Frediani et al. [46] used microwaves to pyrolyse waste tires. Compared with other methods, the pyrolytic carbon black obtained from microwave reactor has less ash content (less than 8 wt.%) and has a BET of 65 m^2/g.

Based on the analysis above, the pyrolytic carbon black cannot be directly applied for reinforcing filler of new rubber products. The reuse of the pyrolytic carbon black as rubber reinforcing filler requires pretreatment. The quality of pyrolytic carbon black can be improved by controlling the pyrolysis temperature and other conditions to reduce carbonaceous deposits. For example, pyrolysis process under vacuum conditions and proper temperature (450-550 °C) can significantly improve the quality of the pyrolytic carbon black. The main compounds of waste tire cannot be separated and remain on the surface of the pyrolytic CB under lower temperature, resulting in low purity. As the heating rate increased, the carbonaceous deposits can be reduced on the surface of the pyrolytic CB [116]. In 1997, Roy et al. proposed that the rubber pyrolysis technology under low pressure improved the value of pyrolytic CB [92]. They found that the surface chemical structures of pyrolytic carbon black became similar to the commercial CB as the pressure decreased. Moreover, it was reported that the BET of the pyrolytic carbon black under vacuum can reach 68 m^2/g and its performance is between N550 and N600 [93].

On the other hand, the pyrolytic carbon black can be modified through post-treatment due to the higher inorganic ash content. Kim et al. [117] found that the inorganic ash content of the pyrolytic carbon black after acid washing dropped below 3 wt.% and carboxyl functional groups increased on the surface, which could improve the dispersion ability of the pyrolytic carbon black in water. However, the carbon black obtained by oxidization needs to be wiped off twice by water in the industry, which caused the problem of a large amount of wastewater.

In recent decades, the recycling of waste rubbers to prepare activated carbon has also been widely studied [118]. The activated carbon obtained from waste tires has also been used in many studies to remove organic pollutants and toxic metals [119-121]. The study of activated carbon from tire pyrolysis in laboratory-scale has proved that steam activated carbon has a good surface area, excellent adsorption of certain harmful chemicals.

The pyrolytic CB with high carbon content has a preliminary porous structure, which can be properly treated for activation. In 1987 [122], some researchers developed a method to produce activated carbon by thermal pyrolysis of waste tire rubbers. In 1996, the activated carbon was obtained from waste tires and then was used for removing methane and SO2. Lehmann et al. [123] also carried out a similar experiment in which the inactivated tire char adsorbed SO2 for cleaning gas.

So far, the application and technologies of pyrolytic carbon black as activated carbon have been developed. Trubetskaya et al. [124] confirmed that a new method of using tire pyrolytic carbon black is to remove toxic chemicals. It was reported that the pyrolytic carbon black can remove phenol and chlorine with a filtration efficiency of 95%, indicating that the pyrolytic carbon black was a promising substitute for wastewater cleaning. Zandi-Atashbar et al. [125] designed a magnetic spinel Fe_2CuO_4/reduced graphene oxide nanoparticle, which was applied for catalyst for the pyrolysis of waste tires. This material can help decompose waste tires into high-value chemicals and effectively remove mercury ions from wastewater. Generally, the BET surface of the activated carbon under the parameters is about 500~2000 m2/g. Under proper activation by acid-washing, the pyrolytic carbon black has about 1000 m^2/g BET surface area and it can effectively absorb mineral salts and tar, thus improving its adsorption performance. These studies [124,125] showed that the application of appropriate activation and modification methods to activate and modify tire pyrolysis carbon is beneficial to transform the pyrolytic carbon black into high prospect activated carbon, by increasing the surface area of the carbon.

Li et al. [82] conducted an activation study on the pyrolytic carbon black obtained from the rotary kiln. By analyzing the pore characteristics of the pyrolytic carbon black and comparing it with commercial activated carbon, they found that the pyrolytic carbon black had less microporosity and had a relatively rich mesoporous structure. After being activated at 950 °C for 4 hours under CO_2 atmosphere, the BET of the activated pyrolytic carbon black reached 306 m^2/g. Thus, activated pyrolytic carbon black can be used as mesoporous activated carbon to adsorb larger molecules in solution.

Lopez et al. [55] studied the effects of different temperatures and activation times on the properties of the pyrolytic carbon black under the condition of water vapor as the activator. The reason for using water vapor as the activator is that it is more active than CO_2. The experimental results showed that the BET of activated pyrolysis carbon black exceeded 500 m^2/g at different activation temperatures of 850 °C and 900 °C. By this way, the activated pyrolytic carbon black had a larger surface area. It is found that the activated pyrolytic carbon black had more mesoporous structure and fewer micropores by the analysis of N2 adsorption curve test, which proved that the activated pyrolytic carbon black is only suitable for adsorbing larger molecular compounds. After being activated, the pyrolytic carbon black had a significant increase in surface area, and its sulfur content was also markedly reduced (less than 1 wt.%). This is beneficial to the industrial application of the pyrolytic carbon black.

Compared with the commercial CB (e.g. N330), the pyrolytic CB usually has lower specific surface area and higher ash depositions, which have negative influence on reinforcing properties to the rubbers. These defects of the pyrolytic CB prevent its large application in the markets. To develop the better properties of the pyrolytic CB, some new methods have been developed. For instance, microwave irradiation has the advantages of fast heating and low heating cost. The pyrolytic CB produced from microwave irradiation has a small ash content in the range of 5.9-9.5 wt.% [48]. Although this method lower production costs in the pyrolysis process, the performance of the pyrolytic CB still cannot meet the standard of commercial carbon black. In addition, there are some difficulties in removing carbon black from chemically cross-linked structures of tire rubbers.

Although the pyrolytic CB contains a high carbon element (>80 wt. %), the performance gap between the pyrolytic CB and the commercial CB is large. The ash content, particle size, chemical composition and surface chemical properties of the pyrolytic CB limit its further application. The pyrolytic CB can be subjected to post-modification treatment such as acid-washing activation, but the energy consumption and the cost are relatively high. Therefore, much research needs to be carried out to overcome these drawbacks in the current pyrolysis process.

2.5 Light pyrolysis of tire rubber

The technology between reclamation and conventional pyrolysis is regarded as light pyrolysis. The cross-linked network of the polymer can be quickly broken down at the elevated temperature during the light pyrolysis process.

2.5.1 Structure evolution of the light pyrolysis

Cheng et al. [126] found that the liquefaction of waste rubber can be performed at a low temperature to decrease the generation of secondary pollutants and obtain liquid rubber with a high percent of sol parts (68.34 wt.%). The carbon black removed from the reactor is aggregated in a size of about 200 nm, and the outer surface of the carbon black is covered with a thin layer of rubber with a size of 2–10 nm, which shows a clear core-shell structure. This mentioned behavior significantly affects the application of the materials.

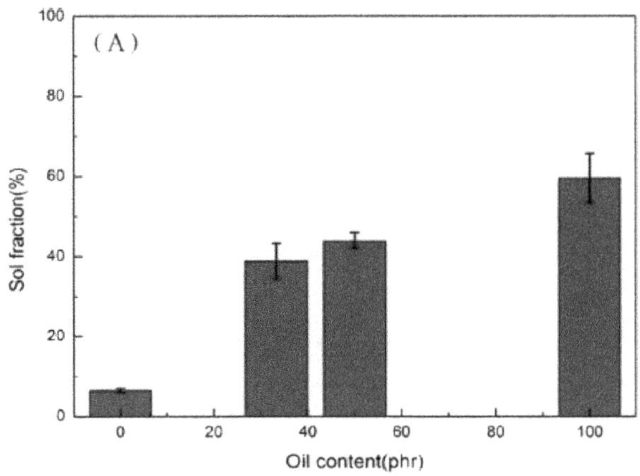

Figure 2.1 *Sol fraction of ground tire rubber (GTR) after liquefaction (Reproduced from Reference 126 with permission from Elsevier).*

Li et al. [127] separated lightly pyrolytic CB from GTR by reactive melt-extrusion process and analyzed the surface chemical composition of the commercial CB (N330), the pyrolytic CB and the light pyrolytic CB by X-ray photoelectron spectroscopy (XPS). They found that the elements C, N and S on the outer surface of the light pyrolytic CB were higher than that of N330. However, the percent of O were lower than N330, meaning that an organic layer was sticked on the surface of the light pyrolytic CB. The bound rubber on the surface of the light pyrolytic CB decreases the expression of O element, while the existence of cross-linking agents and accelerators in the rubber offer the elements of S and N.

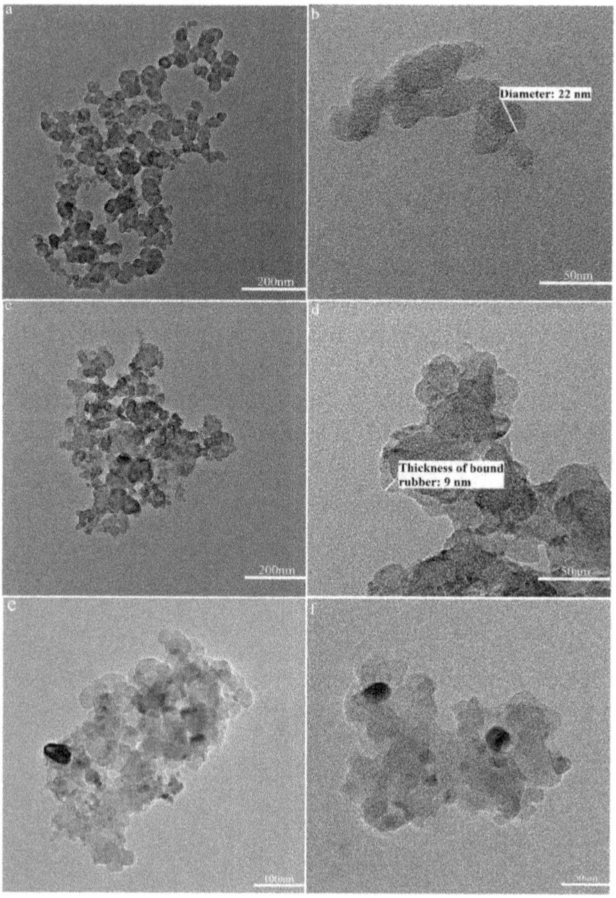

Figure 2.2 *TEM of (a-b) N330, (c-d) the light pyrolytic carbon black and (e-f) the pyrolytic carbon black (Reproduced from Reference 127 with permission from Elsevier).*

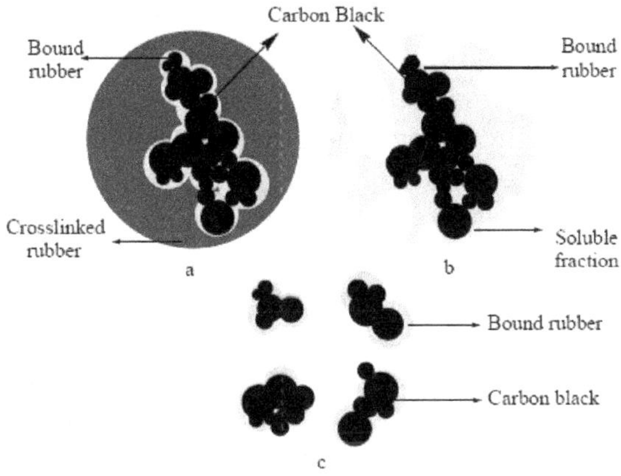

Figure 2.3 *The removal step of carbon black and polymer in light pyrolysis: (a) GTR; (b) lightly pyrolytic rubber; (c) the light pyrolytic carbon black (Reproduced from Reference 127 with permission from Elsevier).*

Compared with the other two CB, the light pyrolysis of tire rubber gives higher percent of N and S on the surface of the light pyrolytic CB from the results obtained by XRD. The light pyrolytic carbon black has the highest concentration of C-C or C-H among these samples, which means that rubber polymer is coated on the surface of the light pyrolytic CB. The characteristic peak of COOH is not found in the commercial carbon black, which may be broken down during the pyrolysis at high temperature. The presence of COOH functional groups makes the light pyrolytic CB more reactive than the commercial carbon black. It was an interesting phenomenon that the oxide peak is only found in the pyrolytic CB and absent in N330 and the light pyrolytic carbon black. The peak intensity of C-O in the pyrolytic carbon black was the highest among these samples due to the oxidative reaction of the pyrolytic carbon. The content of C=O to C-O on the surface of the light pyrolytic carbon black was higher than N330. Moreover, the shake-up peak was not found in the spectrum of N330 and the light pyrolytic CB, meaning that there were no free oxygen molecules close to the aromatic system on the other CB samples. This phenomenon also certified that the chemical structure of the light pyrolytic carbon black is more similar to N330 rather than the pyrolytic carbon black. The average primary particle size of N330, which can be observed by TEM (Transmission Electron Microscope), was estimated to be 21 nm, and the smaller size resulted in higher specific surface area. The presence of bound rubber on the outer surface of the light pyrolytic CB increased the particle size to approximately 29 nm. The particle size and polydispersity index (PDI) of the commercial CB was higher than that of N330 and the light pyrolytic CB because some organic gases released during the pyrolysis process and converted into ash and covered on the surface of the pyrolytic carbon black. The BET surface area of the light pyrolytic carbon black was lower than the pyrolytic carbon black because the the rubber polymer covered on the surface. The cross-linked network was broken down in the pyrolysis process of vulcanized rubber. The interfacial interactions between rubber chains and carbon blacks produces a thin layer of bound rubber at the interface, which further strengthened during the vulcanization reaction. The highly cross-linked bound rubber layers are thermally stable under the treatment at 300 °C.

Song et al. [128] provided a layered exfoliation method to break the cross-linked network and separate bound rubber and encapsulated CB particles from the vulcanized rubber. By this method, soybean oil swelled the rubber network under mild conditions and chain scission of the rubber occurred. A simple method was used to separate bound rubber encapsulated CB particles from the vulcanizates by layer-by-layer. A high sol fraction was obtained after the solvent extraction process because of the high breakdown degree of network. The CB covered bound rubber was subsequently obtained from the degraded vulcanizates. This process makes it possible to exfoliate the bound rubber from the CB surface by layer-by-layer. The gel mainly contained carbon black and a thin layer of loosely bound rubber was found on the surface of the CB. In addition, the bound rubber layer was rougher than that of rubber due to the inhomogeneous degradation. The CB showed similar structure as the original N330, but a thin layer of tightly bound rubber was on the surface. The S, N and Zn elements were observed in the bound rubber. The ZnO was also observed, and the percent of different sulfur bonds was higher than that of mono-sulfide in loosely or tightly bound rubber.

Zhang et al. [129] investigated thermo-oxidative reaction by using a self-designed dynamic recycling reactor. Without any chemical agent, a high sol fraction was obtained in 66.53 wt. % at 200 °C after 20 min from the thermal-oxidative reaction of tire rubber powder. It was concluded that the degraded ground tire rubber containing fillers and sol fraction was easily dispersed into natural rubber matrix.

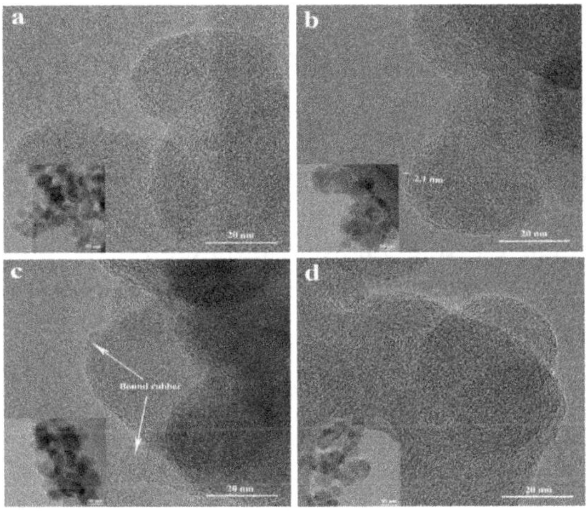

Figure 2.4 *TEM of carbon black surface exfoliated from (a) N330; (b) rubber treated for 4 h; (c) vulcanised rubber treated for 2 h; (d) rubber vulcanizate treated for 4 h (Reproduced from Reference 128 with permission from Elsevier).*

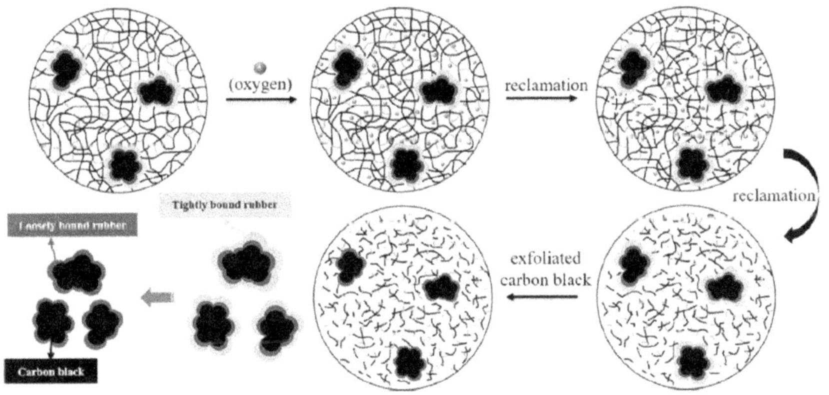

Figure 2.5 *Flow diagram of the effect of oxygen in assisting the dynamic thermal-oxidative reaction. Reproduced from Reference 129 with permission from American Chemical Society).*

2.5.2 Application of the lightly pyrolyzed products

Wu et al. [130] found that waste rubber was lightly pyrolyzed into a liquid rubber by screw extrusion. The liquid rubber obtained at 300

°C (LPR300) significantly promoted the low-temperature rheological properties with the increase of LPR300. LPR with high sol content has potential to be a good alternative of asphalts. The liquid rubber obtained at different reaction temperatures were characterized by gel permeation chromatography, thermogravimetric analysis and extraction analysis. The rheological properties of LPRs and their mixture with asphalt were evaluated by dynamic rheological analysis.

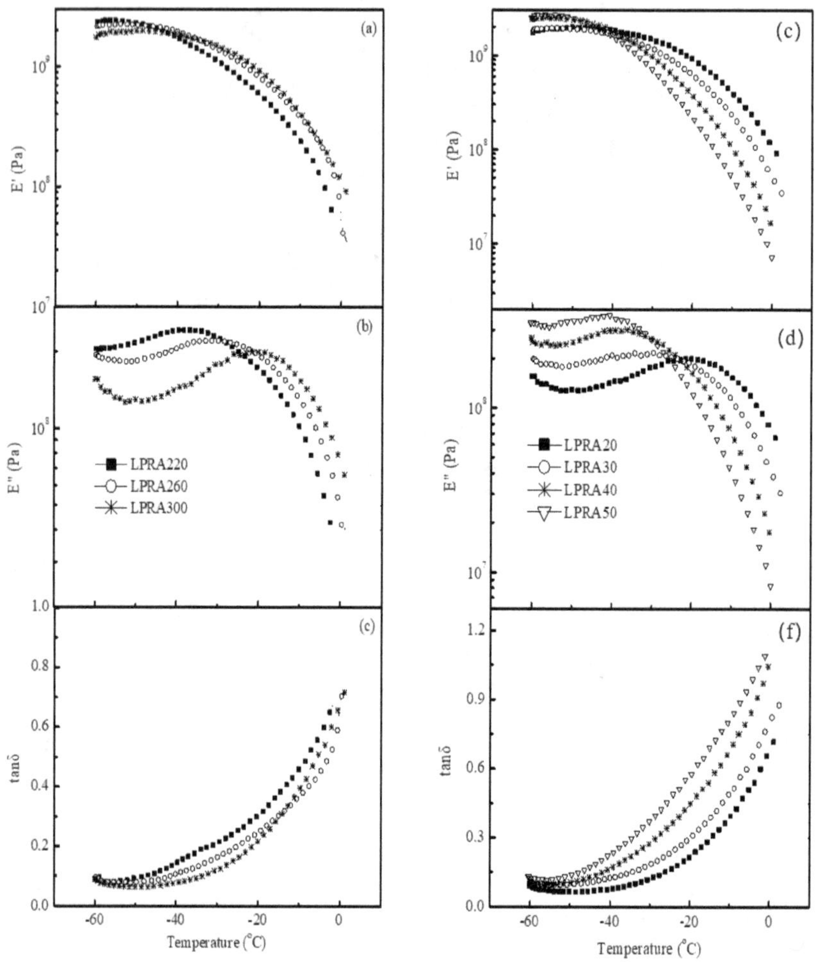

Figure 2.6 *(a)-(c) Low-temperature rheological performances of modified asphalts with different LPRs; (d)-(f) Low-temperature performances of modified asphalts with different percents of LPR300 (Reproduced from Reference 130 with permission from Elsevier).*

LPR with high sol content had typical properties similar to base asphalt, like low viscosity and balanced rheological properties. Figure 2.6(a) showed that the storage modulus of LPR modified asphalts (LPRMAs) extruded at different temperatures were both better than base asphalt (BA). LPRs prepared at higher temperature sped up the decreasing rate of storage modulus over -40 °C, which indicated that adding LPR improved the low-temperature performance and LPR obtained at lower temperature endowed the LPRMA mixture with better low-temperature performance. Figure 2.6(b) showed Tg of LPRMAs were lower than that of BA and increased with extrusion temperatures, which implied that LPR prepared at lower temperature had potential to improve the low-temperature property. Figure 2.6 showed the low-temperature rheological properties of LPRMAs with different contents of LPR300. The storage moduli of LPRMAs were both lower than that of BA over -40 °C. Increasing LPR300 part decreased the storage modulus, indicating that more LPR300 improved low-temperature properties of the LPRMA blends. In addition, loss factor decreased with the increasing content of LPR300, which implied that adding more content of LPR300 makes asphalt stickier and better anti-fatigue. LPR could be applied for replacing asphalt beyond 50%, giving the asphalt with good low-temperature performance.

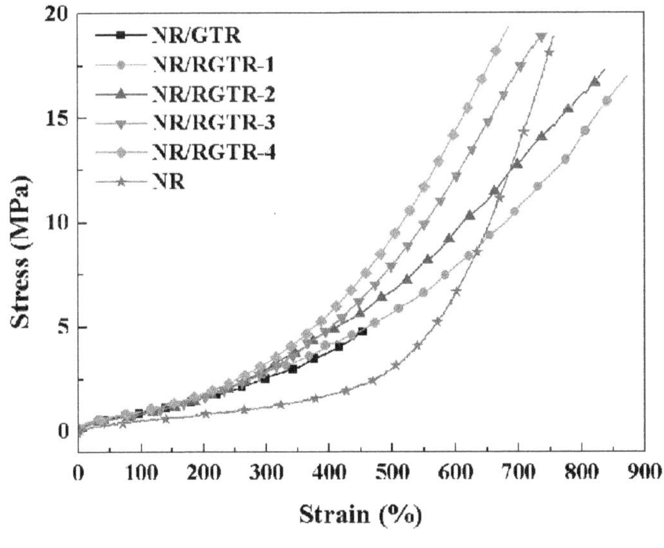

Figure 2.7 *Strain–stress curves for reused NR/RGTRs (Reproduced from Reference 129 with permission from American Chemical Society).*

The physical and aging resistance performances of the modified NR were reinforced by up to 40 wt.% reclaimed ground tire rubber (RGTR) [129]. The CB could be removed from the cross-linked structure of GTR and extracted into the sol fraction because of the high reclaiming extent. The RGTR, which has a high reclaiming extent, exhibited better performance in NR compared with other traditional reclaimed rubber. Adding of 10–40 wt.% of RGTR into NR had no negative influence on the physical performance. Figure 2.7 showed that the highest physical performance were achieved by adding 40 wt.% RGTR. The continuous uniform network formed by the loosely cross-linked network of the gel part of the RGTR and the strong interfacial adhesion between RGTR and NR was beneficial to the mechanical properties of the material. Any improvement of tensile strength beyond that of untreated GTR was due to the improved interfacial adhesion because of network breakdown. The thermal aging resistance of NR was also improved.

2.6 Advantages and disadvantages of tire pyrolysis

The main advantages of tire rubber pyrolysis is that the obtained materials have potential to apply for new products. The liquid product from pyrolysis, usually named as pyrolytic oil, can have 44 MJ/kg of power due to the tire composition and the pyrolysis parameters. The pyrolytic oil has possible application to be alternative fuel. On the other hand, the solid product obtained from pyrolysis is mainly the pyrolytic carbon black with porosity and large surface area. The pyrolytic carbon black shows the possibility to work as the commercial carbon black in the polymer products. Moreover, the pyrolytic CB has also been applied for activated carbon in many aspects to remove organic pollutants and toxic metals.

However, there are some limitations and challenges. Pyrolysis is an energy-consuming process and rubber is a poor conductor of heat, which makes the heat dispersion uneven and causes serious energy loss. In terms of pyrolytic products utilization, the high viscosity of the pyrolysis oil result in problems in the service life, which generate carbon deposit and eventually affect the fuel flow through pipes. The surface of pyrolytic carbon black has a certain number of harmful deposits, thus leading to low quality of renewed products. The material is only considered as a cheap filler but not as high value reinforcing filler and this seriously hinders its application. In addition, the composition of pyrolysis products is very complicated due to the different formula of the tire in the manufactures. Therefore, the modification and application research of pyrolytic products still have a long way to go from the view of high-value application and technological development.

2.7 Conclusions

With the huge demand of rubber, the amount of waste rubber increase year by year. These waste rubbers have been a source of health and environmental concerns. Pyrolysis, as an efficient and sustainable recycling method of waste rubber, is receiving more and more interest and attention due to the great significance to the ecological civilization and sustainable development strategies. This chapter presented the main advances of pyrolysis of waste tires, including different reactors, the important parameters of the pyrolysis process and the pyrolytic products' potential applications.

The pyrolysis of waste rubber is a relatively complicated process, including primary polymer degradation and secondary reactions. From many literatures published in this field, the types of reactors have great impact on the pyrolysis parameters, such as temperature, heating rate and residence time. Moreover, the characteristics and yield of the pyrolytic materials also varied widely due to the different pyrolysis parameters. For pressure and temperature of the pyrolysis process, a pressure increase in the reactor leads to a higher yield of pyrolytic char. The content of the liquid component decreases and the gas component increases as the pyrolysis temperature increases. Vacuum pyrolysis decreases the probability of secondary reactions and the flow of the carrier gases, which makes the pyrolysis process more feasible.

The pyrolytic CB has potential to be a reinforcing filler or pigment of rubber products, which brings considerable benefits to pyrolysis industry. The surface of pyrolytic CB from pyrolysis is mainly contains a certain amount of harmful ash compared with the commercial carbon black and has different micro-structures of particle. This affects the performance of the pyrolytic carbon black as rubber reinforcing fillers and limits its application. Much attempts have been made to improve the quality of the pyrolytic CB by vacuum pyrolysis and modification.

The waste rubber can also be treated by light pyrolysis, resulting in liquid rubber and lightly pyrolytic carbon black. The liquid pyrolyzed rubber (LPR) had some common properties parallel to base asphalt, such as viscosity and rheological properties. LPR can substitute as asphalt over 50 % by weight, which improve the low temperature performance of asphalt. The lightly pyrolytic carbon black obtained by reactive melt-extrusion process, similar to the commercial carbon black (N330), can reinforce the rubber greater than the common pyrolytic CB because of the presence of bound rubber. However, some competing reactions took place simultaneously in the process of the light pyrolysis, resulting in a heterogeneous recycling of waste rubber under different parameters. In addition, the lightly pyrolytic CB was separated from the head product by toluene extracts, which is a complex process for industry.

The recent research studies have gradually moved from the batch small-scale test stage to the continuous pilot test stage, and some technologies have the potential to achieve industrial applications. However, the pyrolysis reaction mechanism is still not clear due to the complex composition of waste rubber. Pyrolysis equipment generally has the problems of high energy-consuming, high investment and complexity of control. Therefore, the future research of waste rubber pyrolysis lies in:

(1) Pyrolysis mechanism, especially reaction kinetics should be further investigated to obtain a clearer understanding of the rubber pyrolysis on a molecular scale;

(2) Reactors needs to be designed and developed for high heat-energy utilization and an easier control of pyrolysis parameters, thereby promoting the cost effectiveness of tire pyrolysis and improving the quality of the pyrolytic products;

(3) More research should be done to focus on the improvement of the quality of the pyrolytic carbon black to achieve its high-value utilization.

Acknowledgements

The authors are grateful for the research foundation provided by International Cooperation Project by MOST, SQ2019YFE012146, Natural Science Foundation of China (No. 51273110) and the Jiangsu Science and Technology Project (BA2016002).

References

[1] Cheng X, Song P, Zhao X, et al (2018) Liquefaction of ground tire rubber at low temperature, Waste Manage 71: 301-310. https://doi.org/10.1016/j.wasman.2017.10.004
[2] Rajan V, Dierkes W, Joseph R, Noordermeer J (2006) Science and technology of rubber reclamation with special attention to NR-based waste latex products, Prog Polym Sci 31(9) 811-834. https://doi.org/10.1016/j.progpolymsci.2006.08.003
[3] Martínez J, Puy N, Murillo R, García T, et al (2013) Waste tyre pyrolysis – A review, Renew Sust Energ Rev 23: 179-213. https://doi.org/10.1016/j.rser.2013.02.038
[4] Beukering P, Janssen M (2001) Trade and recycling of used tyres in Western and Eastern Europe, Resour Conserv Recy 33(4): 235-265. https://doi.org/10.1016/S0921-3449(01)00082-9
[5] Commission E (1999) Landfill of waste directive, Council Directive1999/31/EC. Brussels.
[6] Perkins A, Inayat-Hussain S, Deziel N, et al (2019) Evaluation of potential carcinogenicity of organic chemicals in synthetic turf crumb rubber, Environ Res 169 163-172. https://doi.org/10.1016/j.envres.2018.10.018
[7] Isayev A, (2014) Recycling of natural and synthetic isoprene rubbers, Chemistry, Manufacture and Applications of Natural Rubber: 395-435. https://doi.org/10.1533/9780857096913.3.395
[8] Shulman VL, (2019) Tire Recycling, Waste:489-515. https://doi.org/10.1016/B978-0-12-815060-3.00026-8
[9] Juma M, Koreňová Z, Markoš J, et al (2007) Experimental study of pyrolysis and combustion of scrap tire, Polym Advan Technol 18(2): 144-148. https://doi.org/10.1002/pat.811
[10] Xu J, Yu J, Xu J, et al (2020) High-value utilization of waste tires: A review with focus on modified carbon black from pyrolysis, Sci Total Environ 742: 140235. https://doi.org/10.1016/j.scitotenv.2020.140235
[11] Mouri M, Sato N, Okamoto H, et al (2000) Continuous devulcanisation by shear flow stage reaction control technology for rubber recycling. Part 4. Devulcanisation mechanism for EPDM, Inter Polym Sci Technol 27: T/17-T/22. https://doi.org/10.2324/gomu.72.283
[12] Laird DA, Brown RC, Amonette JE, et al (2009) Review of the pyrolysis platform for coproducing bio-oil and biochar, Biofuel, Bioprod Bior 3(5): 547-562. https://doi.org/10.1002/bbb.169

[13] Kwon E, (2009) Fundamental Understanding of the Thermal Degradation Mechanisms of Waste Tires and Their Air Pollutant Generation in a N_2 Atmosphere, Environ Sci Technol 5996-6002. https://doi.org/10.1021/es900564b

[14] Danon B, Gryp P, Schwarz CE, et al (2015) A review of dipentene (dl-limonene) production from waste tire pyrolysis, J Anal App Pyrol 1121-13. https://doi.org/10.1016/j.jaap.2014.12.025

[15] Antoniou N, Zabaniotou A, (2013) Features of an efficient and environmentally attractive used tyres pyrolysis with energy and material recovery, Renew Sust Energ Rev 20: 539-558. https://doi.org/10.1016/j.rser.2012.12.005

[16] Miranda M, Pinto F, Gulyurtlu I, et al (2013) Pyrolysis of rubber tyre wastes: A kinetic study, Fuel 103542-552. https://doi.org/10.1016/j.fuel.2012.06.114

[17] Diez C, Martinez O, Calvo L, et al (2004) Influence of the final temperature of the process on emissions and the calorific value of the products recovered, Waste Manage 24(5): 463-469. https://doi.org/10.1016/j.wasman.2003.11.006

[18] İlkılıç C, Aydın H, (2011) Fuel production from waste vehicle tires by catalytic pyrolysis and its application in a diesel engine, Fuel Process Technol 92(5): 1129-1135. https://doi.org/10.1016/j.fuproc.2011.01.009

[19] Kordoghli S, Khiari B, Paraschiv M, et al (2017) Impact of different catalysis supported by oyster shells on the pyrolysis of tyre wastes in a single and a double fixed bed reactor, Waste Manage 67: 288-297. https://doi.org/10.1016/j.wasman.2017.06.001

[20] Acevedo B, Barriocanal C, Alvarez R, (2013) Pyrolysis of blends of coal and tyre wastes in a fixed bed reactor and a rotary oven, Fuel 113: 817-825. https://doi.org/10.1016/j.fuel.2012.12.07

[21] Tale D, Hamid HA, Deris RRR, et al (2020) Insights into pyrolysis of waste tire in fixed bed reactor: Thermal behavior, Mater Today: Proceedings 31178-186. https://doi.org/10.1016/j.matpr.2020.01.569

[22] Xu S, Lai D, Zeng X, et al (2018) Pyrolysis characteristics of waste tire particles in fixed-bed reactor with internals, Carbon Resources Conversion 1(3): 228-237. https://doi.org/10.1016/j.crcon.2018.10.001

[23] Laresgoiti MF, Caballero BM, Marco I, et al (2004) Characterization of the liquid products obtained in tyre pyrolysis, J Anal Appl Pyrol 71(2): 917-934. https://doi.org/10.1016/j.jaap.2003.12.003

[24] Rofiqulislam M, Haniu H, Rafiqulalambeg M, (2008) Liquid fuels and chemicals from pyrolysis of motorcycle tire waste: Product yields, compositions and related properties, Fuel 87(13-14): 3112-3122. https://doi.org/10.1016/j.fuel.2008.04.036
[25] Cunliffe AM, Williams PT (1998) Composition of oils derived from the batch pyrolysis of tyres, J Anal Appl Pyrol 44: 131-152. https://doi.org/10.1016/S0165-2370(97)00085-5
[26] Mkhize NM, Gryp P, Danon B, et al (2016) Effect of temperature and heating rate on limonene production from waste tyre pyrolysis, J Anal Appl Pyrol 120: 314-320. https://doi.org/10.1016/j.jaap.2016.04.019
[27] Kaminsky W, Mennerich C, Zhang Z, (2009) Feedstock recycling of synthetic and natural rubber by pyrolysis in a fluidized bed, J Anal Appl Pyrol 85(1-2): 334-337. https://doi.org/10.1016/j.jaap.2008.11.012
[28] Edwin Raj R, Robert Kennedy Z, et al (2013) Optimization of process parameters in flash pyrolysis of waste tyres to liquid and gaseous fuel in a fluidized bed reactor, Energ Convers Manage 67: 145-151. https://doi.org/10.1016/j.enconman.2012.11.012
[29] Walter Kaminsky CM, (2001) Pyrolysis of synthetic tire rubber in a fluidised-bed reactor to yield 1,3-butadiene, styrene and carbon black, J Anal Appl Pyrol 58-59: 803-811. https://doi.org/10.1016/S0165-2370(00)00129-7
[30] Dai X, Wu C, Zhang We, et al (2001) Pyrolysis of waste tires in a circulating fluidized-bed reactor, Energy 26385-399. https://doi.org/10.1016/S0360-5442(01)00003-2
[31] Dai X, Yin X, Wu C, et al (2001) Pyrolysis of waste tires in a circulating fluidized-bed reactor, Energy 26(4): 385-399. https://doi.org/10.1016/S0360-5442(01)00003-2
[32] Yazdani E, Hashemabadi SH, Taghizadeh A (2019) Study of waste tire pyrolysis in a rotary kiln reactor in a wide range of pyrolysis temperature, Waste Manage 85: 195-201. https://doi.org/10.1016/j.wasman.2018.12.020
[33] Antoniou N, Zabaniotou A (2018) Re-designing a viable ELTs depolymerization in circular economy: Pyrolysis prototype demonstration at TRL 7, with energy optimization and carbonaceous materials production, J Clean Prod 174: 74-86. https://doi.org/10.1016/j.jclepro.2017.10.319
[34] Syamsiro M, Dwicahyo MS, Sulistiawati Y, et al (2019) Development of a rotary kiln reactor for pyrolytic oil production from waste

tire in Indonesia, IOP Conference Series: Earth Environ Sci 245. https://doi.org/10.1088/1755-1315/245/1/012044

[35] Ayanoğlu A, Yumrutaş R (2016) Rotary kiln and batch pyrolysis of waste tire to produce gasoline and diesel like fuels, Energ Convers Manage 111: 261-270. https://doi.org/10.1016/j.enconman.2015.12.070

[36] Banar M, Akyıldız V, Özkan A, et al (2012) Characterization of pyrolytic oil obtained from pyrolysis of TDF (Tire Derived Fuel), Energ Convers Manage 62: 22-30. https://doi.org/10.1016/j.enconman.2012.03.019

[37] Li S (2004) Pilot-Scale Pyrolysis of Scrap Tires in a Continuous Rotary Kiln Reactor, Ind Eng Chem Res 43(17): 5133-5145. https://doi.org/10.1021/ie030115m

[38] Luo S, Feng Y (2017) The production of fuel oil and combustible gas by catalytic pyrolysis of waste tire using waste heat of blast-furnace slag, Energ Convers Manage 136: 27-35. https://doi.org/10.1016/j.enconman.2016.12.076

[39] Song Z, Yang Y, Sun J, et al (2017) Effect of power level on the microwave pyrolysis of tire powder, Energy 127: 571-580. https://doi.org/10.1016/j.energy.2017.03.150

[40] Dai L, Fan L, Duan D, et al (2017) Microwave-assisted catalytic fast co-pyrolysis of soapstock and waste tire for bio-oil production, J Anal Appl Pyrol 125: 304-309. https://doi.org/10.1016/j.jaap.2017.03.012

[41] Wang Y, Dai L, Fan L, et al (2017) Microwave-assisted catalytic fast co-pyrolysis of bamboo sawdust and waste tire for bio-oil production, J Anal Appl Pyrol 123: 224-228. https://doi.org/10.1016/j.jaap.2016.11.025

[42] Formela K, Hejna A, Zedler L, et al (2019) Microwave treatment in waste rubber recycling - recent advances and limitations, Express Polym Lett 13(6): 565-588.

[43] Russell AD, Antreou EI, Lam SS, et al (2012) Microwave-assisted pyrolysis of HDPE using an activated carbon bed, RSC Adv 2(17): 6756-6760. https://doi.org/10.1039/c2ra20859h

[44] Song Z, Yan Y, Xie M, et al (2018) Effect of Steel Wires on the Microwave Pyrolysis of Tire Powders, ACS Sust Chem Eng 6(10): 13443-13453. https://doi.org/10.1021/acssuschemeng.8b03337

[45] Undri A, Meini S, Rosi L, et al (2013) Microwave pyrolysis of polymeric materials: Waste tires treatment and characterization of the value-added products, Journal of Analytical and Applied Pyrolysis 103 149-158. https://doi.org/10.1016/j.jaap.2012.11.011

[46] Undri A, Rosi L, Frediani M, et al (2014) Upgraded fuel from microwave assisted pyrolysis of waste tire, Fuel 115: 600-608. https://doi.org/10.1016/j.fuel.2013.07.058

[47] Xie Q, Addy M, Liu S, et al (2015) Fast microwave-assisted catalytic co-pyrolysis of microalgae and scum for bio-oil production, Fuel 160: 577-582. https://doi.org/10.1016/j.fuel.2015.08.020

[48] Rosi L, Bartoli M, Frediani M (2018) Microwave assisted pyrolysis of halogenated plastics recovered from waste computers, Waste Manag 73: 511-522. https://doi.org/10.1016/j.wasman.2017.04.037

[49] Suriapparao DV, Vinu R, (2015) Resource recovery from synthetic polymers via microwave pyrolysis using different susceptors, J Anal Appl Pyrol 113: 701-712. https://doi.org/10.1016/j.jaap.2015.04.021

[50] Lopez G, Alvarez J, Amutio M, et al (2017) Waste truck-tyre processing by flash pyrolysis in a conical spouted bed reactor, Energ Convers Manage 142523-532. https://doi.org/10.1016/j.enconman.2017.03.051

[51] Ding K, Zhong Z, Zhang B, et al (2016) Catalytic pyrolysis of waste tire to produce valuable aromatic hydrocarbons: An analytical Py-GC/MS study, J Anal Appl Pyrol 122: 55-63. https://doi.org/10.1016/j.jaap.2016.10.023

[52] Olazar M, Arabiourrutia M, López G, et al (2008) Effect of acid catalysts on scrap tyre pyrolysis under fast heating conditions, J Anal Appl Pyrol 82(2): 199-204. https://doi.org/10.1016/j.jaap.2008.03.006

[53] Miandad R, Barakat MA, Rehan M, et al (2018) Effect of advanced catalysts on tire waste pyrolysis oil, Process Saf Environ 116: 542-552. https://doi.org/10.1016/j.psep.2018.03.024

[54] Niksiar A, Sohrabi M, Rahimi A (2013) A correction on a published kinetic model for tyre pyrolysis in a conical spouted bed reactor, J Anal Appl Pyrol 104: 707-709. https://doi.org/10.1016/j.jaap.2013.06.013

[55] Lopez G, Olazar M, Amutio M, et al (2009) Influence of Tire Formulation on the Products of Continuous Pyrolysis in a Conical Spouted Bed Reactor, Energ Fuel 23(11): 5423-5431. https://doi.org/10.1016/j.jaap.2013.06.013

[56] Gartzen L, Aguado R, Elordi G, et al (2010) Vacuum Pyrolysis of Waste Tires by Continuously Feeding into a Conical Spouted Bed Reactor, Ind Eng Chem Res 49: 8990-8997. https://doi.org/10.1021/ie1000604

[57] Zhang X, Wang T, Ma L, et al (2008) Vacuum pyrolysis of waste tires with basic additives, Waste Management 28: 2301-2310. https://doi.org/10.1016/j.wasman.2007.10.009

[58] Juan RF, Conesa A, Marcilla A (1996) Gas from the Pyrolysis of Scrap Tires in a Fluidized Bed Reactor, Energ Fuel 10: 134-140. https://doi.org/10.1021/ef950152t

[59] Mkhize NM, Danon B, Alvarez J, et al (2019) Influence of reactor and condensation system design on tyre pyrolysis products yields, J Anal Appl Pyrol 143: 104683.1-104683.10. https://doi.org/10.1016/j.jaap.2019.104683

[60] Williams PT, Brindle AJ (2003) Fluidised bed pyrolysis and catalytic pyrolysis of scrap tyres, Environ Technol 24(7): 921-929. https://doi.org/10.1080/09593330309385629

[61] Hu H, Fang Y, Liu H, et al (2014) The fate of sulfur during rapid pyrolysis of scrap tires, Chemosphere 97: 102-107. https://doi.org/10.1016/j.chemosphere.2013.10.037

[62] Aylon E, Fernandez-Colino A, Murillo R, et al (2010) Valorisation of waste tyre by pyrolysis in a moving bed reactor, Waste Manage 30(7): 1220-1224. https://doi.org/10.1016/j.wasman.2009.10.001

[63] Abnisa F, Wan WMA, Ramalingam S, et al (2013) Co-pyrolysis of palm shell and polystyrene waste mixtures to synthesis liquid fuel, Fuel 108311-318. https://doi.org/10.1016/j.fuel.2013.02.013

[64] Unapumnuk K, Keener TC, Lu M, et al (2006) Pyrolysis behavior of tire-derived fuels at different temperatures and heating rates, J Air Waste Manage 56(5): 618-627. https://doi.org/10.1080/10473289.2006.10464481

[65] Ding S, Dasgupta R, Lee C, et al (2015) New bulk sulfur measurements of Martian meteorites and modeling the fate of sulfur during melting and crystallization – Implications for sulfur transfer from Martian mantle to crust–atmosphere system, Earth Planet Sci Lett 409: 157-167. https://doi.org/10.1016/j.epsl.2014.10.046

[66] Jasminská N, Brestovič T, Čarnogurská M, (2013) The Effect of Temperature Pyrolysis Process of Used Tires on the Quality of Output Products, Acta Mechanica et Autom 7(1): 20-25. https://doi.org/10.2478/ama-2013-0004

[67] Chen R, Li Q, Zhang Y, et al (2019) Pyrolysis kinetics and mechanism of typical industrial non-tyre rubber wastes by peak-differentiating analysis and multi kinetics methods, Fuel 2351224-1237. https://doi.org/10.1016/j.fuel.2018.08.121

[68] Williams PT, Besler S (1995) Pyrolysis-thermogravimetric analysis of tyres and tyre components., Fuel 14: 1277-1283. https://doi.org/10.1016/0016-2361(95)00083-H

[69] Singh RK, Ruj B, Jana A, et al (2018) Pyrolysis of three different categories of automotive tyre wastes: Product yield analysis and characterization, J Anal Appl Pyrol 135: 379-389. https://doi.org/10.1016/j.jaap.2018.08.011

[70] Ucar S, Karagoz S, Ozkan AR, Yanik J (2005) Evaluation of two different scrap tires as hydrocarbon source by pyrolysis, Fuel 84(14-15): 1884-1892. https://doi.org/10.1016/j.fuel.2005.04.002

[71] Islam MR, Joardder MU, Hasan, SM, et al (2011) Feasibility study for thermal treatment of solid tire wastes in Bangladesh by using pyrolysis technology, Waste Manage 31(9-10): 2142-2149. https://doi.org/10.1016/j.wasman.2011.04.017

[72] Mohammed Kyari AC, Williams PT, (2005) Characterization of Oils, Gases, and Char in Relation to the Pyrolysis of Different Brands of Scrap Automotive Tires, Energ Fuel 19: 1165-1173. https://doi.org/10.1021/ef049686x

[73] Elbaba IF, Wu C, Williams PT, (2010) Catalytic Pyrolysis-Gasification of Waste Tire and Tire Elastomers for Hydrogen Production, Energ Fuel 24(7): 3928-3935. https://doi.org/10.1021/ef100317b

[74] Cataldo F (1998) Thermal depolymerization and pyrolysis of cis-1,4-polyisoprene: preparation of liquid polyisoprene and terpene resin, J Anal Appl Pyrol 44(2): 121-130. https://doi.org/10.1016/S0165-2370(97)00081-8

[75] Kan T, Strezov V, Evans T, (2017) Fuel production from pyrolysis of natural and synthetic rubbers, Fuel 191403-410. https://doi.org/10.1016/j.fuel.2016.11.100

[76] Seidelt S, Müller-Hagedorn M, Bockhorn H (2006) Description of tire pyrolysis by thermal degradation behaviour of main components, J Anal Appl Pyrol 75(1): 11-18. https://doi.org/10.1016/j.jaap.2005.03.002

[77] Cheung KY, Lee KL, Lam KL, et al (2011) Integrated kinetics and heat flow modelling to optimise waste tyre pyrolysis at different heating rates, Fuel Process Technol 92(5): 856-863. https://doi.org/10.1016/j.fuproc.2010.11.028

[78] Kuz'mina RI, Romadenkina SB, Mikhel' AA, et al (2016) Rubber waste processing by high-temperature pyrolysis, Solid Fuel Chem 50(4): 260-264. https://doi.org/10.3103/S036152191604008X

[79] Nisar J, Ali G, Ullah N, et al (2018) Pyrolysis of waste tire rubber: Influence of temperature on pyrolysates yield, J Environ Chem Eng 6: 3469-3473. https://doi.org/10.1016/j.jece.2018.05.021

[80] Efika CE, Onwudili JA, Williams PT, (2018) Influence of heating rates on the products of high-temperature pyrolysis of waste wood pellets and biomass model compounds, Waste Manage 76: 497-506. https://doi.org/10.1016/j.wasman.2018.03.021

[81] Choi GG, Jung SH (2014) Total utilization of waste tire rubber through pyrolysis to obtain oils and CO2 activation of pyrolysis char, Fuel Process Technol 12357-64. https://doi.org/10.1016/j.fuproc.2014.02.007

[82] Li W, Huang C, Li D, et al (2016) Derived oil production by catalytic pyrolysis of scrap tires, Chinese J Catal 37(4): 526-532. https://doi.org/10.1016/S1872-2067(15)60998-6

[83] Nisar J, Ali G, Ullah N, et al (2018) Pyrolysis of waste tire rubber: Influence of temperature on pyrolysates yield, J Environ Chem Eng 6(2): 3469-3473. https://doi.org/10.1016/j.jece.2018.05.021

[84] Isabel MFL, Cabrero MA,et al (2001) Pyrolysis of scrap tyres, Fuel Process Technol 72: 9-22. https://doi.org/10.1016/S0378-3820(01)00174-6

[85] Rodríguez E, Gutiérrez A, Palos R, et al (2019) Cracking of Scrap Tires Pyrolysis Oil in a Fluidized Bed Reactor under Catalytic Cracking Unit Conditions. Effects of Operating Conditions, Energ Fuel 33(4): 3133-3143. https://doi.org/10.1021/acs.energyfuels.9b00292

[86] López G, Olazar M, Aguado R,et al (2010) Continuous pyrolysis of waste tyres in a conical spouted bed reactor, Fuel 89(8): 1946-1952. https://doi.org/10.1016/j.fuel.2010.03.029

[87] Hita I, Arabiourrutia M, Olazar M, et al (2016) Opportunities and barriers for producing high quality fuels from the pyrolysis of scrap tires, Renew Sust Energ Rev 56: 745-759. https://doi.org/10.1016/j.rser.2015.11.081

[88] Danon B, Görgens J, (2015) Determining rubber composition of waste tyres using devolatilisation kinetics, Thermochim Acta 621: 56-60. https://doi.org/10.1016/j.tca.2015.10.008

[89] Quek A, Balasubramanian R, (2013) Liquefaction of waste tires by pyrolysis for oil and chemicals—A review, J Anal Appl Pyrol 1011-16. https://doi.org/10.1016/j.jaap.2013.02.016

[90] Quek A, Balasubramanian R (2012) Mathematical modeling of rubber tire pyrolysis, J Anal Appl Pyrol 95: 1-13. https://doi.org/10.1016/j.jaap.2012.01.012

[91] Miller SJ, Shah N, Huffman GP, (2006) Feedstock Recycling and Pyrolysis of Waste Plastics: Converting Waste Plastics into Diesel and Other Fuels. New York: John Wiley and Sons; 345-355 https://doi.org/10.1002/0470021543

[92]. Roy ACC, Darmstadt H (1999) The vacuum pyrolysis of used tires End-uses for oil and carbon black products, J Anal Appl Pyrol 51: 201-221.

[93] Zhang X, Wang T, Ma L, et al (2008) Vacuum pyrolysis of waste tires with basic additives, Waste Manage 28(11): 2301-2310. https://doi.org/10.1016/j. wasman.2007.10.009

[94] Betancur M, Martinez JD, Murillo R (2009) Production of activated carbon by waste tire thermochemical degradation with CO2, J Hazard Mater 168(2-3): 882-887. https://doi.org/10.1016/ j.jhazmat.2009.02.167

[95] Williams PT, Bottrill RP, Cunliffe AM (1998) Combustion of Tyre Pyrolysis Oil, Process Saf Environ 76(4): 291-301. https://doi.org/10.1205/095758298529650

[96] Pakdel H, Népo Murwanashyaka J, Roy C (2002) Extraction of Betulin by Vacuum Pyrolysis of Birch Bark, J Wood Chem Technol 22(2-3): 147-155. https://doi.org/10.1081/WCT-120013359

[97] Sadao Ogasawara MK, Wakao N (1987) Preparation of Activated Carbon by Thermal Decomposition of Used Automotive Tires, Ind Eng Chem Res 26(12): 2552-2556. https://doi.org/10.1021/ie00072a030

[98] Zhang X, Zhong L, Zeng G, et al (2018) Process intensification of honeycomb fractal micro-reactor for the direct production of lower olefins from syngas, Chem Eng J 351: 12-21. https://doi.org/10.1016/j.cej.2018.06.078

[99] Martínez JD, Rodríguez-Fernández J, Sánchez-Valdepeñas J, et al (2014) Performance and emissions of an automotive diesel engine using a tire pyrolysis liquid blend, Fuel 115: 490-499. https://doi.org/10.1016/j.fuel.2013.07.051

[100] Czajczyńska D, Krzyżyńska R, Jouhara H, et al (2017) Use of pyrolytic gas from waste tire as a fuel: A review, Energy 134: 1121-1131. https://doi.org/10.1016/j. energy2017.05.042

[101] Juan JME., Gonza´lez F, Canito JL, et al (2001) Pyrolysis of automobile tyre waste. Influence of operating variables and kinetics study, J Anal Appl Pyrol 58: 667-683. https://doi.org/10.1016/S0165-2370(00)00201-1

[102] Hita I, Rodríguez E, Olazar M, et al (2015) Prospects for Obtaining High Quality Fuels from the Hydrocracking of a Hydrotreated

Scrap Tires Pyrolysis Oil, Energ Fuel 29(8): 5458-5466. https://doi.org/10.1021/acs.energyfuels.5b01181

[103] Galvagno SCS, Casabianca T, Calabrese A, et al (2002) Pyrolysis process for the treatment of scrap tyres: preliminary experimental results, Waste Manage 22: 917-923. https://doi.org/10.1016/S0956-053X（02）00083-1

[104] Vichaphund S, Aht-ong D, Sricharoenchaikul V, et al (2017) Effect of CV-ZSM-5, Ni-ZSM-5 and FA-ZSM-5 catalysts for selective aromatic formation from pyrolytic vapors of rubber wastes, J Anal Appl Pyrol 124: 733-741. https://doi.org/10.1016/j.jaap.2016.11.011

[105] Kar Y, (2011) Catalytic pyrolysis of car tire waste using expanded perlite, Waste Manage 31(8): 1772-82. https://doi.org/10.1016/j.wasman.2011.04.005

[106] Mastral AM, Murillo R, Callen MS, et al (2000) Influence of process variables on oils from tire pyrolysis and hydropyrolysis in a swept fixed bed reactor, Energ Fuel 14(4): 739-744. https://doi.org/10.1021/ef990183e

[107] Ayanoğlu A, Yumrutaş R (2016) Production of gasoline and diesel like fuels from waste tire oil by using catalytic pyrolysis, Energy 103: 456-468. https://doi.org/10.1016/ j.energy.2016.02.155

[108] Machin EB, Pedroso DT, de Carvalho JAJ, (2017) Technical assessment of discarded tires gasification as alternative technology for electricity generation, Waste Manage 68: 412-420. https://doi.org/10.1016/ j.wasman.2017.07.004

[109] Acosta R, Fierro V, Martinez de Yuso A, et al (2016) Tetracycline adsorption onto activated carbons produced by KOH activation of tyre pyrolysis char, Chemosphere 149: 168-76. https://doi.org/10.1016/j.chemosphere.2016.01.093

[110] Mouzourakis E, Georgiou Y, Louloudi M, et al (2017) Recycled-tire pyrolytic carbon made functional: A high-arsenite [As(III)] uptake material PyrC350((R)), J Hazard Mater 326: 177-186. https://doi.org/10.1016/ j.jhazmat.2016.12.027

[111] Anjum A, Ramani B, Bramer E (2019) Role of Recovered Carbon Black Ash Content Composition on In-Rubber Performance, International Rubber Conference, IRC, 2019.

[112] Saleh TA, Gupta VK (2014) Processing methods, characteristics and adsorption behavior of tire derived carbons: a review, Adv Colloid Interfac 21193-101. https://doi.org/10.1016/j.cis.2014.06.006

[113] Sagar M, Nibedita K, Manohar N, et al (2018) A potential utilization of end-of-life tyres as recycled carbon black in EPDM rubber,

Waste Manage 74110-122. https://doi.org/10.1016/j. wasman2018.01.003

[114] Hans Darmstadt CR, Kaliaguine S, (1995) Charteriazation of pyrolytic carbon blacks from commercial tire pyrolysis plants, Carbon 33: 1449-1455. https://doi.org/10.1016/0008-6223(95)00096-V

[115] Rajapaksha AU, Chen SS, Tsang DC, et al (2016) Engineered/designer biochar for contaminant removal/immobilization from soil and water: Potential and implication of biochar modification, Chemosphere 148276-91. https://doi.org/10.1016/j.chemosphere.2016.01.043

[116] Quek A, Balasubramanian R, (2009) An algorithm for the kinetics of tire pyrolysis under different heating rates, J Hazard Mater 166(1): 126-32. https://doi.org/10.1016/j.jhazmat.2008.11.034

[117] Choi GG, Oh SJ, Kim JS, (2016) Scrap tire pyrolysis using a new type two-stage pyrolyzer: Effects of dolomite and olivine on producing a low-sulfur pyrolysis oil, Energy 114: 457-464. https://doi.org/10.1016/j.energy.2016.08.020

[118] Ariyadejwanicha WTP, Nakagawa K, (2003) Preparation and characterization of mesoporous activated carbon from waste tires, Carbon 41: 157-164. https://doi.org/10.1016/S0008-6223(02)00267-1

[119] Saleh TA, Danmaliki GI, (2016) Influence of acidic and basic treatments of activated carbon derived from waste rubber tires on adsorptive desulfurization of thiophenes, J Taiwan Institute of Chem E 60460-468. https://doi.org/10.1016/j.jtice.2015.11.008

[120] Gupta VK, Gupta B, Rastogi A, et al (2011) Pesticides removal from waste water by activated carbon prepared from waste rubber tire, Water Res 45(13): 4047-55. https://doi.org/10.1016/j.watres.2011.05.016

[121] Gupta VK, Nayak A, Agarwal S, et al (2014) Potential of activated carbon from waste rubber tire for the adsorption of phenolics: effect of pre-treatment conditions, J Colloid Interf Sci 417: 420-30. https://doi.org/10.1016/j.jcis.2013.11.067

[122] Brady TA, Rostam-Abadi M, Rood MJ, (1996) Applications for activated carbons from waste tires: natural gas storage and air pollution control, Gas Separation & Purification 1097-102. https://doi.org/10.1016/0950-4214(96)00007-2

[123] Abadi MR, Christopher MB, Rood MJ, et al (1998) Reprocessing and Reuse of Waste Tire Rubber to Solve Air-Quality Related

Problems, Energ Fuel 12: 1095-1099. https://doi.org/10.1021/ef9801120

[124] Trubetskaya A, Kling J, Ershag O, et al (2019) Removal of phenol and chlorine from wastewater using steam activated biomass soot and tire carbon black, J Hazard Mater 365: 846-856. https://doi.org/10.1016/j.jhazmat.2018.09.061

[125] Hooshmand Ahoor A, Zandi-Atashbar N, (2014) Fuel production based on catalytic pyrolysis of waste tires as an optimized model, Energ Convers Manage 87: 653-669. https://doi.org/10.1016/j.enconman.2014.07.033

[126] Cheng X, Song P, Zhao XY, et al (2018) Liquefaction of ground tire rubber at low temperature, Waste Manage 71: 301-310. https://doi.org/10.1016/j.wasman.2017.10.004

[127] Li S, Wan C, Wu X, et al (2016) Core-shell structured carbon nanoparticles derived from light pyrolysis of waste tires, Polym Degrad Stab 129192-198. https://doi.org/10.1016/j.polymdegradstab.2016.04.013

[128] Song P, Zhao X, Cheng X, et al (2018) Recycling the nanostructured carbon from waste tires, Compos Commun 7: 12-15. https://doi.org/10.1016/j.coco.2017.12.001

[129] Zhang Y, Zhang Z, Wemyss AM, et al (2020) Effective Thermal-Oxidative Reclamation of Waste Tire Rubbers for Producing High-Performance Rubber Composites, ACS Sust Chem Eng 8(24): 9079-9087. https://doi.org/10.1021/acssuschemeng.0c02292

[130] Wu X, Wang S, Dong R (2016) Lightly pyrolyzed tire rubber used as potential asphalt alternative, Constr Build Mater 112: 623-628. https://doi.org/10.1016/j.conbuildmat.2016.02.208

[131] Li C, Zhang C, Zhang L, et al (2020)"Catalytic pyrolysis of tire waste: Impacts of biochar catalyst on product evolution." Waste Manage 116:9-21. https://doi.org/10.1016/j.wasman.2020.07.045

3

Chemical Devulcanization of Waste Rubber

Ricky Saputra[a,*], **Rashmi Walvekar**[a] and **Mohammad Khalid**[b]

[a]*School of Computer Science and Engineering, Faculty of Innovation and Technology, Taylor's University Lakeside Campus, No. 1 Jalan Taylor's, 47500 Subang Jaya, Selangor, Malaysia.*
ricky.saputra@sd.taylors.edu.my
[b]*Graphene & Advanced 2D Materials Research Group (GAMRG), School of Science and Technology, Sunway University, No. 5, Jalan Universiti, Bandar Sunway, 47500 Subang Jaya, Selangor, Malaysia.*

3.1 Introduction

Selective cleavage on vulcanizates' the three-dimensional sulphur crosslink, particularly mono-, di- and polysulfidic bonds, is one of the biggest challenges faced not just by academics and researchers but also industries. Such effort is still vastly being explored up to this point since the invention of vulcanization process by Charles Goodyear, to effectively recycle waste rubber without the significant lost or depreciation of its properties. It is commonly known that the diminishing properties of vulcanizates after treatment is caused by the scission of carbon backbone of rubber [1]. Typically, such phenomenon is caused, for instance, by the introduction of intensive physical forces and/or extreme exposure to severely high temperature setting.

Amongst the known state of the art devulcanization processes, chemical devulcanization in particular is one of the most interesting approaches explored as they are able to provide various pathways for faster, efficient and possibly selective crosslink cleavage [2]. The concept of chemical devulcanization of rubber simply defined as a method to selectively cleave the three-dimensional crosslink network with reduced degradation of polymer, through the assistance of suitable chemical agents at a certain pressure and temperature setting [1]. For the purpose of enhancing the rate of the reaction, this particular technique is not limited to a simple batch process where the vulcanizates are imbibed with chemicals, but most commonly coupled with thermal [3] and mechanical [4] energy leading to the emergence of thermochemical and mechanochemical devulcanization processes, respectively. Additionally, recent developments have combined suitable chemicals or devulcanizing agents with microwave irradiation [5] as well as ultrasonic devulcanization [6].

Such versatile characteristic of incorporating suitable devulcanizing agent in devulcanizing techniques makes chemical-based devulcanization process the most employed approaches in rubber reclaiming industries [7], despite the fact that some of the used chemicals involve a certain degree of toxicity [8]. However, investigation on a more sustainable and economical agents are also being explored since the concept of green chemistry was publicized. The often used devulcanizing chemicals can be classified as organic sulfides and thiols [9], amino compounds [10] as well as sulphur-free organic and inorganic compounds [11]. It is interesting to observe that, although the number of devulcanizing agents used for the delink phenomena are quite limitless, their devulcanization reaction mechanisms follow a certain chemistry, according to the type of agent used. In this chapter, the authors have identified four common devulcanization reaction mechanisms which includes inorganic catalytic, free radical, nucleophilic scission and delink by swelling, whereby their occurrences are thoroughly discussed.

3.2 Devulcanization by Inorganic Catalytic System

This particular devulcanization system mainly involves two different types of chemical probes, namely metallic salt-based and alkali-based. The utilization of these particular chemicals is due to their ability to attack and possibly break the crosslink network in a vulcanizate through phase transfer catalytic system, that makes the reaction between two immiscible phases possible. The advantage of using such particular system is their catalytic or semi-catalytic behavior which allows the devulcanizing agent to get regenerated and hence, economically attractive. On the other hand, their major drawbacks are the safety, toxicity and no ecofriendly aspect.

3.2.1 Salt-based Substance

Devulcanization through this group of substance is generally performed under room temperature and pressure conditions, whereby the process commonly combined with the addition of H_2O_2 as the initiator for reduction-oxidation reaction. Particularly, phenylhydrazine-iron (II) or Iron (III) chloride were used to catalyzed the oxidation of the crosslink network within the rubber. The procedure of this devulcanization pathway follows that reported by S. Yamashita *et al.* [12, 13], whereby vulcanizate (100 g) was first imbibed with a dropwise amount of H_2O_2, phenylhydrazine (0.5 g) and iron (II) chloride (0.25 g) in 5 ml of methanol and 10 ml of benzene. The mixture was then left for a day. A reaction scheme illustrated in Figure 3.1 was proposed. Based on Figure 3.1, the initial process that happened was the production of radicals by $FeCl_2$ and atmospheric O_2. It must be noted that the produced radical(s) can be phenyl-hydrazine, phenyl and/or peroxyl. The catalyst was also oxidized along with the production of H_2O_2.

After either of the radical(s) formed, the process proceeded to stage 2 which is Bolland oxidative degradation [14]. For simplicity purposes, the mechanism reported in Figure 3.1 only shows the reaction that includes peroxy radical, whilst other radicals follow the scheme similarly. Initially, the produced peroxy radical entered rubber network and instead of producing the preferred Zaitsev's product, the majority of products were those of Hoffmann's (attacking main chain). In other words, depolymerization was preferred over devulcanization. The attack of peroxy radical onto the C-H bond in the polymeric backbone resulted in the production of alkylperoxy benzene and polymeric radical. According to M. Myhre *et al.* [2], such phenomena is due to the excessive production of H_2O_2 side product in the first step which has strong influence to negate the production of Zaitsev's product. This influence is exceptionally common in many organic syntheses, such as that of anti-Markovnikov's reactions [15].

The reaction continued through the generation of alkoxy radical and phenolate radicals which were used to excessively cleave C-C bond within the polymeric network, generating various products that are shown in Figure 3.1 (alkyl-phenolate, polyether compound, polymer chain elongation produced from two polymeric radicals, etc.). Hence, this scheme is a better fit to be called depolymerization or reclamation process rather than devulcanization as claimed [16]. Similar study was performed by A. A. Yehia *et al.* [17], whereby the authors used three different metal halide catalyst, namely $FeCl_2$, $FeCl_3$ and $ZnCl_2$, along with phenylhydrazine at differ-

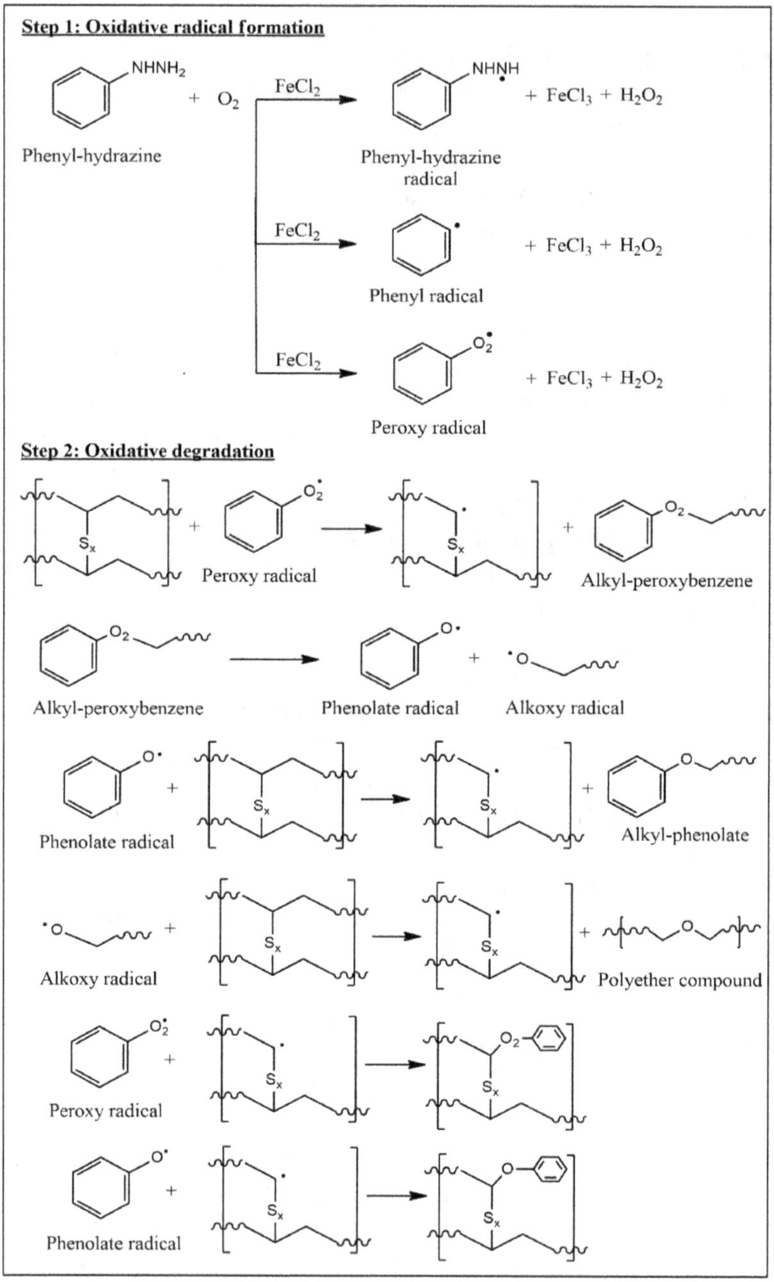

Figure 3.1 *Oxidative degradation by FeCl2 catalyst. Reproduced from ref. [14] with permission from Rubber Research Institute of India*

ent concentration. Devulcanization assessment was performed by the general gel-sol content analysis. The author reported that the best catalytic agent for devulcanization was $ZnCl_2$ at 1.00 phr, resulting in a sol content of 63.10%, while the optimal concentration for the phenylhydrazine was 1.5 phr, giving a sol content of 59.10%. Two years later, the same authors revulcanized the devulcanizate (treated at the obtained optimal conditions), with high-abrasion furnace carbon as the filler, mixed with natural rubber. The result obtained shows that the retention of the tensile strength and elongation at break of the revulcanizate was up to 50%.

An attempt to enhance the metal halide-based devulcanization treatment was carried out by K. Deivasigamani *et al.* [18] by means of ethanol, o-xylene, $CuCl_2$ and nano-$CuCl_2$ due to their active capability in transforming C-S bond to C-C and/or C-H bond(s). The mixture of these substances was then refluxed at 90°C for 60 mins. Successful devulcanization was studied by means of carbon, hydrogen, nitrogen and sulphur analyzer (CHNS) and Fourier transform infrared spectroscopr (FTIR). The earlier showed that treatment with $CuCl_2$ and nano-$CuCl_2$ reduced the sulphur percentage from 2.01% to 0.83-1.14%. On the other hand, FTIR investigation revealed the existence of C=S and S=O bonds at a bandwidth around 1250 and 1050 cm^{-1} respectively, indicating changes experienced by the crosslink network. The authors explained that such changes were due to the modification, grafting or attachment done by the devulcanizing agent onto the cleaved sulfidic bonds, as there was no liberation of gaseous H_2S observed.

3.2.2 Alkaline-based Chemical

Under this umbrella of devulcanizing agent, the initial principal that must be understood is that there are three phases involved within the system. The first being the waste vulcanizate acting as the viscous organic phase, NaOH solution as both of the aqueous phase as well as the devulcanizing agent and finally a quaternary ammonium salt (QAS) behaves as the catalyst. The most important substance in this system is the OH^- present in aqueous NaOH solution, due to the fact that this anion serves as the agent that is actually attacking and cleaving the sulphur crosslink network within the polymeric network [19]. However, simple mixing between the viscous and aqueous phase does not allow the migration of OH^- into the viscous organic phase, as the occurrence is at an exceptionally slow rate with small probability of reaction to get initiated. This is due to the large difference in the solubility between NaOH and vulcanizate [16]. Such large difference in solubilities was fortunately overcame upon the introduction of QAS (acting as catalyst) into the system, increasing the rate of OH^- transfer from aqueous to viscous phase and enabling the initiation of crosslink scission. In other words, the transfer of the hard anion (hydroxide ion) into the polymeric network is now possible. This interesting process was first patented in 1979 [20] and published in 1982 by P. Nicholas [21] which is commonly known as phase transfer catalytic system, whereby the QAS acting as the phase transfer catalyst (PTC). In his publication, P. Nicholas reported the procedure to conduct such devulcanization technique. The first step reported was refluxing the powdered vulcanizate with aromatic solvent (i.e. benzene or toluene) for 6 hours. The produced vulcanizate (10 g) was then mixed and refluxed with 24 cm^3 of aqueous sodium hydroxide at 2.08 mol/dm^3 and a QAS catalyst (i.e., N-methyl-N,N,N-tri-n-alkylammonium chlorides) at temperature below 100^oC for 120 mins. Next, the devulcanizate produced was filtered, dried and the crosslink density was measured. The author reported that 70-75% degree of devulcanization was achieved which majorly consisted of di- and polysulfidic bonds, while monosulfidic scission was little to not being experienced [21]. Additionally, the author also acknowledged that this particular system was exothermic and semi-catalytic as the QAS used was identified to experience degradation after the treatment process and suggesting the use of QAS with longer alkyl chain to enhance the degree of devulcanization. Prior to understand the mechanism of this process, it is important to note that the diffusion or transfer of the hydroxide ion is only onto the surface of the vulcanizate, allowing the hydroxide ion to selectively and readily attack the crosslink network [16]. Figure 3.2 illustrates the step by step mechanism for the major reaction, occurred for di- and polysulfide scission.

In the first step, QAS catalyst would exchange its one halide ion with a hydroxide ion from NaOH, forming sodium halide salt and quaternary ammonium hydroxide. The earlier would remain in the aqueous phase, while the other moved the OH^- ion to the rubber phase. The exchange process would continue until the system's saturation point or depletion of the QAS' halide ion. Once it is in the rubber phase, the OH^- ion is then ready to attack either the di- and/or polysulfide bond by donating its electron pair (Lewis' base), as shown in step 2(a). The heterogeneous scission experienced by the sulphide bond would then produce a sulfenic acid and thiolate ion. However, it must be noted that sulfenic acid is a tautomer-active molecule and able to arrange itself to form alkyl-hydrosulfinyl compound as shown in step 2(b). When the alkyl-hydrosulfinyl produced is exposed to another molecule of sulfenic acid that eventually formed, the hydroxyl group from the sulfenic acid will deprotonate the alkyl-hydrosulfinyl forming sulfinothioate compound and water. Finally, the formed sulfinothioate compound would react with one hydroxide ion transferred by QAS to form a stable sulfinic acid and thiolate ion. With respect to the reaction described, the deprotonation of alkyl-hydrosulfinyl indicates the formation of water molecule which marks the safety of this scheme and further waste treatment is not required.

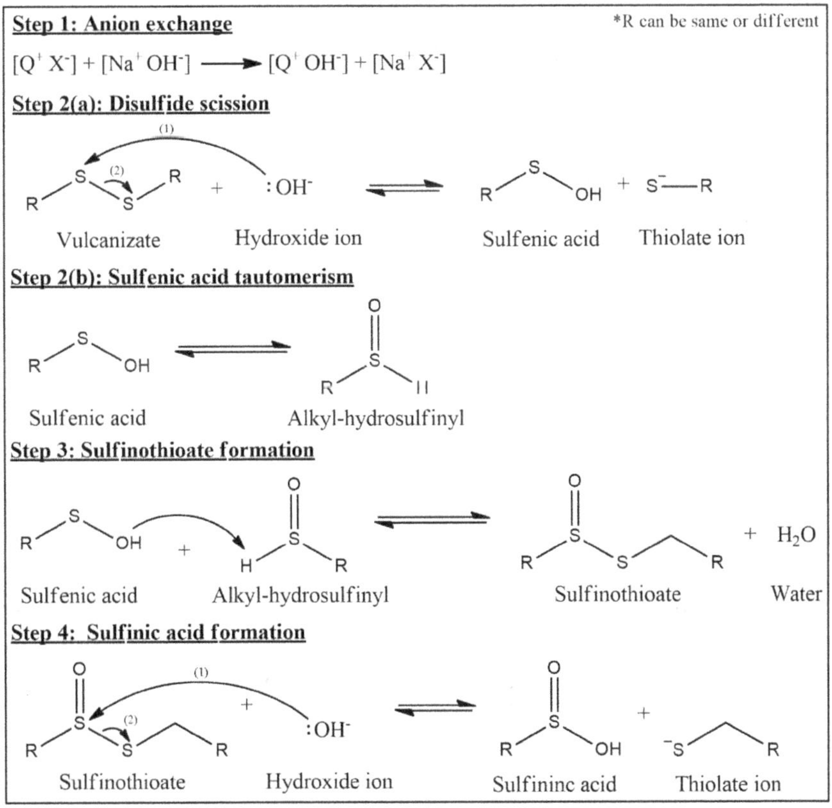

Figure 3.2 *Devulcanization chemistry of QAS-based PTC system. Reproduced from ref. [21] with permission from American Chemical Society.*

Observing the PTC mechanism described in Figure 3.2, it is clear that the rubber crosslink experiences oxidation reaction through functional group modification by the addition of oxygen atoms. This in turn also increased the polarity of the produced devulcanizate as well as improving its hydrophilicity and, hence, better adhesion upon revulcanization with filler or to be reproduced as composite [22]. After such interesting finding, M. Nuzaimah *et al.* [11] further the study using different concentration of NaOH. The authors found that treatment with 7.0 and 10.0% of NaOH gave the best degree of wettability and hydrophilicity of devulcanizate, seen though the reduction of surface contact angle of filler towards revulcanizate by about 27% and the highest value of tensile strength. Additionally, such decrease alleviated the hydrophobicity of the devulcanizate and enabled better interfacial adhesion with its composite constituents when it undergoes revulcanization, alongside with the increase of their bonding strength [23]. However, decrement in both tensile strength as well as elongation at break of revulcanizate from original vulcanizate were also reported which was caused by the damaged received by the vulcanizate upon treatment. This finding corroborates with the recent publication reported by A. Kashani *et al.* [24], where the authors also concluded that prior treatment using NaOH allowed the reduction of contact angle from the initial vulcanizate due to a better interfacial adhesion.

3.3 Devulcanization by Free Radical Mechanism

The background of this devulcanization process came from the understanding that when vulcanizate is subjected to high mechanical shearing or exposed to high enough temperature, the crosslink network would then be homolytically cleaved and fragmentized to form sulfide radicals [25]. Nevertheless, it is to be noted that only di- and polysulfidic bonds are often cleaved due to their lower thermal stability, while monosulfidic bond still remain intact. Increasing the temperature and/or shear rate beyond critical levels would only enhance depolymerization which leads to the significant decrement of polymer's mechanical and physical properties [14]. Additionally, the free radicals formed most often recombine and regenerate the sulphur bridge. Hence, experiencing lower degree of devulcanization. Such is the basic occurrence for the typical thermal, mechanical and thermomechanical devulcanization processes. To prevent the recombination of these reactive sulfide radicals (which typically occurs in thermal, mechanical and thermomechanical devulcanization), an introduction of specific agent(s) is necessary to graft itself onto these sulfide radicals. Hence, preventing these radicals to be reactive. Such agents are commonly known as radical scavenger, whereby disulphides, benzoyl peroxide and benzoic acid are most often used.

3.3.1 Disulphides-based Chemicals as Radical Scavenger

Chemicals fall under this group of devulcanizing agent are considered to be the most reported within the literature. For instance, S. O. Movahed *et al.* [26] reported the assistance of 2-mercaptobenzothiazole-disulfide (MBTS) in thermomechanical devulcanization of ethylene-propylene-diene rubber (EPDM). The authors reported that the devulcanization process occurred at 220°C with screw speed of 180-200 rpm, and the produced devulcanizate was compounded with pure EPDM. The result showed that when the blend was vulcanized at 10, 20 40 and 50 wt% of EPDM devulcanizate, there was no change in tensile strength, compression set, Mooney viscosity and hardness respectively. Additionally, the author also acknowledged that since the properties of vulcanizate blends produced were highly dependent on the amount of EPDM devulcanizate added, adjustments need to be made in accordance to ones' needs.

S. Ghorai *et al.* [27] employed bis(3-triethoxysilyl propyl) tetrasulfide (TESPT) to devulcanize ground natural rubber vulcanizate (GNRV). Their initial procedure was soaking the GNRV with 10 mL of spindle oil and TESPT at various amount (2-8 mL) for a period of 24 hr. Next, the mixture was ambiently introduced into a two-roll mill at variable time and a friction ratio of 1.2. The investigation showed that treatment at 60 min gave the highest soluble content, but the maximum positive gradient was only experienced by treatment at 20 min as seen in Figure 3.3 (i). Such indicating the dependency of sol content on devulcanization time. Additionally, it was also reported that the optimal amount of TESTP used was 6 mL, while at 8 mL a decreasing trend was seen such as that illustrated in Figure 3.3 (ii). In a different study, X. Zhang *et al.* [9] utilized a different concentration of novel disulphides based chemical, 4,4'-dithiobis(2,6-di-t-butylphenol) or thiobisphenols, to chemo-thermochemically devulcanize waste rubber powder (WRP). The process was carried out at 180-200°C in an internal mixer and the authors suggests that higher amount of thiobisphenols as well as temperature were promoting the decrement of crosslink density with some main chain scission. Nevertheless, treatment with 0.5 gr of thiobisphenols at an operating condition of 200°C was the optimal setting.

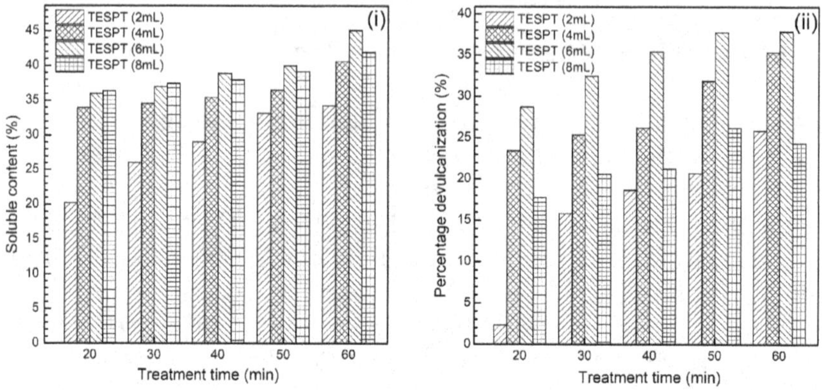

Figure 3.3 *Effect of TESPT on (i) soluble fraction and (ii) devulcanization efficiency.*

The utilization of disulphides such as MBTS and TESPT are just some of the chemicals used. Other disulphides chemicals such as tetramethylthiuram di-sulfide (TMTD) [28], diphenyl disulphides (DD) [29], 2,2-dibenzamidodiphenyldisulphide (DBDD) [30] and 2,2'-di-benzo-thiazole-disulfide (DBTS) [4] were also employed, and some degree of inferiority between the revulcanizate blend with the original vulcanizate was observed. Their measured chemical as well as physical properties are compiled and shown in both Table 3.1 and Table 3.2. Interestingly, many authors stated that these agents have symmetrical geometry [28–31], which is the main factor for the initiation of free radical mechanism, proven through the identical mechanism reported. For simplicity purposes, only mechanism reported by X. Zhang *et al.* [9] is explained and illustrated in Figure 3.4.

***Table 3.1** Summary of chemical properties of disulphides-treated devulcanizate*

Disulphide	Amount of agent used	Temperature (°C)	Soluble content (%)	Crosslink density (mol/cm³)	Percentage devulcanization (%)	Ref.
DBTS	0.50 phr	110.0	66.90	3.10×10^{-5}	53.00	[4]
Thiobisphenol	0.50 gr	200.0	19.00	9.50×10^{-5}	-	[9]
MBTS	6.00 phr	220.0	17.60	1.45×10^{-5}	92.19	[26]
TESPT	6.00 mL	Ambient	42.50	-	37.80	[27]
TMTD	2.75 gr	Ambient	-	5.14×10^{-5}	-	[28]
DD	1.00 phr	110.0	-	4.21×10^{-5}	38.33	[29]
DBDD	1.00 phr	150.0	31.00	2.22×10^{-5}	85.00	[30]

As seen in Figure 3.4 step 1(a), the first step involved homolytic cleavage of the devulcanizing agent due to mechanical shearing and/or thermal exposure, producing two sulfide radicals. This is possible as the resonance structure between the benzene ring and S-S bond were substantially disrupted and weakened. Such thermomechanical exposure, simultaneously caused the scission of both crosslink and carbon backbone of polymer, producing their respective radicals as shown in step 1(b) and (c) respectively. Next, due to the formation of various free radicals, three different reactions may occur which are crosslink capping, main chain capping and polymeric chain extension as shown in step 2(a), (b) and (c) respectively.

Table 3.2 *Summary of mechanical properties of disulphides-treated devulcanizate*

Disulphide	Blend composition (phr)	Tensile strength (MPa)	Elongation at break (%)	Tear Strength (N/mm)	Hardness (shore A)	Ref.
DBTS	NR(60)/WRP(40)	16.87	447.00	29.42	50.00	[4]
Thiobisphenol	Pure WRP	5.80	320.00	-	-	[9]
MBTS	Pure EPDM	3.00	60.00	-	71.00	[26]
TESPT	Pure GNRV	7.79	387.04	-	67.10	[27]
TMTD	Pure GNRV	14.17	1240.12	-	40.00	[28]
DD	Pure GNRV	9.40	1006.00	18.07	38.00	[29]
DBDD	NR(30)/dNR(70)	-	450.00	50.00	-	[30]

Referring to Figure 3.4 step 2(a), thiobisphenols radicals were highly attracted to the crosslink radicals which resulted in the formation of thiobisphenol terminal. This, however, is the preferred reaction pathway as the reactivity of the crosslink radical was negated and resulted in the reduction of crosslink density. On the contrary, reaction following step 2(b) and (c) were not preferred as it led to no change in crosslink density, seen through the undisrupted crosslink network by the thiobisphenol radical. Instead, capping at unwanted terminal and chain extension was experienced. The possible occurrence of step 2(b) and (c) was expected due to the insufficient amount of thiobisphenol used in the devulcanization process, implying the significant importance in studying the most optimal amount of devulcanizing agent used. In the review published by L. Asaro *et al.* [8], the amount of devulcanizing agent that should be added is about 0.5-10 wt%.

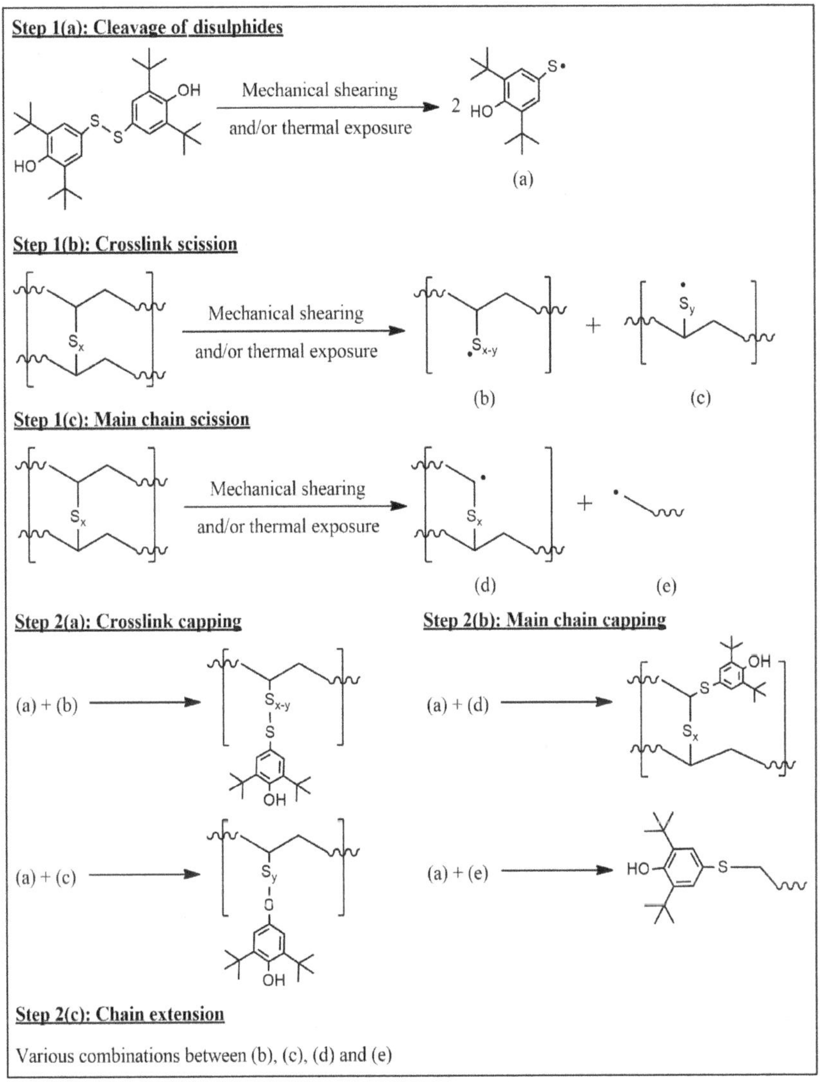

Figure 3.4 *Devulcanization mechanism by thiobisphenols.*

3.3.2 Benzoyl Peroxide as Radical Scavenger

Other than by means of disulphides-based chemical, devulcanization through free radical mechanism was also seen through the utilization of benzoyl peroxide (BPO) as the devulcanizing agent. This particular pathway was first reported by Rooj *et al.* in 2011

[32], whereby its mechanism is somewhat similar to that of disulphides but with the possibility of cleaving the monosulfidic bond. The authors evaluated the impact of both chemical and mechanochemical-based devulcanizations on natural rubber vulcanizate as well as the concentration of the BPO used. In the earlier investigation, the authors observed a decrement in volume fraction of devulcanizate compared to that of the original sample, which is a common indicator for the successful cleavage of crosslink scission. This is due to the fact that the formation of crosslink always forms a compact molecular environment which resulted in the increase of polymeric volume fraction. Additionally, the authors observed the declination in crosslink density of devulcanizate, i.e., the increase in percentage of devulcanization, as treatment time was increased from 2-6 hrs. The author also expected the crosslink scission experienced was a mixture of mono-, di- and polysulfide due to the decrease in tensile strength as well as moduli.

A detail description on the devulcanization performed by Rooj *et al.* [32] is shown in Table 3.3. The table indicates that there is no significant difference in terms of physical properties and percentage of devulcanization between the control sample and its revulcanizates that were treated mechanochemically. However, it is interesting to find that the treatment time was needed to be decreased from 2-4 hrs to 10-30 mins to achieve the same degree of devulcanization as those which were only treated chemically. Such is obvious as high shear force was introduced. Their study was then continued by C. Temram *et al.* [33] using various concentration of BPO to devulcanize GNRV (Table 3.3). The authors found that the optimum concentration of BPO was 5 phr, seen through the negligible increase in the percentage of devulcanization when BPO concentration was increased to 10 phr. Such giving a maximum degree of devulcanization of 64.5%. Additionally, the effect of devulcanization temperature (80-105°C) was investigated. However, there was no appreciable change observed in the crosslink density, percentage devulcanization as well as soluble content. To further understand the chemistry, we proposed the reaction scheme illustrated in Figure 3.5. The first initiation in step 1(a) involves the symmetrical and homolytic cleavage of BPO leading to the formation of benzoyl radicals. It is to be noted that the chemistry of benzoyl radical is highly unstable and was further broken down to form a highly reactive phenyl radical, while releasing CO_2 as shown in step 1(b). The generated phenyl radical would attack either the weakest S-S bond or C-S bond within the polymeric network according to step 2(a) and 2(b) respectively, producing sulfide radicals. Moving on to step 3 of Figure 3.5, the presence of phenyl radical would further initiate C-S bond cleavage in accordance to counter cure phenomenon developed by ESCHER Holding Corp, New York [32]. This, in turn, resulted in a formation of carbon radical within the polymeric chain as well as cyclo-sulphur compounds (typically hexa- or octagonal sulphur). We suggest that carbon radicals formed earlier are expected to either capped by the phenyl radical or experience chain extension. The earlier is suspected to be the Zaitsev's product due to the nucleophilicity of the phenyl. On the other hand, cyclo-sulphur was reported to experience the typical oxidation process due to the atmospheric treatment environment where O_2 was readily supplied [14].

Through FTIR characterization, Rooj *et al.* showed that the crosslink network peak at 556 cm^{-1} showed a decrement in intensity as treatment time was increased. Such inferring to the breakdown of S-S linkage [32]. M. Sabzekat *et al.* [34] whom also employed BPO in GNRV thermochemical devulcanization, through their Horikx analysis, reported that lower BPO concentration and shorter treatment time resulted in a better selectivity towards cleaving the crosslink network. Furthermore, their investigation on revulcanizate's tensile properties showed no adverse effect when the 40 phr of NR devulcanizate was compounded with virgin rubber.

BPO is considered as a hazardous chemical which may cause explosion at high concentration. As an effort to mitigate its usage, this particular chemical had recently been replaced with a chemical mixture of quinone with benzoic acid in devulcanization application. Substitution of BPO with quinone-benzoic acid (QBA) mixture, was studied by M. M. Hassan *et al.* [35] and K. F. Al-Nemr *et al.* [36]. The result obtained showed comparable values in terms of devulcanizates' gel content. For instance, their study reported a gel content between 68.80-79.90%, compared to that reported by Rooj *et al.* [32] which was between 60-70%. The approximately nine percent difference was expected due to a slightly higher operating temperature in BPO-based devulcanization. The mechanism scheme for devulcanization by means of QBA is practically similar to BPO's which was illustrated in Figure 3.4 earlier. The only difference is their reactant for benzoyl radical formation. Unlike in BPO, where two benzoyl radicals were easily formed through homolytic scission, the benzoyl radical in QBA was only able to be formed through the presence of quinone-based compounds. Furthermore, only 1 benzoyl radical could be formed per-QBA. A greater number of phenyl radicals formed per-BPO compared to per-QBA, was also expected to be one of the factors that affected the difference in their reported degree of devulcanizations.

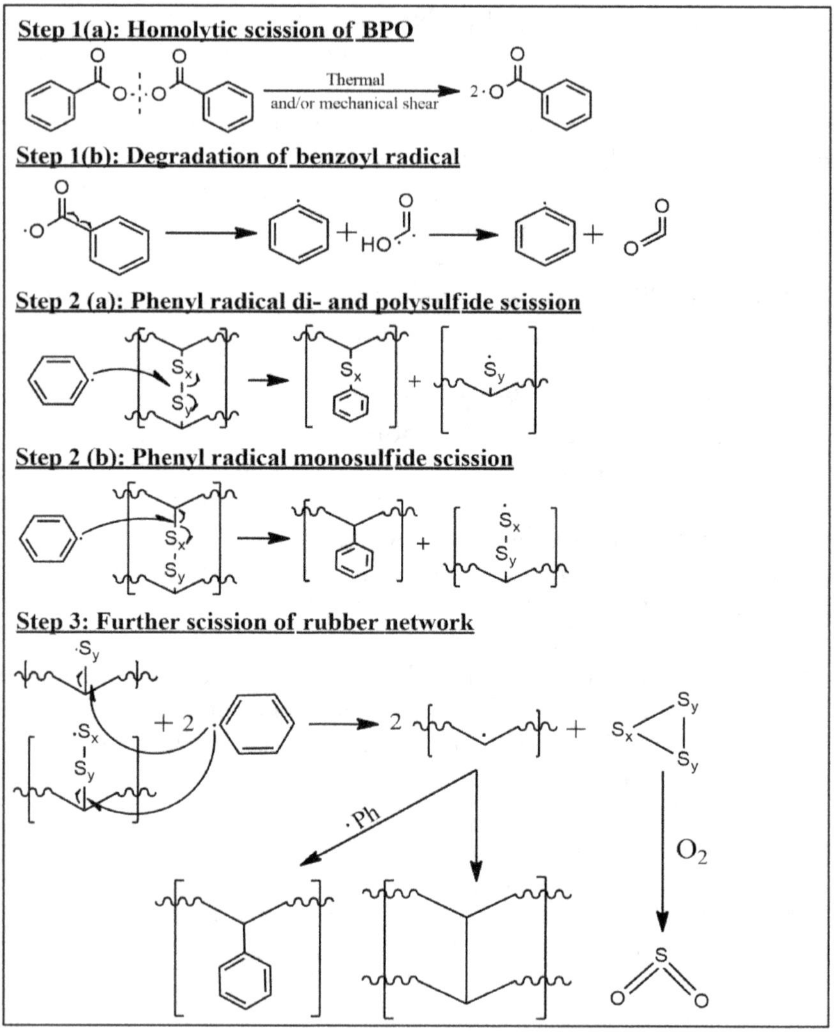

Figure 3.5 *Mechanism for BPO devulcanization reaction.*

Table 3.3 *BPO-based chemical and mechanochemical devulcanization comparison*

Reference	M. Sabzekat *et al.* [34]				C. Temram *et al.* [33]				Rooj *et al.* [32]						
Formulation	**C***	**1**	**2**	**3**	**C***	**1**	**2**	**3**	**C***	**1**	**2**	**3**	**4**	**5**	**6**
Temperature (°C)	-	80	80	110	-	90	105	105	-	80	80	80	80	80	80
Time (min)	-	120	240	240	-	55	55	55	-	120	240	360	10	20	30
Screw speed (rpm)	-	-	-	-	-	25.0	25	25	-	-	-	-	60	60	60
Hydrocarbon oil	-	100 mL	100 mL	100 mL	-	10 phr	10 phr	10 phr	-	400 mL	400 mL	400 mL	400 mL	400 mL	400 mL
BPO conc. (phr)	-	6.0	6.0	6.0	-	1.0	5.0	10.0	-	2.0	2.0	2.0	2.0	2.0	2.0
Volume fraction	-	0.2	0.2	0.2	-	-	-	-	0.2	0.2	0.1	0.1	0.2	0.2	0.1
Sol fraction (%)	-	1.2	12.5	12.8	30.3	38.1	39.1	39.7	-	-	-	-	-	-	-
Crosslink density x10^{-4} (mol/cm^3)	2.6	2.0	1.1	0.9	9.9	7.5	3.9	3.8	7.3	6.4	4.1	3.9	6.3	5.1	4.0
%Devulcanization	-	23.0	57.0	65.0	0.0	29.0	64.5	64.1	-	10.5	43.5	46.8	13.5	29.6	45.2
200% moduli (MPa)	4.1	4.6	4.8	4.9	-	-	-	-	1.1	0.9	0.9	0.9	1.0	1.0	0.9
†T_s (MPa)	22.0	12.0	16.5	16.0	-	-	-	-	15.0	13.7	4.2	1.8	11.2	7.1	1.8
•E_b (%)	447	389	408	400	-	-	-	-	1200	1151	618.0	370.0	1101	612	340
Hardness (Shore A)	50.5	56.6	55.8	54.5	-	-	-	-	-	-	-	-	-	-	-

*C : control or untreated rubber
†T_s : tensile strength
•E_b : elongation at break
\- : not reported and/or not available

3.4 Nucleophile-assisted Devulcanization Scheme

Nucleophile defined as a compound or ion that has and able to donate a pair of electrons (or acting as an agent in proton transfer) to the neighboring compound for initiating a certain type of reaction. In this case, devulcanization reaction. Various studies had reported their findings in the utilization of thiols and amine [37–39] as the devulcanising agent through this particular scheme. Moreover, between the two agents, amine-based chemicals are particularly known to be the strongest nucleophile as they are able to cleave the sulfidic bond that presents in the general cyclo-octasulfur geometry.

3.4.1 Amino Derivative Compounds

Amine's degree of nucleophilicity decreases from quaternary to primary amine and it was expected that the better the degree of nucleophilicity of a devulcanizing agent, higher degree of devulcanization would be achieved. However, it was previously declared that the different types of amine, its degree of basicity, and number of hydrogen attached to the nitrogen atom, did not influence the devulcanization process in any ways [2]. Instead, the presence and the number of α-hydrogen within the amine molecule is the primary factor that strongly and directly influence the degree of devulcanization as described in a 2003 patent which was filed by M. van Duin *et al.* [40]. The patent also noted that the presence of any substituent within the amine does not disrupt the reactivity of the α-hydrogen, while J. K. Premachandra *et al.* [37] recommended the use of aliphatic amine that has 10-14 carbon atoms for the best devulcanization treatment. In choosing the length of carbon chain present in the amine, we suggest a prior economic evaluation to be done, as longer chain compounds are typically much more expensive.

α-hydrogen mentioned earlier defined as hydrogen atom(s) that is attached to an α-carbon atom, while α-carbon is referring to a carbon next to the nitrogen atom within the amine molecular structure. Figure 3.6 illustrates the four different types of amines along with their numbers of α-hydrogen. The unique characteristic of this particular α-hydrogen was explained by B. Chan *et al.* [41], whereby it exhibits an abstraction effect towards the crosslink network that resulted in the promotion of amine compound's α-carbon and sulfide fragmentations. The α-carbon radical of amine that was formed would then be stabilized by its internal captodative effect. On the other hand, the sulfide radical formed attracted the α-hydrogen radical released by the α-carbon.

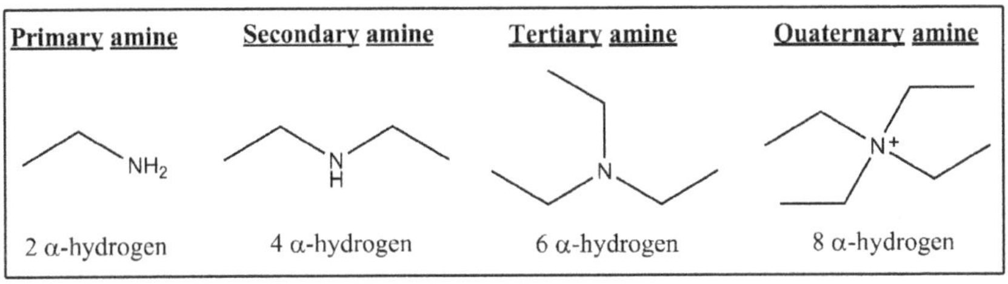

Figure 3.6 *Types of amines and their α-hydrogen.*

C. Jiang *et al.* [39] reported their findings on the utilization of tetraethylenepentamine (TEPA) as the delinking agent in mechanochemical devulcanization of ground tire rubber (GTR), whereby amine based compounds was observed to be able to graft itself onto the fragmented crosslink chain due to the abstraction and captodative effect described earlier. The devulcanization of GTR was carried out in a two-roll mill through the addition of 5 wt% of TEPA at ambient condition. Interestingly, the authors found that an FTIR peak at about 3100-3400 cm^{-1} in the sol fraction appeared, which indicating the successful in-situ grafting of TEPA onto the GTR resulting in its dissolution in xylene. The author also reported that the optimal tensile properties achieved by revulcanizing devulcanized GTR-high density polyethylene (HDPE) rubber blend was 18.30 MPa, 100.40 kN/m and 287% for tensile strength, tear strength and elongation at break respectively. The mechanism reported by the author is shown in Figure 3.7.

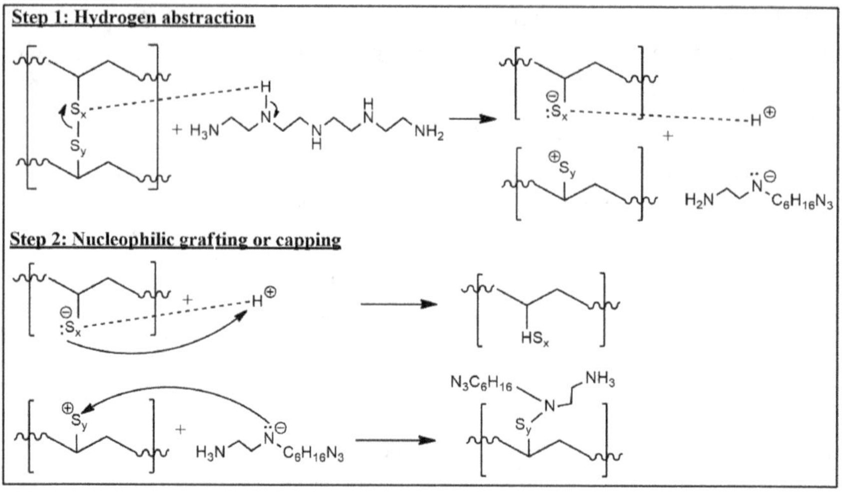

Figure 3.7 *TEPA grafting on GTR.*

K. A. J. Dijkhuis *et al.* [42] performed their thermomechanical devulcanization of EPDM rubber at temperature range of 225-275°C through the utilization of hexadecylamine (HDA), whereby the author reported a maximum degree of devulcanization of about 50% at HDA content of 2.5 phr. The author also reported the type of crosslink which was broken in their investigation were only di- and polysulfide, while monosulfidic bonds were not successfully cleaved. Despite some crosslinks were successfully cleaved, significant depolymerization as well as reclamation was observed from the Horikx analysis. We expect such outcome due to the overly high operating temperature leads to the simultaneous cleavage of the C-C backbone of the polymer. Nevertheless, the revulcanizates' tensile strength, modulus and hardness were concluded to be at satisfactory. In a different study, GTR at 40-mesh was mechanochemically devulcanized by means of undisclosed amino compound (5-20 gr) at temperature range of 30-70°C in a two-roll mill [37]. The author reported that optimal amount of amino compound used was 6 phr, characterized through the lowest value of crosslink density and the degree of devulcanization valued at about 80 % which confirms the suitability of amino compound to be used in devulcanization process as reported by A. Fazli *et al.* [43]

3.4.2 Aliphatic and Aromatic Thiols or Mercaptans

The second devulcanizing agent that can cleave the sulphur crosslink bond through nucleophilic scission is mercaptans or thiols which has the functional group of -SH. This particular agent is the next most effective nucleophile after amino compounds especially when they are attached to an aromatic ring instead of aliphatic [19]. However, the study on both aliphatic and aromatic-based thiols had been explored and it is interesting to observe that in almost all of the cases, thiol-based agents were mixed with piperidine that was acting as a base [14]. For instance, A. M. Joseph *et al.* [44] studied the influence of chain length and curing system used in thiol-based NR mechanochemical devulcanization. The first devulcanizing agent that they used was a mixture of short chain thiol (propane-2-thiol) and 0.40 mol/dm^3 piperidine in toluene, while the other was a mixture of 1.00 mol/dm^3 of hexane-1-thiol in toluene. Such forming thiol-amine. The authors reported that treatment with propane-2-thiol resulted in a higher polysulfidic cleavage, while the other resulted in a mixture of di- and polysulfidic cleavage (i.e., the latter being higher). The achievable percentage of devulcanization was only between 20-40%, however, retention of their mechanical properties was at 60-70%, except for tear strength which was at 30-40%. Such was due to the fact that some main chain scission occurred as proven by means of Horikx curve.

To prove the fact that aromatic thiol works better in devulcanizing elastomeric vulcanizate, P. Thaicharoen *et al.* [45] employed thiosalicylic acid as the devulcanizing agent for natural rubber vulcanizate. The process occurred at 140°C for 30 mins in a two-roll mill. Their investigation found that the maximum achievable degree of devulcanization was 74.83% at 2 phr of thiosalicylic acid, determined through the significant decrement in devulcanizate crosslink density. However, utilizing thiosalicylic acid at 1 phr resulted in 73.47% of devulcanization. This particular observation clearly indicates that there was no reason to use thiosalicylic acid at an amount greater than 1 phr, as there was only a 1.36% of increase at the exact same treatment condition. Additionally, the authors also revulcanizing the produced devulcanizate with virgin NR at different ratio and their mechanical properties were determined. Surprisingly, the tensile strength as well as elongation at break were negligibly influenced by the ratio of the blend, while the optimal amount of thiosalicylic acid was again observed to be 1 phr. The author suspected that increasing the amount of thiosalicylic acid resulted in an increase chance for depolymerization. Figure 3.8 illustrates the nucleophilic bond scission mechanism that happened.

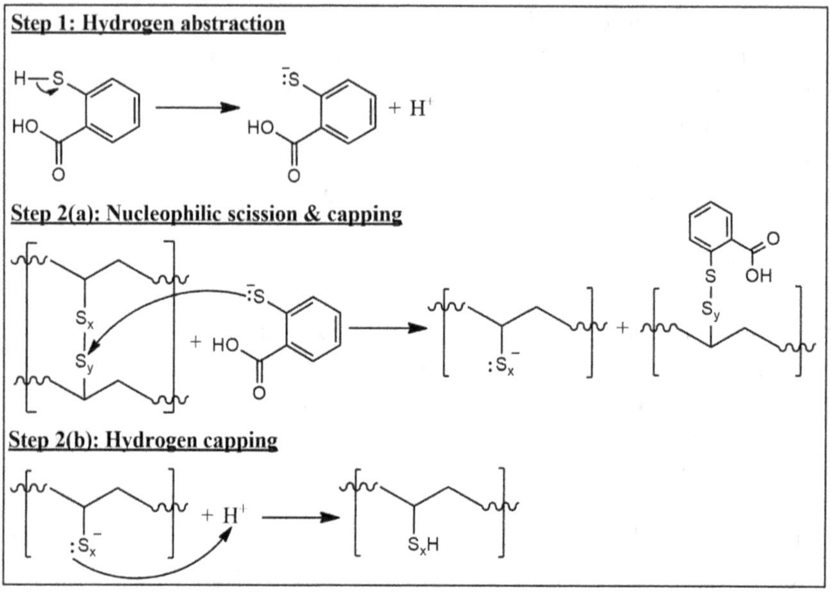

Figure 3.8 Possible thiosalicylic acid devulcanization mechanism. Reproduced from ref. [14] with permission from Rubber Research Institute of India.

The first step of Figure 3.8 shows that in this scheme, hydrogen was abstracted from the thiol group similar to the abstraction of hydrogen from an amino compound described in section 4.1. Such is due to the nucleophilicity of the compounds. It was earlier mentioned that amino group is more nucleophilic than thiol, such is due to the presence of free pair of electrons within the amino molecule. Moreover, aromatic-thiol is more nucleophilic than aliphatic-thiol, this resulted from the high degree of nucleophilicity of the benzene ring. Due to the fact that this particular mechanism is initiated by the attack of nucleophile, stronger nucleophile is expected to result in a higher degree of devulcanization. Such has been seen by comparing the study performed by A. M. Joseph *et al.* [44] and P. Thaicharoen *et al.* [45] described earlier. Continue on to step 2 (a) of the mechanism illustrated in Figure 3.8, the [:S$^-$] nucleophile would then readily attack one of the crosslink sulfurs resulted in their heterolytic cleavage. Sulphur from the crosslink cleavage that did not receive an electron would combine with the deprotonated thiosalicylic acid. On the contrary, sulphur from the crosslink that received a pair of electrons would act as nucleophile and attack the free [H^+], forming a thiol group onto the cleaved crosslink chain. However, it should be noted that this sulphur may experience oxidation at atmospheric oxygen and form a more stable C=S bond instead of C-SH.

4.4.3 Ionic Liquids and Eutectic Solvents

Various published studies in regards to lignin depolymerization by means of ionic liquids (IL) are kept on increasing annually due to the amount of essential products they could generate [46, 47]. However, limited studies were found of rubber devulcanization by means of this intriguing solvent that has the capability to have tailored properties. Until 2015 that is, where S. Seghar *et al.* [48] pioneered the first microwave devulcanization study of styrene butadiene rubber (SBR) through the utilization of pyrrolidinium hydrogen sulfate (PHS) IL amounted at 10 wt%. The concept behind their idea was the fact that IL is able to absorb microwave radiation and translate that to heat which can penetrate into the core of the grounded SBR. Their sample formulation is shown in Table 3.4 and the results from the devulcanization study are illustrated in Figure 3.9.

***Table 3.4** SBR-PHS formulation in IL-assisted microwave devulcanization*

SBR without PHS	SBR with PHS	Magneton power (W)	Treatment time (s)	Energy (Wh/kg)
SBR	SBR-P	0	0	0
SBR-110	SBR-P-110	656	30	110
SBR-220	SBR-P-220	656	60	220
SBR-330	SBR-P-330	656	90	330
SBR-440	SBR-P-440	656	120	440

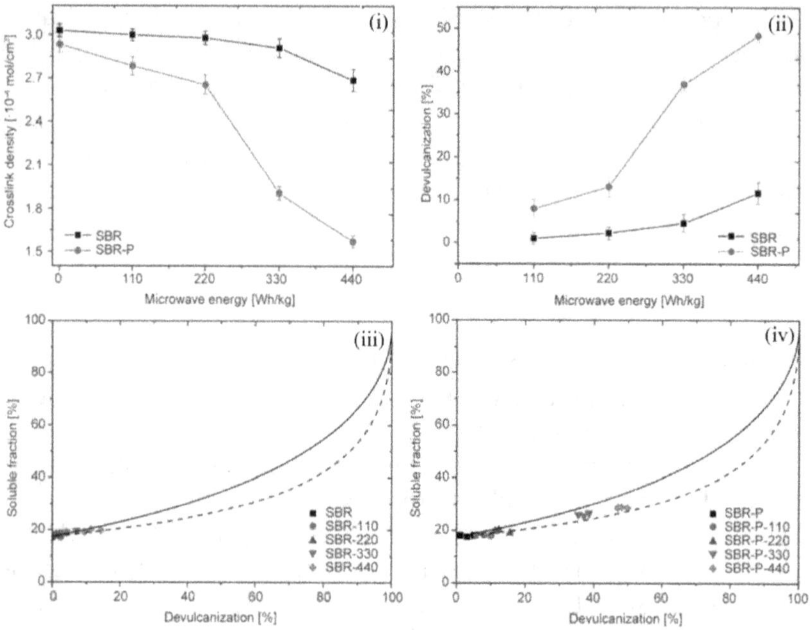

Figure 3.9 *PHS-treated SBR (i) crosslink density and (ii) percentage devulcanization, along with Horikx analysis for (iii) non-PHS and (iv) PHS treated samples. Reproduced from ref. [48] with permission from Express Polymer Letters.*

With respect to Figure 3.9 (i), it is clearly seen that the degree of devulcanization between PHS treated and non-PHS treated samples was only evident when the microwave energy was at 220 Wh/kg or higher. Furthermore, the effectiveness of IL could also be observed from Figure 3.9 (i) whereby it shows that samples that was imbibed by PHS shows a significant negative gradient in terms of crosslink density. Such indicates the fact that a certain degree of sulfides bonds was cleaved. The inverse similarity of the reported Figure 3.9 (ii) towards Figure 3.9 (i), demonstrates the inversely proportional relationship between crosslink density and percentage devulcanization as described in Flory-Rehner equation [49]. Despite the favorable degree of devulcanization achieved, both Horikx curves display in Figure 3.9 (iii) and (iv) concluded that the crosslink scission experienced was accompanied by polymeric cleavage. The mechanical properties of revulcanizate and their mechanism were not reported, however, similar scheme as that illustrated in Figure 3.7 is expected. Nevertheless, S. Seghar *et al.* [48] have proven the positive role of IL as a potential devulcanizing agent. We expected such outcome was due to the excessive increase in soluble content which may be due to the high temperature exposure and prolonged treatment time. Such insight is in agreement with the recent depolymerization (or metathetic degradation) study of NR by means of hydrophobic IL, reported by A. Mouawia *et al.* [50]. For a certain period of time, ILs were considered as a potential and attractive solvent due to their tailorable properties. However, they are commonly known for being toxic, non-biodegradable, high in viscosity, corrosive, expensive and difficult to be synthesized [51]. Such is true as IL generally need to be made from exceptionally pure constituents, costing 5 to 20 times higher [52].

Recently, its alternative, referred to as deep eutectic solvent (DES), emerged with much more sustainable properties compared to that of IL, while still possessing the desired tailorable properties [53]. The first invented and most studied as well as applied DES up to the current times are those of choline chloride (ChCl)-based [54]. As a replacement to IL, the study of DES as a devulcanizing agent was pioneered by R. Saputra *et al.* [55] using the combination of ChCl, $ZnCl_2$ and urea in their GTR thermosonic devulcanization, at 37 kHz and 180°C. The authors reported that a significant shift in FTIR sulphidic bandwidths was observed alongside with the decrement of crosslink density which led to the high degree of devulcanization, ranging from 93-97%. To verify the selective cleavage, the authors fitted their devulcanization data onto the Horikx curve and the result indeed only showed crosslink scission. Moreover, their publication further proved that soluble content assessment without crosslink density evaluation will not give an accurate overview of devulcanization phenomena. In the next publication by the same authors, the influence of water dilution on DES was also evaluated and as expected, the increase in percentage dilution proportionally decrease the degree of devulcanization experienced [56]. In a different set of study published by R. Saputra *et al.* [6], ammonium-based ternary deep eutectic solvents (TDES) were used as a devulcanizing agent and the operating parameter was optimized. The author, through statistical analysis, reported that ultrasonic time and frequency as well as heating time and temperature were significant variables, whereby their optimal value were at 60 min, 37 kHz, 55 min and 180°C respectively. Horikx curve fitting was also made, showing only crosslink scission was experienced and proving the positive potential of DES as a devulcanizing agent. Figure 3.10 illustrates the proposed mechanism by the authors.

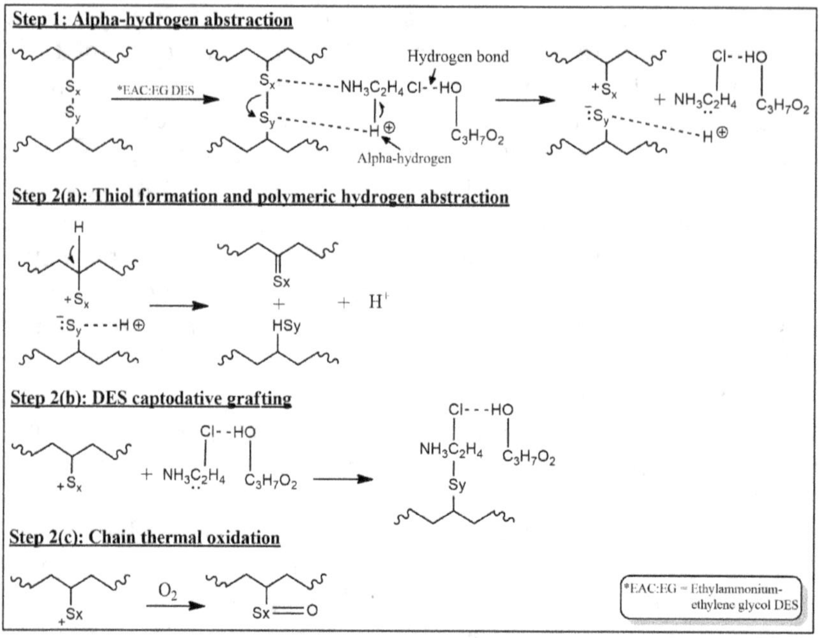

Figure 3.10 *Mechanism of DES in devulcanization. Reproduced from ref. [6] with permission from Elsevier.*

Based on Figure 3.10, the first step of the mechanism shows the behavior of a highly reactive α-H that is influenced by the nucleophilic characteristic of the sulphur atoms and/bonds that are present in the crosslink. This, in turn, resulted in the cleavage or opening of the sulfidic bond through heterolytic scission, producing a positive and negative terminal sulfide. However, due to the presence of hydrogen bond within the DES, abstraction of α-H is further enhanced due to the strong pulling force towards the involved electronegative atoms. Proceeding from this, two possible reactions were proposed:

the first described in step 2 (a):

- The negative terminal sulfide produced earlier would use its free pair of electrons to generate a thiol with the abstracted α-H.
- The positive terminal sulfide is highly unstable and prompts the abstraction of the adjacent H atom by cleaving the C-H bond. This in turn produces a more stable C=S bond.

- This particular scheme was proposed as there was an increase in thermal stability of devulcanizate (observed by TGA) compared to the untreated GTR.

the second described in step 2 (b) and 2 (c):

- The DES produced from the abstraction of α-H in step 1 earlier, was carrying a pair of free electrons – a nucleophile.
- Such nucleophile was expected to attack/donate its electron pair to the positive terminal sulfide to dismiss its instability and resulted in a possible hydrogen bond formation.
- Hence, DES was observed to somewhat grafted onto the negative terminal sulfide.
- This scheme was proposed as N-H bond in FTIR signal, about the bandwidth of 3250 cm^{-1}, was observed on devulcanizate which indicates grafting [39].
- Continue on to step 2 (c), on the other hand, the positive terminal sulfide generated from step 1 was expected to experience thermal oxidation by means of ambient O_2.
- Such, in turn, producing a sulfoxide terminal within the cleaved crosslink chain.

Another DES that was reported by the authors was diethylethanolammonium chloride-based, which has an oxygen atom attached with two pairs of free electrons. However, their investigation showed that it gave the lowest degree of devulcanization (~60%) compared to DESs that do not have oxygen atom attached (~65 to 75%). The author believed that such outcome could be due to the influence of the free electron pairs possessed by the oxygen atom which may slightly create competition or disrupt the crosslinking scission performed by the α-H. Nevertheless, the difference of percentage of devulcanization caused by such phenomena was not significantly large.

3.5 De-crosslinking by Swelling

Swelling of elastomeric vulcanizate defined as a polymeric enlargement phenomenon due to the expansion of its molecular network. It is generally caused or induced by the introduction of a certain type of fluid in polymer. It is further expected that the swelling of vulcanizate's matrix will open up both of the polymeric as well as the crosslink networks that are tightly entangled between each other. Hence, allowing an easier solvent (devulcanizing agent) penetration and equivalent dispersion into the matrix which eventually increase the efficacy of crosslink scission. Fluids that are used for swelling and had been studied up to this point can be categorized into two groups, which are unsaturated compounds (i.e. various aliphatic, aromatic and paraffinic oils) and supercritical (or subcritical) fluids.

3.5.1 Unsaturated Compounds

This particular group of swelling agents are generally those of hydrocarbon-based with a number of hydrogen deficiency index of, at least, one within its molecular structure. This allows the unsaturated carbons to be highly active. S. O. Movahed *et al.* [26] reported that such compounds were used as they are not just able to increase the degree of plasticity of devulcanizates, but also possesses an oxidative acceleration capability and can act as a radical acceptor. For instance, the mentioned author utilized twin-screw extruder for devulcanization of EPDM rubber, by mixing it with 6 phr of MBTS and 15 phr mixture of aromatic, aliphatic and paraffinic oils at 220°C. It was interestingly found that, the degree of devulcanization achieved was ranging from 91.85 to 92.19%, with a soluble content of 13.10 to 17.60%. Despite Horikx analysis was not reported, we believe that such high degree of devulcanization with small percentage of sol content is a good indicator for selective crosslink scission. The produced EPDM devulcanizate was also revulcanized in a blend with virgin EPDM at variable ratios and investigation showed that up to 20 wt% of EPDM devulcanizate added resulted in unchanged mechanical properties.

In a different study, S. Saiwari *et al.* [57] also utilized an unsaturated hydrocarbon compound, treated distillate aromatic extract (TDAE), as a swelling agent to assist the devulcanization by means of DD dispersion. TDAE swelling for 30 mins was found to be more selective compared to the sample that was swelled for 60 min. Additionally, the author reported that when swelling is done with 15% of TDAE, optimal condition was achieved. Above 15%, the author explained that ineffective shearing occurred within the internal batch mixer due to the lubricative property of the oil. Other observation showed that heterogeneous bond cleavage was experienced and the average decrement in crosslink density was seen to be 70%. Similar observation was also reported by Y. Li *et al.* [58] where the authors performed thermomechanical devulcanization of GTR by means of waste engine oil (WEO). They showed that competition between decrosslinking and secondary crosslinking took place. Furthermore, the presence of WEO allowed GTR to swell much more effectively which leads the breakage of partial sulfidic bonds. The effect of WEO was seen to be beneficial up to 45 wt%, while addition at an excessive amount was declared to decrease the degree of devulcanization due to its lubricity characteristic. Recently, the use of these unsaturated hydrocarbon compounds as a swelling agent in devulcanization were hardly reported due to their flammability and flashpoints that are considerable close to the operating temperature [48]. As an effort to substitute their usage, the emergence of soybean (or vegetable) oil as a swelling agent had been introduced in a number of publications due to their synergistic effect and the occurrence of controlled oxidation [59–61].

The comparison study on the effective SBR rubber devulcanization (at 150°C) between the typical hydrocarbon-based and vegetable-based oils as the swelling agent, was recently studied and reported by Z. Cheng *et al.* [60]. The authors used 10-mesh of SBR which was mixed with soybean oil at various ratios for an imbibition period of 2 hrs, in order to optimize the swelling. The result obtained was, as usual, reported in terms of Flory-Rehner analysis, soluble content and Horikx curve. In accordance to the reported soluble content, the authors showed that the addition of either soybean or paraffinic oil significantly enhanced the soluble content of devulcanizate. Moreover, it is also declared that the soybean-swelled samples are far more superior compare to those swelled by paraffinic oil with respect to soluble fraction that is about doubled. The authors suspected such occurrence was experienced due to the higher efficacy of soybean oil in weakening the intermolecular forces experienced by the polymeric network. This is caused by the higher degree of unsaturation of the soybean oil which resulted it to be much more prone to oxidation. The higher capability of soybean oil in swelling GTR, in turn, allowed the availability of larger free molecular environment for the ambient oxygen to attack either the S-S or C-C bond. Furthermore, at the same time, the oxidized soybean oil was also expected to induce the cleavage of these bonds. The latter, however, much more preferred due to the lower energy requirement. Moreover, it was apparent that the highest soluble content observed was 47% at (optimal) soybean oil content of 200 phr, while further increase resulted in the decrease of soluble content. Such was suspected due to the shielding effect of the soybean oil on the rubber which preventing it from experiencing further oxidation. The fate of the devulcanized GTR by means of Horikx theory showed that all of the treated sample lies above the main chain scission. Such indicates that this particular treatment mainly cleaves the C-C bond. The degree of this cleavage, however, is much more significant at higher soybean oil content. The effect of treatment time in regards to soluble content also was reported. Observation indicates that 4 hours was the optimal treatment time, while increasing the exposure time would instead decrease the soluble fraction. This result explains the possible intensification of re-crosslinking phenomena in comparison toward de-crosslinking.

In a different study, P. Song *et al.* [61] also conducted their investigation of the usage of soybean oil as a green plasticizers in a low temperature devulcanization process of GTR. The authors also stressed the importance of oil in stretching the crosslink chains which weaken the intermolecular attraction. Two important conclusions were reported. The first being the fact that the integration of the remaining gel content after devulcanization was strengthened due to the increase of crosslink density resulted from recombination. The other was a grafting occurrence of soybean oil onto the soluble fraction of the devulcanizate, determined by means of FTIR analysis and evidenced through the increase in the molecular weight observed. The results obtained were reproduced as Figure 3.11. In accordance to Figure 3.11 (a), it is clear that the optimal soybean oil content was 100 phr, while at higher amount did not result in an appreciable change. Furthermore, this is also in concurrence with the reported physical properties shown in Figure 3.11 (b), which corroborates with Z. Cheng *et al.* [60]. Nevertheless, the embedment of GTR devulcanizate into the blend resulted in the improvement of tensile and elongation at break compared to the embedment of untreated GTR. Due to such interesting phenomena, the author proposed a series of mechanism for the occurrence of this event, and it is illustrated in Figure 3.12.

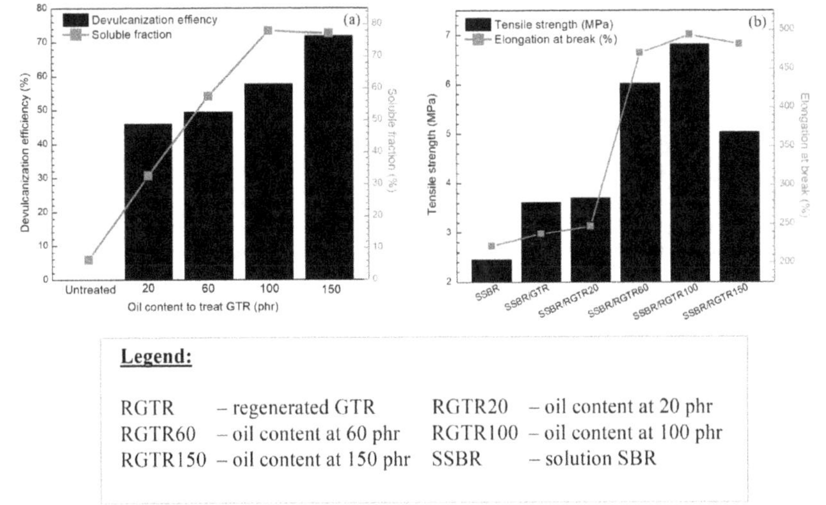

Figure 3.11 *Investigation on soybean oil-induced devulcanization at various content with respect to (a) soluble content, (b) Horikx curve and (c) mechanical properties.*

With respect to step 1 of Figure 3.12, it is clearly seen that the initial phenomenon that happened was the thermal scission of the GTR. The author reported that the scission which happened at 150°C, produced three different radicals which were C•, C-S• and •S• radicals. The next step can be divided into three different schemes, as shown in Figure 3.12 step 2(a)-(c). In the step 2(a), it was seen that the C• was thermally oxidized by accepting an electron pair from the incoming oxygen atom, while at the same time homolytically repelling its hydrogen. Such produced a H• and a carbonyl (C=O) terminal within the polymeric network that was expected to have a higher thermal stability. Referring to step 2(b), the C-S• formed earlier underwent further reaction with the unsaturated carbon from the soybean oil. This caused the π-bond present in the soybean oil to homolytically cleaved and attached onto the C-S•. Additionally, the adjacent tertiary C• produced due to such attachment was protonated by the H• produced in step 2(a) earlier. Lastly, in step 2(c), the crosslink radicals formed from the thermal scission in step 1 could experience crosslink recombination as well as chain extension which further strengthening the crosslink network.

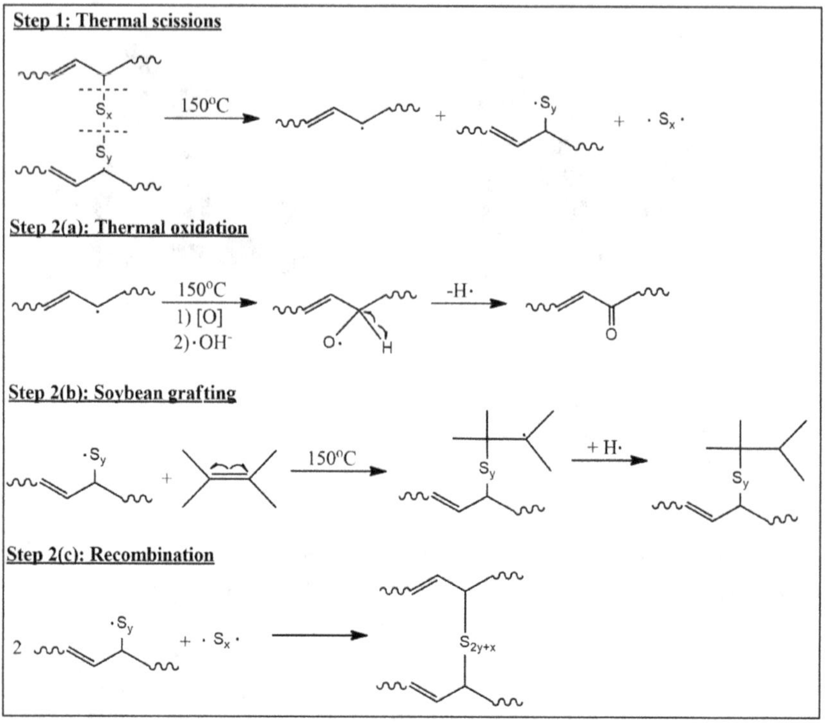

Figure 3.12 *Soybean-assisted devulcanization. Reproduced from ref. [62] with permission from American Chemical Society.*

C. C. Pegollo *et al.* [59] also reported their study on the use of oxidized soybean oil in rubber devulcanization as a swelling agent through a 2^2 factorial design. The effect of treatment time (20 and 40 mins) and soybean content (6 and 10 phr) were selected as the independent variables with respect to the responses, which are crosslink density and soluble content measurements. The authors reported that at lower bound of treatment time as well as soybean oil content, devulcanization process is much more superior. The value of crosslink density was found to drop by 71.10% from 2.1×10^{-4} to 6.2×10^{-5} mol/cm^3, while the soluble content was found to increase by 228.2% from 3.1 to 10.20%. Such corroborates with the previously discussed studies by Z. Cheng *et al.* [60] and P. Song *et al.* [61], whereby it is evident that there is a saturation point on the amount of swelling agent added and higher than that, would lead to ineffective treatment.

3.5.2 Supercritical and/or Subcritical Fluids

Other than swelling through the utilization of hydrocarbon oils or sustainably green vegetable oils, as described in the previous subsection, vulcanizates swelling by the assistance of supercritical or subcritical fluid is another popular technique that is widely reported in literature to aid the devulcanization process. Supercritical fluid defined as a fluid that is in a condition above of its critical point, whereby there is no clear distinction between its liquid/gas phase. Additionally, the gas that is above its critical temperature could not be compressed by any amount of pressure and due to indistinguishable liquid/gas phase, supercritical fluid generally does not exhibit any surface tension, but having a transitional viscosity of which value between its liquid and gas values. Despite being in such a state, supercritical fluid can dissolve solutes just like a liquid but also has a diffusivity capability similar to that of gas [8]. Hence, they are expected to be able to penetrate porous network of a solid which could enhance mass transport phenomena.

Supercritical ethanol (ScEtOH) utilization as an effort to swell and devulcanize waste sidewall rubber at variable temperature (200-260°C), time (30-120 mins) and pressure (8-12 MPa) was reported by X. Li *et al.* [63], accompanied by DD agent. Their full factorial investigation showed that only temperature was proven to be a significant variable, while pressure and time did not have any effect on devulcanization process. Moreover, the author also reported that gradual increase in soluble fraction (as well as decrease in crosslink density) was observed from 210 to 240°C and dramatic increase in soluble fraction (as well as increase in crosslink density) was observed from 240 to 270°C. Further increase, however, did not produce any appreciable enhancement in soluble fraction that was valued at 56%. The authors related the dramatical increase of soluble fraction between 240-270°C to the fact that ethanol was within its supercritical state. While it is in such state, the fluid is able to enhance the devulcanization process mainly through a more effective permeation of the fluid deep into the rubber network. Thus, targeting the inner crosslink networks (instead of just surface devulcanization) which result in a higher degree of devulcanization. Horikx study was also performed, showing treatment at 250-270°C mainly favored main chain scission, while treatment at 210°C equally favored main chain and crosslink scission. The outcome of this study was then compared to subcritical ethanol-assisted devulcanization reported by T. Zhang *et al.* [64]. In this study, the process occurred at 200°C and the highest resulted soluble content was about 72%. Higher degree of soluble fraction achieved in this study compared to the ScEtOH discussed earlier was expected due to the usage of an extruder that introduced a mechanical shearing effect onto the rubber, promoting more scission occurrences.

Aside from sub-/supercritical ethanol-assisted devulcanization, and mostly due to its flammable property, a differently modern as well as sustainably green swelling agent was proposed [65–68]. This particular agent is supercritical carbon dioxide ($ScCO_2$), with the properties of being non-toxic, inert, economical and non-flammable. In the early 2004, M. Kojima *et al.* [69] reported their brief study on NR devulcanization using $ScCO_2$ alongside with DD. The authors concluded that the amount of carbon black present in the NR vulcanizate did not affect the devulcanization process and the achievable soluble content was ranging from 20-40%. Furthermore, revulcanizing virgin rubber with devulcanizate produced (up to 40 phr) resulted in only 10% depreciation on the tensile strength.

In a different study, DD and $ScCO_2$-assisted devulcanization study of sulfur-cured butyl rubber (BR) vulcanizate was also published K. Jiang *et al.* [65]. The authors conducted the treatment in a high-pressure reactor where 100 g of solid CO_2, 50 g of BR vulcanizate and 0-5 g of DD were added together. The reactor was then sealed and allowed the mixture to be heated at thermal gradient of 3°C/min from the temperature of 140-190°C, whereas treatment time was varied between 0-240 mins. With respect to the exerted pressure within the chamber, the author reported that pressure was in linear relationship with soluble fraction. Pressure <7.38 MPa, however, only give rise to a low soluble fraction, whereas pressure ≥7.38 MPa resulted in a significant increase in devulcanizates' soluble fraction. Such is due to the fact that CO_2 was at its supercritical state when the pressure was ≥7.38 MPa which resulted in a proper and better swelling phenomenon. This, in turn, allowed the penetration of DD into the rubber network to have higher efficacy. On the other hand, reaction time, DD content and treatment temperature seemed to give negligible change in terms soluble content (~98.50%) when they were valued at ≥120 mins, 8 phr and 180°C respectively. Devulcanizates produced at such optimal conditions were then revulcanized with virgin BR at variable ratios and their mechanical properties were measured. It was observed that tear strength and hardness decreased from 14.8 kN/m and 34 shore A to

10.8 kN/m and 29 shore A when the BR content introduced was 30%. Further increase in BR content to 50%, proportionally decrease them both to 8.90 kN/m and 23.2 shore A respectively. Similar trend was also observed for 100% as well as 300% moduli. Such was reported due to the decrease amount of crosslink density. On the contrary, tensile strength and elongation at break increased with BR devulcanizate content. For instance, at 30% and 50% BR content, tensile strength was seen to increase to 6.8 and 5.5 MPa from 1.7 MPa, while elongation at break was observed to reach the value of 890 and 1010% from 400% respectively. Nevertheless, the author concluded that the revulcanization process that uses the formulation which contains 30% of BR devulcanizate is better than the formulation that uses 50% of BR devulcanizate.

$ScCO_2$-assisted devulcanization study was also reported by Z. Liu *et al.* [67] who conducted the experiment through a 2^2-full factorial design to statistically evaluate the influence of various parameters on the values of soluble fraction and crosslink density of waste treat rubber (mainly consisted of SBR and BR). These parameters were reaction temperature, pressure, time and DD content. With respect to their analysis of variance (ANOVA) investigation for both responses (soluble fraction and crosslink density), the authors reported that pressure was the only insignificant variable whereas temperature, time and DD content were significant. The latter being the most significant variable. Furthermore, the interaction between these three variables were also found to be significant and reported in their article.

The response in terms of soluble fraction and crosslink density as a function of temperature and DD content is showing that the increase in DD content significantly increase the soluble fraction and decreasing the crosslink density. However, steeper gradient for both responses with respect to DD content than that of temperature is observed. Such proven the earlier statement declaring DD was the most significant factor.

The effect of temperature and treatment time were then plotted by the authors onto the Horikx curve. The fitted data points for treatment ≥200°C were described to be more preferential towards main chain scission rather than treatment <200°C, despite heterogeneous scission were still being experienced. This suggests a lower temperature treatment is more beneficial in persevering the polymeric backbone. Similarly, shorter treatment time was also reported to be more effective, although the experienced phenomena was reclamation. Similar factorial study reported by Z. Liu *et al.* [67] was also performed by I. Mangili *et al.* [66]. The result of their study corroborates with each other, whereby treatment temperature, DD content and their interaction were the significant variables. Within the same year, the same author also published another article in regards to DD and $ScCO_2$ utilization for devulcanization, complete with a reaction scheme as shown in Figure 3.13.

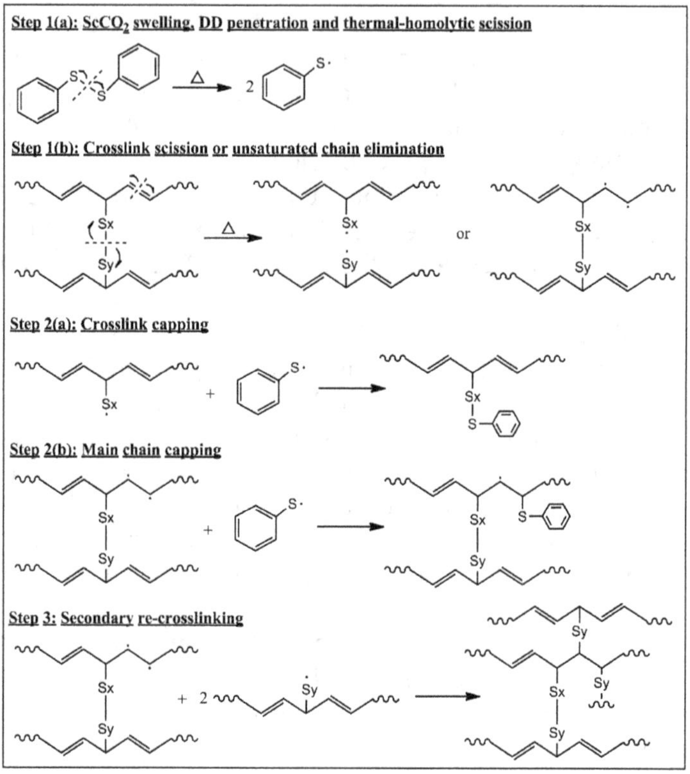

Figure 3.13 *$ScCO_2$ reaction scheme.*

With respect to Figure 3.13, two initial steps were expected to occur. The first, as shown in step 1 (a), is the swelling of rubber network by $ScCO_2$, allowing easy penetration of DD in which it experienced homolytic scission producing two Ph-S• radicals. Simultaneously, at the time of network expansion, crosslink network was expected to cleave alongside with the possible elimination of the adjacent unsaturated C=C bond(s). Such producing RS• and R• radicals. The next step would be the involvement of these radicals in the suspected radical capping. In step 2 (a), the earlier generated RS• would combine with the Ph-S• which prevent the recombination of RS• radicals to form crosslinks. On the other hand, Ph-S• could also combine with R• and grafted on the main chain as shown in step 2(b). The author acknowledge that this particular process mainly attacks the crosslink network instead of the carbon main chain. Despite the absence of Horikx analysis to confirm this, a certain probability of secondary re-crosslinking could happen as described in step 3. Lastly, the author also stressed that the amount of DD added during the devulcanization process would affect the revulcanizate curing as well as tensile properties. For instance, in their reported study, increase in elongation at break and decrement in moduli was experienced.

3.6 Concluding Remarks

State-of-the-art devulcanization by means of chemical or devulcanizing agent have been thoroughly compiled and discussed within this chapter. It was interesting to find that depending on the devulcanizing agent used, a certain reaction scheme could be followed. Throughout the recent 10 years of chemical-based devulcanization, four different schemes have been identified which are inorganic catalytic, free radical, nucleophilic and swelling systems. In the earlier days chemical devulcanization was seen to just behave as an imbibition agent in a batch process, however, its versatile use has been proven in a variety of processes described in this short chapter. Moreover, due to its flexibility and versatility, it is a common choice for industries to employ chemical devulcanization despite the certain degree of toxicity the agents possess. On that note, more sustainable agents had also been explored through the usage of the recently emerged ionic liquids, eutectic solvents and supercritical CO_2. Intriguingly, a comparable as well as more superior properties exhibited by the obtained devulcanizate and revulcanizates were observed, indicating the exceptionally high potential of these solvents to be acting as a new generation of devulcanising agent in the future, despite further extensive research studies required.

References

1. Akca E, Gursel A, Sen N (2018) A review on devulcanization of waste tire rubber. Period Eng Nat Sci 6:154–160. https://doi.org/10.21533/pen.v6i1.167

2. Myhre M, Saiwari S, Dierkes W, Noordermeer J (2012) Rubber Recycling: Chemistry, processing, and applications. Rubber Chem Technol 85:408–449. https://doi.org/10.5254/rct.12.87973
3. Worlee A, Saiwari S, Dierkes W, et al (2019) Influence of filler network on thermo-chemical de-vulcanization efficiency of carbon black filled natural rubber. J Met Mater Miner 29:76–81. https://doi.org/10.14456/jmmm.2019.38
4. Liu HL, Wang XP, Jia DM (2019) Recycling of waste rubber powder by mechano-chemical modification. J Clean Prod 118716. https://doi.org/10.1016/j.jclepro.2019.118716
5. Zhang Y, Zhang Z, Wemyss AM, et al (2020) Effective thermal-oxidative reclamation of waste tire rubbers for producing high-performance rubber composites. ACS Sustain Chem Eng 8:9079–9087. https://doi.org/10.1021/acssuschemeng.0c02292
6. Saputra R, Walvekar R, Khalid M, et al (2020) Devulcanisation of ground rubber tyre by novel ternary deep eutectic solvents. J Mol Liq. https://doi.org/https://doi.org/10.1016/j.molliq.2020.112913
7. Bockstal L, Berchem T, Schmetz Q, Richel A (2019) Devulcanisation and reclaiming of tires and rubber by physical and chemical processes: A review. J Clean Prod 236:117574. https://doi.org/10.1016/j.jclepro.2019.07.049
8. Asaro L, Gratton M, Seghar S, Aït Hocine N (2018) Recycling of rubber wastes by devulcanization. Resour Conserv Recycl 133:250–262. https://doi.org/10.1016/j.resconrec.2018.02.016
9. Zhang X, Saha P, Cao L, et al (2018) Devulcanization of waste rubber powder using thiobisphenols as novel reclaiming agent. 78:980–991. https://doi.org/10.1016/j.wasman.2018.07.016
10. Markl E, Lackner M (2020) Devulcanization technologies for recycling of tire-derived rubber: A review. Materials (Basel) 13:. https://doi.org/10.3390/ma13051246
11. Nuzaimah M, Sapuan SM, Nadlene R, Jawaid M (2020) Sodium hydroxide treatment ofwaste rubber crumb and its effects on properties of unsaturated polyester composites. Appl Sci 10:. https://doi.org/10.3390/app10113913
12. Yamashita S, Kawabata N, Sagan S, Hayashi K (1977) Recla-

mation of vulcanized rubbers by chemical degradation. V. Degradation of vulcanized synthetic isoprene rubber by the phenylhydrazine–ferrous chloride system. J Appl Polym Sci 21:2201–2209. https://doi.org/10.1002/app.1977.070210817

13. Yamashita S, Kato S, Kawabata N, Okamoto T (1978) Reclamation of vulcanized rubbers by chemical degradation. VIII. Absorption of oxygen and degradation of cis-1,4-polyisoprene by phenylhydrazine–iron(II) chloride system. J Appl Polym Sci 22:353–360. https://doi.org/10.1002/app.1978.070220204
14. Joseph A, George B, Madhusoodanan K, Alex R (2015) Current status of sulphur vulcanization and devulcanization chemistry: process of vulcanization. Rubber Sci 28:82–121
15. Dutta Chowdhury A, Kumar Lahiri G (2012) A generalized approach for iron catalyzed chemo- and regioselective formation of anti-Markovnikov acetals from styrene derivatives. Chem Commun 48:3448–3450. https://doi.org/10.1039/c2cc17889c
16. Langer E, Bortel K, Waskiewicz S, Lenartowicz-Klik M (2020) Methods of PET Recycling. In: Langer E, Bortel K, Lenartowicz-Klik M (eds) Plasticizers Derived from Post-Consumer PET, First. Matthew Dean, Cambridge, pp 127–171
17. Yehia A, Abdelbary EM, Mull M, et al (2012) New trends for utilization of rubber wastes. Macromol Symp 320:5–14. https://doi.org/10.1002/masy.201251001
18. Deivasigamani K, Nanjan J, Mani HP (2017) Recycling of waste gasket rubber granules by bulk CuCl2 and nano CuCl2: Removal of Hg(II) ions by recycled rubber Granules. Proc R Soc A Math Phys Eng Sci 473:. https://doi.org/10.1098/rspa.2016.0771
19. K. De S, I. Isayev A, Klementina K (2005) Rubber Recycling. CRC Press - Taylor & Francis Group, Boca Raton
20. Nicholas PP (1979) Devulcanized Rubber Composition and Process for Preparing Same (U.S. Patent No. 4161464). U.S. Patent and Trademark Office. https://patents.google.com/patent/US4161464A/en.
21. Nicholas PP (1987) Phase-Transfer Catalysts. ACS Symp Ser 326:155–168
22. Safan M, Eid FM, Awad M (2017) Enhanced properties of crumb rubber and its application in rubberized concrete. Int

J Curr Eng Technol 7:1784–1790

23. Guo S, Dai Q, Si R, et al (2017) Evaluation of properties and performance of rubber-modified concrete for recycling of waste scrap tire. J Clean Prod 148:681–689. https://doi.org/10.1016/j.jclepro.2017.02.046
24. Kashani A, Ngo TD, Hemachandra P, Hajimohammadi A (2018) Effects of surface treatments of recycled tyre crumb on cement-rubber bonding in concrete composite foam. Constr Build Mater 171:467–473. https://doi.org/10.1016/j.conbuildmat.2018.03.163
25. Joseph AM, Madhusoodanan KN, Alex R, George B (2019) Stable free radical assisted mechanical devulcanisation: Devulcanisation of NR/BR blends. Rubber Chem Technol 32:68–82
26. Ostad Movahed S, Ansarifar A, Karbalaee S, Athary Far S (2015) Devulcanization and recycling of waste automotive EPDM rubber powder by using shearing action and chemical additive. Prog Rubber, Plast Recycl Technol 31:87–116. https://doi.org/10.1177/147776061503100202
27. Ghorai S, Bhunia S, Roy M, De D (2016) Mechanochemical devulcanization of natural rubber vulcanizate by dual function disulfide chemicals. Polym Degrad Stab 129:34–46. https://doi.org/10.1016/j.polymdegradstab.2016.03.024
28. De D, De D, Singharoy GM (2007) Reclaiming of Ground Rubber Tire by a Novel Reclaiming Agent. I. Virgin Natural/Reclaimed GRT Vulcanizates. Polym Eng Sci 47:1091–1100. https://doi.org/10.1002/pen
29. Jana GK, Das CK (2005) Devulcanization of natural rubber vulcanizates by mechanochemical process. Polym - Plast Technol Eng 44:1399–1412. https://doi.org/10.1080/03602550500209853
30. Rajan V V., Dierkes WK, Joseph R, Noordermeer JWM (2006) Recycling of NR based cured latex material reclaimed with 2,2′-dibenzamidodiphenyldisulphide in a truck tire tread compound. J Appl Polym Sci 102:4194–4206. https://doi.org/10.1002/app.24563
31. Molanorouzi M, Mohaved SO (2016) Reclaiming waste tire rubber by an irradiation technique. Polym Degrad Stab 128:115–125. https://doi.org/10.1016/j.polymdegradstab.2016.03.009
32. Rooj S, Basak GC, Maji PK, Bhowmick AK (2011) New route

for devulcanization of natural rubber and the properties of devulcanized rubber. J Polym Environ 19:382–390. https://doi.org/10.1007/s10924-011-0293-5

33. Temram C, Wattanakul K (2012) Investigation the degradation and devulcanization reaction of thermoplastic vulcanizate using peroxide compound. In: AIP Conference Proceedings. pp 246–251
34. Sabzekar M, Chenar MP, Mortazavi SM, et al (2015) Influence of process variables on chemical devulcanization of sulfur-cured natural rubber. Polym Degrad Stab 118:88–95. https://doi.org/10.1016/j.polymdegradstab.2015.04.013
35. Hassan MM, Aly RO, Abdel Aal SE, et al (2013) Mechanochemical devulcanization and gamma irradiation of devulcanized waste rubber/high density polyethylene thermoplastic elastomer. J Ind Eng Chem 19:1722–1729. https://doi.org/10.1016/j.jiec.2013.02.012
36. El-Nemr KF, Khalil AM, Fathy ES (2018) Thermoplastic elastomers based on waste rubber and expanded polystyrene: Role of devulcanization and ionizing radiation. Int J Polym Anal Charact 23:58–69. https://doi.org/10.1080/1023666X.2017.1387447
37. Premachandra JK, Edirisinghe DG, De Silva MIA (2011) A novel reclaiming agent for ground rubber tyre (GRT). Part 1: Property evaluation of virgin natural rubber (NR)/novel reclaimed GRT blend compounds. Prog Rubber, Plast Recycl Technol 27:31–48. https://doi.org/10.1177/147776061102700103
38. Sutanto P, Laksmana FL, Picchioni F, Janssen LPBM (2006) Modeling on the kinetics of an EPDM devulcanization in an internal batch mixer using an amine as the devulcanizing agent. Chem Eng Sci 61:6442–6453. https://doi.org/10.1016/j.ces.2006.05.024
39. Jiang C, Zhang Y, Ma L, et al (2018) Tailoring the properties of ground tire rubber/high-density polyethylene blends by combining surface devulcanization and in-situ grafting technology. Mater Chem Phys 220:161–170. https://doi.org/10.1016/j.matchemphys.2018.08.040
40. Van Duin M, Noordermeer JWM, Verbruggen MAL, & Van Der Does L (2003) Method for Devulcanizing Rubber with an Amine (U.S. Patent No. 0013776). U.S. Patent and Trademark Office https://patents.google.com/patent/US6956065B2/en.

41. Chan B, Karton A, Easton CJ, Radom L (2016) α-Hydrogen abstraction by •OH and •SH radicals from amino acids and their peptide derivatives. J Chem Theory Comput 12:1606–1613. https://doi.org/10.1021/acs.jctc.6b00007
42. Dukhuis KAJ, Babu I, Lopulissa JS, et al (2008) A mechanistic approach to EPDM devulcanization. Rubber Chem Technol 81:190–208. https://doi.org/10.5254/1.3548204
43. Fazli A, Rodrigue D (2020) Waste rubber recycling: A review on the evolution and properties of thermoplastic elastomers. Materials (Basel) 13:. https://doi.org/10.3390/ma13030782
44. Joseph AM, George B, Madhusoodanan KN, Alex R (2016) Effect of devulcanization on crosslink densityand crosslink distribution of carbon black filled natural rubber Vulcanizates. Rubber Chem Technol 89:653–670. https://doi.org/10.5254/rct.16.84819
45. Thaicharoen P, Thamyongkit P, Poompradub S (2010) Thiosalicylic acid as a devulcanizing agent for mechano-chemical devulcanization. Korean J Chem Eng 27:1177–1183. https://doi.org/10.1007/s11814-010-0168-9
46. Das L, Xu S, Shi J (2017) Catalytic Oxidation and Depolymerization of Lignin in Aqueous Ionic Liquid. Front Energy Res 5:1–12. https://doi.org/10.3389/fenrg.2017.00021
47. Zhang T, Zhang Y, Wang Y, et al (2019) Theoretical insights into the depolymerization mechanism of lignin to methyl p-hydroxycinnamate by [Bmim][FeCl4] ionic liquid. Front Chem 7:1–12. https://doi.org/10.3389/fchem.2019.00446
48. Seghar S, Aït Hocine N, Mittal V, et al (2015) Devulcanization of styrene butadiene rubber by microwave energy: Effect of the presence of ionic liquid. Express Polym Lett 9:1076–1086. https://doi.org/10.3144/expresspolymlett.2015.97
49. ASTM Standard D6814–02 (2014) Standard Test Method for Determination of Percent Devulcanization of Crumb Rubber Based on Crosslink Density. Annu B ASTM Stand 2:2013–2015. https://doi.org/10.1520/D6814-02R13.2
50. Mouawia A, Nourry A, Gaumont AC, et al (2017) Controlled metathetic depolymerization of natural rubber in ionic liquids: From waste tires to telechelic polyisoprene oligomers. ACS Sustain Chem Eng 5:696–700. https://doi.org/10.1021/acssuschemeng.6b01777
51. García G, Aparicio S, Ullah R, Atilhan M (2015) Deep eutectic solvents: Physicochemical properties and gas separation ap-

plications. Energy & Fuels 29:2616–2644. https://doi.org/10.1021/ef5028873

52. Kunz W, Häckl K (2016) The hype with ionic liquids as solvents. Chem Phys Lett 661:6–12. https://doi.org/10.1016/j.cplett.2016.07.044
53. Perna FM, Vitale P, Capriati V (2020) Deep eutectic solvents and their applications as green solvents. Curr Opin Green Sustain Chem 21:27–33. https://doi.org/10.1016/j.cogsc.2019.09.004
54. Abbott AP, Capper G, Davies DL, et al (2001) Preparation of novel, moisture-stable, lewis-acidic ionic liquids containing quaternary ammonium salts with functional side chains. Chem Commun 1:2010–2011. https://doi.org/10.1039/b106357j
55. Saputra R, Walvekar R, Khalid M, et al (2019) Effective devulcanization of ground tire rubber using choline chloride-based deep eutectic solvents. J Environ Chem Eng 7:103151. https://doi.org/10.1016/j.jece.2019.103151
56. Saputra R, Walvekar R, Jayaweera U, & Khalid M (2019). Dilution and ZnCl2 impact on eutectic solvents as devulcanizing reagent in de-linking phenomena of waste ground rubber tire. AIP Conference Proceedings, 2137(1), 20013. https://doi.org/10.1063/1.5120989
57. Saiwari S, Dierkes WK, Noordermeer JWM (2013) Devulcanization of whole passenger car tire material. KGK Kautschuk Gummi Kunststoffe 66:20–25
58. Li Y, Shen A, Lyu Z, et al (2019) Ground tire rubber thermo-mechanically devulcanized in the presence of waste engine oil as asphalt modifier. Constr Build Mater 222:588–600. https://doi.org/10.1016/j.conbuildmat.2019.06.162
59. Pegollo CC, Pajarito BB (2019) Feasibility of oxidized soybean oil for rubber devulcanization. IOP Conf Ser Mater Sci Eng 634:. https://doi.org/10.1088/1757-899X/634/1/012036
60. Zhang Z, Wan C, Song P, et al (2020) Soybean oil induced efficient thermal–oxidative degradation of covalently cross-linked styrene butadiene rubber. J Appl Polym Sci 137:1–11. https://doi.org/10.1002/app.48935
61. Song P, Wan C, Xie Y, et al (2018) Vegetable derived-oil facilitating carbon black migration from waste tire rubbers and its reinforcement effect. Waste Manag 78:238–248.

https://doi.org/10.1016/j.wasman.2018.05.054
62. Hassan AA, Formela K, Wang S (2019) Reclaimed rubber in situ grafted with soybean oil as a novel green reactive plasticizer in SBR/silica compounds. ACS Sustain Chem Eng 7:14991–15001. https://doi.org/10.1021/acssuschemeng.9b03339
63. Li X, Deng XQ, Dong C (2018) Effect of temperature on devulcanization of waste sidewall rubber by supercritical ethanol. J Braz Chem Soc 29:2169–2179. https://doi.org/10.21577/0103-5053.20180093
64. Zhang T, Xing Z, Zhang L, Zhang Y (2018) Recycling of resin cured IIR-based ground bladder rubber with the assistance of subcritical fluids. J Elastomers Plast 50:677–696. https://doi.org/10.1177/0095244318757836
65. Jiang K, Shi J, Ge Y, et al (2013) Complete devulcanization of sulfur-cured butyl rubber by using supercritical carbon dioxide. J Appl Polym Sci 127:2397–2406. https://doi.org/10.1002/app.37542
66. Mangili I, Oliveri M, Anzano M, et al (2014) Full factorial experimental design to study the devulcanization of ground tire rubber in supercritical carbon dioxide. J Supercrit Fluids 92:249–256. https://doi.org/10.1016/j.supflu.2014.06.001
67. Liu Z, Li X, Xu X, et al (2015) Devulcanizaiton of waste tread rubber in supercritical carbon dioxide: Operating parameters and product characterization. Polym Degrad Stab 119:198–207. https://doi.org/10.1016/j.polymdegradstab.2015.05.017
68. Li X, Xu X, Liu Z (2020) Cryogenic grinding performance of scrap tire rubber by devulcanization treatment with ScCO2. Powder Technol 374:609–617. https://doi.org/10.1016/j.powtec.2020.07.026
69. Kojima M, Tosaka M, Ikeda Y, Kohjiya S (2005) Devulcanization of carbon black filled natural rubber using supercritical carbon dioxide. J Appl Polym Sci 95:137–143. https://doi.org/10.1002/app.20806
70. Mangili I, Collina E, Anzano M, et al (2014) Characterization and supercritical CO2 devulcanization of cryo-ground tire rubber: Influence of devulcanization process on reclaimed material. Polym Degrad Stab 102:15–24. https://doi.org/10.1016/j.polymdegradstab.2014.02.017

4

Mechano-chemical Devulcanization of Rubber Parts

Mohammad Karrabi and **Leila Bazli**
Rubber Group, Iran Polymer & Petrochemical Institute, Tehran 1497713115, Iran

4.1 Introduction

In recent years, one of the major social and environmental issues is the recycling of waste rubbers and elastomers. Waste rubber stockpiles, by-products of rubber factories, and used tires are increasing the complexity of their disposal. The amount of stockpiled scrap tires in the U.S. alone is more than 2.5 billion. These discarded materials can be considered as raw materials for the rubber industry, however, the recycled part of the waste rubbers is small [1].

Rubber-like materials possess a crosslinked molecular structure preventing them from melting as well as subsequent processing. On the other hand, the waste materials are usually a combination of different rubbers containing carbon black, additives, and oil. Rubber devulcanization is one of the main routes for rubber recycling, in which the crosslinks are broken by, for instance, chemical [2], ultrasonic [3], and microwave treatments [4]. As a result of severe rubber degradation during the devulcanization process and high energy consumption, there is no preferred method for rubber recycling. Each method is not applicable to all rubber types. Reducing the size of waste rubbers is another approach to rubber recycling. The obtained powder is reused as raw materials or fillers for the production of new articles. The preliminary step in all recycling processes is the size reduction of rubber materials from a coarse-size to a medium-size. Consequently, the surface area and the resultant surface activity increase. Additionally, the size reduction does not cause severe degradation of rubber [5].

The process of size reduction can be carried out at ambient temperature or under cryogenic conditions. A toothed-wheel mill is usually employed in ambient grinding for the production of large particles in the range of 0.5 to 1.5 mm. The grinding can be performed at or above ambient temperatures [6]. As a result of the high amount of large particles, the fabricated powder is limited to applications that require low strength. In the cryogenic grinding process, using liquid nitrogen, the temperature of rubber reaches values below its glass transition, and the cooled rubber is crushed into small grains by high-strain-rate milling. The particle size between 75 to 600 μm can be obtained through this process [7]. The angular particles that are produced by cryogenic grinding possess a low specific area, therefore, ambient milling is used subsequently to roughen the ground particles and increase their surface area [8]. Cryogenic grinding is an expensive route because around 1.5 kg liquid N_2 is required for grinding 1 kg of rubber [9]. Also, the capital cost of cooling equipment for maintaining the temperature below room temperature imposes high costs. Thus, it is highly desirable to develop an economically feasible approach to fabricate relatively small particles at temperatures close to room temperature [10].

Russian researchers first developed a novel concept of polymer noncryogenic pulverization [11-13]. In this process, compressive shear is imposed on polymeric materials by a twin-screw extruder at determined temperatures and polymer particles are produced without the help of cryogenic fluid to cool material down. Researchers at the Center of Excellence in Polymer Science and Engineering at ILLINOIS TECH, Chicago, U.S. suggested solid-state shear extrusion (SSSE) as an promoted version of the twin-screw extruder process [14]. A less costly version of the SSSE process by the single-screw extruder was also developed for the pulverization of rubber into finer particles [15].

4.2 Brief review of rubber parts recycling processes

End-of-life-tires (ELTs) should be granulated for material recovery. Grinding approaches include ambient, cryogenic, and wet grindings [16]. In most tire recycling technologies, textile (cord) and metallic materials are separated and the tire dimensions are reduced significantly through a grinding process. Granulates of a few millimeters or below are generated by the grinding process in which the processing temperature can be at room temperature or below the glass transition temperature of the contributing polymers in the tires. The obtained powder can be incorporated in new tire compounds as a filler but the addition should be a few amounts at a time. The compatibility between ELTs or their powder and the new rubber compound can be enhanced by the surface modification of the ELTs or devulcanization and breaking the dimensional cross-linking network

[17, 18]. In contrast to thermoplastics, elastomers cannot simply be recycled due to the cross-linking network [19]. The aim of the devulcanization process is the selective breakage of the C-S bonds rather than C-C bonds. The three-dimensional cross-linking network that is formed during vulcanization is totally or partially broken up by energy applied during the devulcanization of waste rubber [20]. The breaking energies of the C-S and S-S bonds are 273 kJ/mol and 227 kJ/mol, respectively. This value has been calculated to be 348 kJ/mol for the C-C bonds which is close to the values for the S-S and C-S bonds revealing that selectivity in devulcanization is hard to be achieved [21]. Higher selectivity leads to better mechanical properties of materials. In a vulcanized rubber network, the network breakdown mechanism can be investigated by a tool developed by Horikx [22]. The theory considers the soluble (sol) fraction increase rate in the rubber (devulcanized fraction) as a function of the remaining insoluble (gel) fraction (the measured cross-link density). According to this theory, this value is different for the breakage of C-C bonds and C-S bonds. According to Edwards et al. [23], the dominant mechanism of network cleavage can be indicated by measurements of cross-linking density and sol fraction in devulcanized rubber samples.

About 70% of the rubber produced globally is allocated to tires, which are composed of up to 60% of synthetic rubber and natural rubber [24]. Thus, the major resource for recycling and reclaiming is waste tires.

In devulcanization processes, ELT materials are first transformed into a ground tire rubber (GTR). In an ill-controlled manner, waste tires are usually burnt or landfilled, while GTR can be used in cement, bitumen, asphalt, tiles, concrete, acoustic and thermal isolation, and other products. Nevertheless, if untreated GTR is simply mixed with an elastomeric matrix, the mechanical properties of the matrix will be decreased significantly. This is the result of the interfacial adhesion between GTR and the matrix as well as the poor dispersion of GTR. The investigations on the improvement of the interfacial adhesion by devulcanization have been conducted for more than five decades [25].

The main factor determining the process success is the temperature of the rubber mixture. In order to prevent the main chain degradation, it is required to avoid too high temperatures [26].

There have been extensive studies on different devulcanization methods including biological [27], microwave-assisted [28], ultrasonic-based [29], chemical [30], and thermomechanical [31] devulcanization methods. An irradiation technique has been proposed by Molanorouzi and Mohaved [32]. Thermomechanical devulcanization using a twin-screw extruder is most practical, because it can be scaled to industrial volume by commonly-used extrusion machinery [24].

4.3 Mechano-chemical processes

Upon the devulcanization process, the intermolecular C-S and/or S-S bonds are broken and polymer chains with shorter lengths are generated leading to weaker rubber properties. Generally, physical, biotechnological, or chemical processes are employed for rubber waste devulcanization. The physical process involves devulcanization by applying external energy such as ultrasonic, microwave, mechanical, and cryomechanical energy. However, upon these processes, polymeric main chain linkage bonds and crosslinks are randomly cleaved, which results in the deterioration of rubber quality after revulcanization [33].

Different chemical agents have been utilized in the chemical devulcanization process including mercaptans or organic disulfides [34]. Research investigations have also reported the application of inorganic compounds [35]. In the chemical process, common organic solvents such as cyclohexane, benzene, and toluene are first used to swell the waste rubber. Nonetheless, the removal of such solvents after the reaction is an arduous task [2, 36]. It has been reported that supercritical toluene and n-butanol could be applied for the used rubber decomposition into fuel oil [37]. But high environmental and economic costs are the main problem of chemical devulcanization.
In the biotechnological process, different species of Nacardia, Acidithiobacillus, Thiobacillus, and so on are used for oxidation-reduction reactions with the rubber powder resulting in biodegradation of the rubber waste. Particle sizes were found to influence the rate and efficiency of devulcanization. Thus, the biotechnological process is limited to the elastomers' surface [27].

In terms of mechanical properties, the mechano-chemical process has been found to be more efficient. In this process, a mechanical force is applied to the rubber in the presence of devulcanizing agents [38].

4.3.1 Chemical devulcanization

As mentioned previously, in the chemical devulcanization approach, C-S or S-S bonds are broken by chemical agents. For acceleration of the devulcanization treatment, this method is generally accompanied by thermal and mechanical energy. Chlorinated hydrocarbons, hydroxide, thiol-amine reagents, disulfides mainly thiophenols, diphenyl disulfide, and their zinc salts are used as devulcanizing chemical agents [39]. The concentrations of these chemicals typically vary between 0.5 to 10 wt.% [24].

The chemical devulcanization is usually carried out in batch processes i.e., at controlled pressure and temperature. The chemical agents are mixed with the waste rubber in a mixer. Toxicity is the main drawback of this technique [40]. Understanding the chemical reaction taking place upon devulcanization is difficult due to its complexity. However, some mechanisms have been proposed for the high temperature devulcanization with diphenyl disulfide (DD) [17, 41, 42]. Oxidative breakdowns of the sulfur crosslinks are initiated by DD and it reacts with the radicals. It has been thought that it might break some main chains resulting in rubber degradation. Some products, such as thiols (RSH) and hydrogen sulfide (H_2S) are produced during the reaction between the rubber and DD. The reaction steps are presented in Figure 4.1 and are as follows [24, 42]:

Step I: Two reactive radicals are generated by the thermal degradation of DD

Step II: The crosslinking sites react with the formed radicals resulting in the breakage of S-S bonds. These chemical bonds can be opened by nucleophilic reagents, rearrangement, stress shear, or temperature. The benzene-sulfide radical subtracts the rubber allylic hydrogen because this is a relatively simple reaction. The activation of the protons existed in an allylic position is carried out by the double bonds, and the generated carbon radicals are resonance stabilized. The production of a radical and thiophenol (C_6H_5SH) is the result of this step.

Step III: A new active crosslinking site is created by another sulfide radical that subtracted an atom of hydrogen from the main chain of the polymer. The following revulcanization process continues by the newly created active sites.

Step IV: Depending on the proposed use, the devulcanized rubber could be blended with virgin polymer matrices or revulcanized.

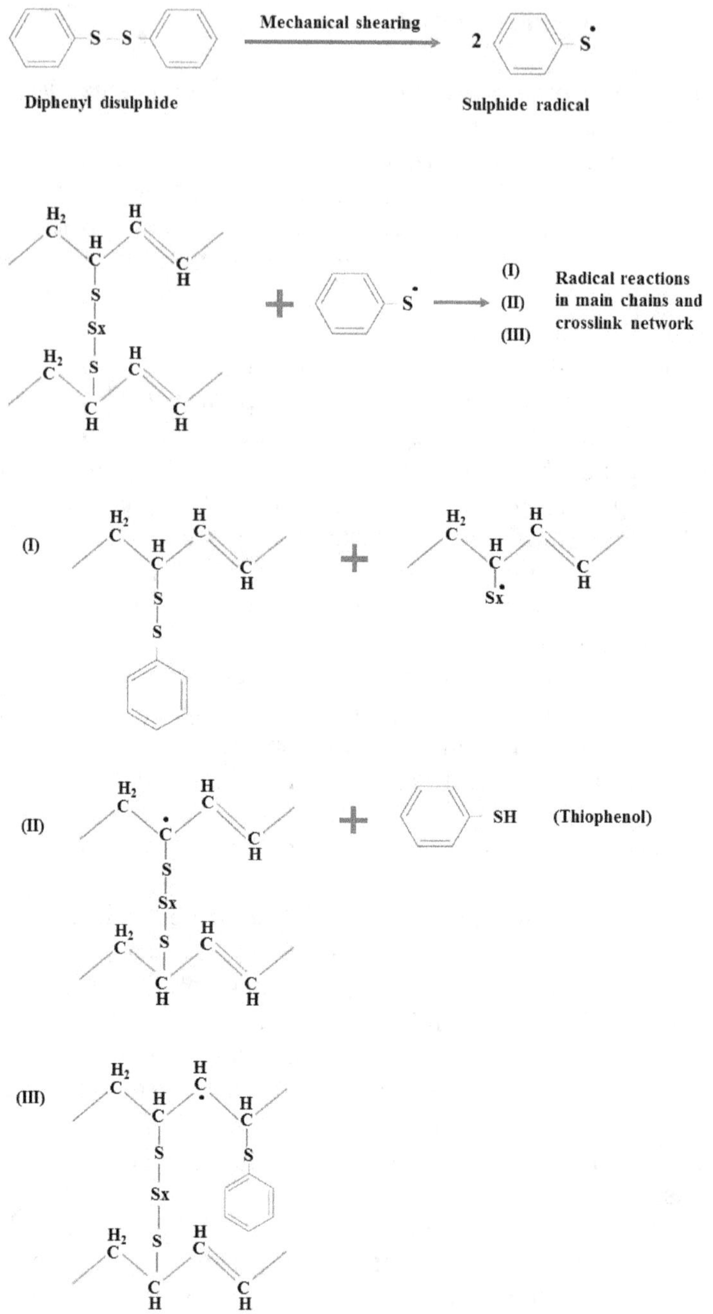

Figure 4.1 *Chemical devulcanization mechanism by diphenyl disulfide.*

4.3.2 Solid-state shear extrusion (SSSE)

A new concept of polymer grinding and pulverization was introduced in Russia. In this approach, compressive shear is applied by a twin-screw extruder to the polymer at selected temperatures and produces polymer powder [12, 13]. Later on, researchers at the Center of Excellence in Polymer Science and Engineering at IIT improved the process and introduced solid-state shear extrusion (SSSE) [14, 15].

It has been found that the extruder is another approach to grind and devulcanize rubber fragments. Extrusion technology has been widely used in the rubber and plastics industries for a long time. In this technology, tire fragments are first fed into the extruder through the feed hopper and subsequently, the tire chips are pulverized and transported down a long barrel by a screw. In a shearing action that depends on the screw design, intermeshing of the tire fragments occurs. As a result of this process, heat is produced which should be taken out from the system to avoid damage to rubber particles. It has been reported that particles with very large surface areas are possible to be achieved by this method. More information about the operating of these systems for the rubber particle production is forthcoming [21, 43].

It is claimed that chemical activation or alteration of the rubber occurs while the rubber passes through the extruder chamber. The extruder consists of a set of twin screws for the pulverization of the rubber wastes. The extruder concept and screw nomenclature are illustrated in Figure 4.2 and Figure 4.3. In these units, horsepower is a direct function of output, and they operate more efficiently with time. As the clearances between the screw and the barrel wear, efficiency and production are enhanced [43].

It is believed that due to excessive shear stresses and high strain energy applied to the rubber granulate, fracture initiation and pulverization occurs. As a result of a decrement in channel depth and shear stresses resulting from the screw rotation, radial compression is applied to the rubber granulates [44]. It has been shown that by superposition of compressive stresses on torsional deformation, much higher strain energy can be stored. Furthermore, critical compressive stress exists which below this value no powder is produced. Due to large cohesive or frictional forces between the fractured surfaces, the fractured rubber granulates are held together by the compressive stresses. Therefore, shearing stresses are transferred from each material section to the next without the need for relative movement along the fracture surfaces. Consequently, a vast network of fractures is formed by the enhancement of stored strain energy and the shear stresses. There is a threshold in the stored energy at which the granulate volume is reduced to the critical value so that they cannot be held together by frictional forces and the granulate begins to pulverize [45].

To break bonds in natural and synthetic rubbers, the applied stress should be two orders of magnitude larger than the value applied macroscopically for the deformation of rubber. Moreover, the apparent crosslink density increases, and the extensibility of chain segments between the crosslinks is reduced due to the presence of carbon black, additives, and crosslinks. Consequently, the ultimate forces that are transmitted to the polymeric chain segments as well as the forces acting at a given strain in the rubber network are enhanced. Additionally, carbon black and crosslinks create non-homogeneous local stress fields that act as potential defect sites in the rubber matrix. Because the bond strength of sulfur bonds is lower than that of cabon-carbon bonds, the majority of the bond cleavage is attributed to the sulfur bonds. As a result of this, the reduction of the crosslink density is greater than the gel fraction after the SSSE processing of rubber granulates [45].

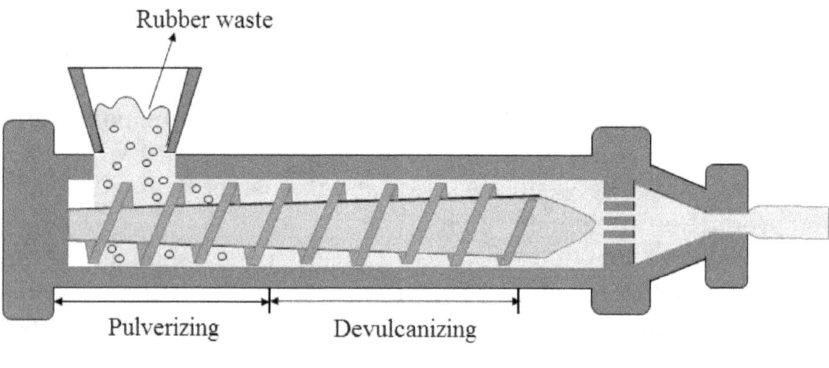

Figure 4.2 *Schematic illustration of an extruder.*

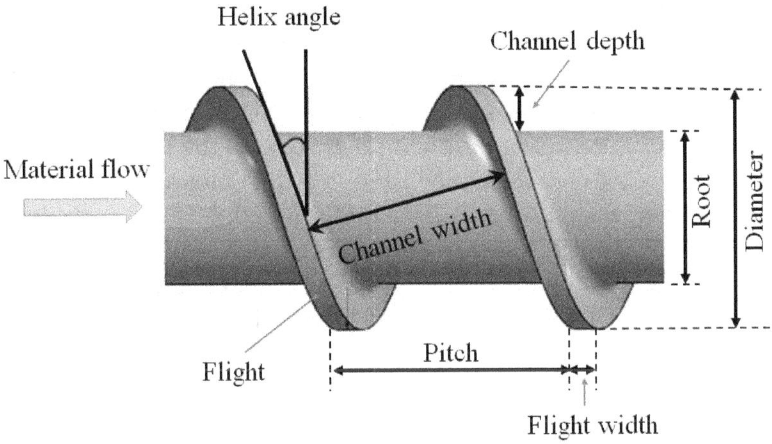

Figure 4.3 *Illustration of screw parameters.*

4.3.3 Screw configurations for low, medium and high shear rate

The processing conditions including screw configuration, screw rotation speed, and temperature profile influences the quality of rubber devulcanized by the SSSE method [46]. Figure 4.4 shows the most influential parameters in the rubber devulcanization process using twin-screw extruders. One effective parameter that provides more accurate knowledge about the devulcanization operating principles in a twin-screw extruder is residence time. The residence time is affected by different factors, namely, screw speed, screw geometry, as well as output rate. The design of screw configuration can be in a way to create different shear rates including low, medium, and high shear rates [47, 48].

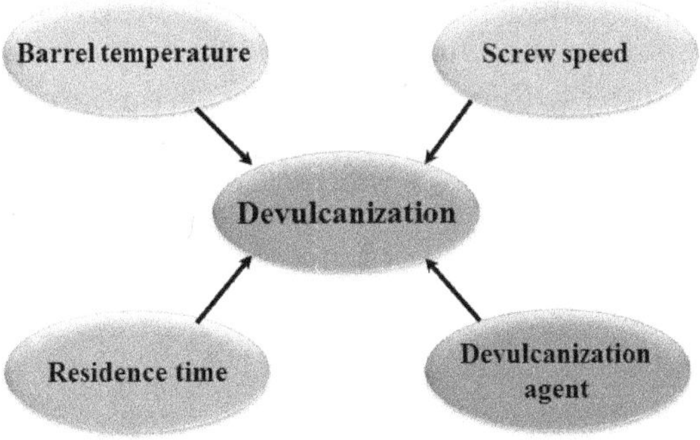

Figure 4.4 *Parameters affecting waste tire devulcanization by an extruder.*

In an extruder, the material is conveyed forward to a downstream feeding or mixing section after melting by conveying elements. A possible screw element configuration for material conveying is shown in Figure 4.5. In this conveying section, large-pitch elements are used to reduce the pressure and minimizing the work on the melt.

Twin-screw extruders involve two types of mixing: distributive mixing and dispersing mixing. Distributive mixing is conducted under a process with a low shear rate and is used for uniform distribution of particles throughout the melt by breaking the melt stream and its recombining. Distributive mixing is normally utilized for distributing fillers with high aspect ratios and fibrous materials (fiberglass, carbon fiber, aramid fiber, etc.) and mixing polymers and additives that are shear sensitive [49].

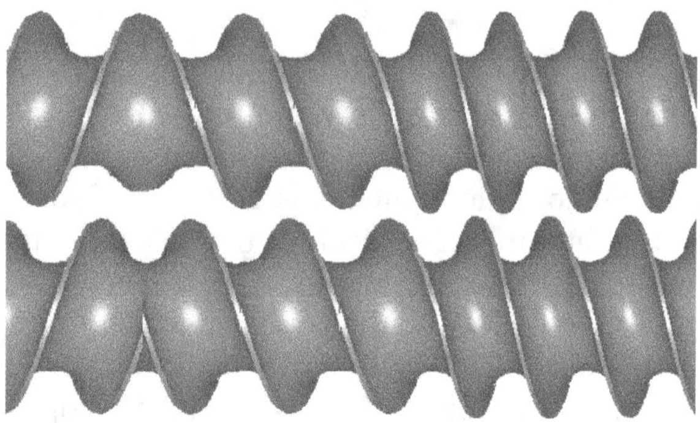

Figure 4.5 *Conveying element.*

Dispersive mixing operates at a high-shear rate in which breaking down of large particles to finer particles occurs, and the finer particles are dispersed within the melt. This type of mixing is used for the dispersion of non-reinforcing fillers, liquid additives, and disperse pigments. The schematic illustrations of these two types of mixing are shown in Figure 4.6. For achieving each type of mixing, different screw geometries and mixing elements are used. Effectiveness or level of mixing is affected by screw geometry, temperature, screw channel fill percent, and screw speed [50].

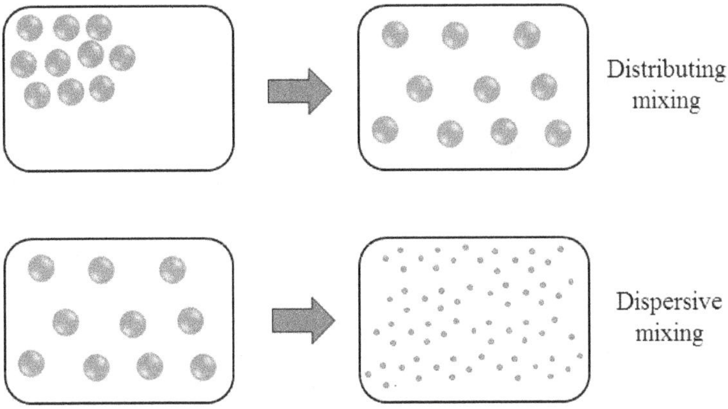

Figure 4.6 *Comparison of dispersive and distributive mixing.*

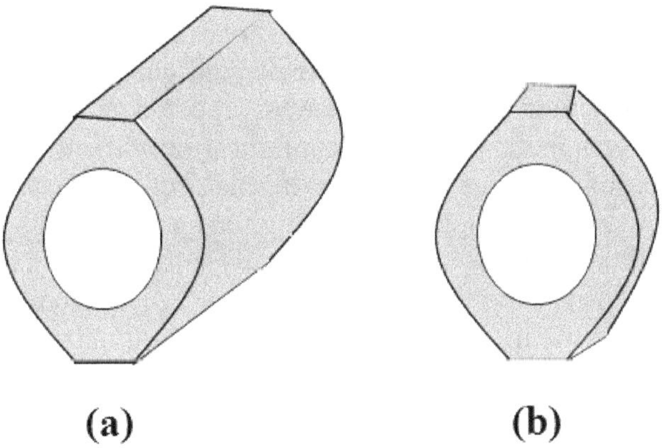

Figure 4.7 *Kneading blocks: a) wide dispersive mixing b) narrow distributive mixing.*

Good distributive mixing is provided by narrow kneading blocks as material flow around and in the blocks. However, good distributive mixing is not achieved as materials flows between the barrel wall and the block. There are various kinds of kneading blocks including narrow blocks, wide blocks, neutral blocks (nonconveying blocks), and forward or rearward conveying. In the case of dispersive mixing, wider kneading blocks are employed. In the wider disks exhibited in Figure 4.7, additives and molten polymer flow around and over the top of the disks conducting dispersive mixing, while materials flow around the narrow disks providing more distributive mixing. Based on the width of each kneading block, neutral kneading blocks can act as dispersive or distributive mixing. Neutral kneading blocks do not provide forward or rearward conveying [51].

Kneading blocks generally consist of three or five blocks and come in various lengths, which are denoted by numbers. As an example, a 90/3/15 kneading block in a 40-mm twin-screw extruder represents that there are three blocks with 5 mm widths and the total length of the blocks is 15 mm, and the blocks are positioned 90 degrees from each other. Good distributive mixing is provided by this element. More dispersive mixing can be achieved by a 90/5/60 kneading block, in which the position degree between the blocks, the number of blocks, the total length, and the wide of each block are 90, 5, 60 mm, and 12 mm, respectively. When the individual disk is wider, more dispersive mixing is obtained. For the processing of a specific formulation, one should combine various elements in order to obtain specific mixing capabilities. Figure 4.8 shows examples of screw configurations for low, medium, and high shear rates [52]. Yazdani et al. [47] investigated the influence of screw configuration on the devulcanization behavior of SBR/NR. It was reported that a longer residence time was achieved in the configuration with the high shear rate because of a stagger angle of 90 in this configuration. Moreover, in the screw configuration with the high shear rate, the sol fraction of the reclaimed rubber was lower compared to the two other configurations. The reason was stated to be the higher viscous shear energy created in the high-shear configuration leading to higher volatilization of low-molecular-weight components.

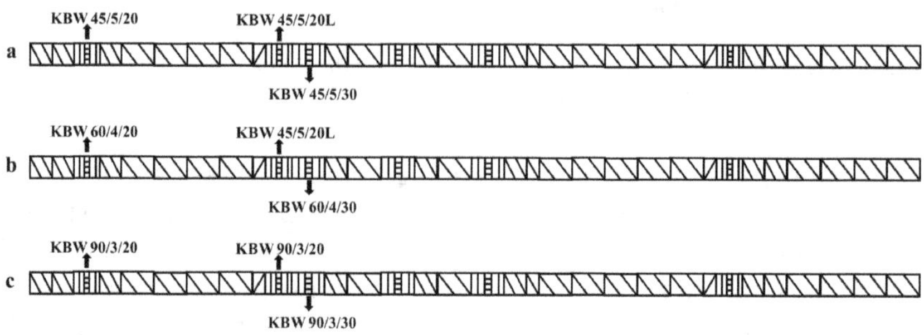

Figure 4.8 *Screw configurations for (a) low shear rate, (b) medium shear rate, and (c) high shear rate.*

4.3.4 Co-rotating twin-screw extruder

Various twin-screw extruders are commercially accessible, and these extruders are selected based on the end-use application. In all models, there are two screw shafts parallel to each other with different distances between the screw shafts. These shafts can rotate in opposite directions (counterrotating) or rotate in the same direction (corotating). In intermeshing screws, the distance between the centerline of the shafts is lower than the screw diameter, while when this value is equal to the screw diameter the screws are called non-intermeshing [53]. A short segment of fully intermeshing screws with parallel corotating and counter-rotating shafts is illustrated in Figure 4.9. A corotating, high-speed twin-screw extruder is utilized to compound resin with additives like stabilizers, reinforcements, colorants, flame retardants, fillers, devolatilization for the removal of solvents, and in situ chemical reactions in the extruder. Counterrotating and corotating extruders with low speeds are used for the fabrication of pipes and profiles. Compounding is usually carried out by counterrotating twin screws. Non-intermeshing, counterrotating extruders are typically employed for devolatilization as well as chemical reactions such as functionalization, controlled crosslinking, addition, polycondensation, and grafting. Owing to the possibility of compounding different formulations in the same extrusion, twin-screw extruders are being employed for the extrusion of films and sheets. Consequently, for the production of a final part, compounding and re-extruding are not needed [54, 55].

Using twin-screw extruders for rubber devulcanization has been considered by researchers recently. Compared to other devulcanization techniques, this approach has some advantages such as continuity, higher efficiency, and high speed [47].

4.4 Devulcanized rubber in polymer blends

Polymers are divided into three material groups: thermoplastics, thermosets, and rubbers. GTR and RTR (reclaimed tire rubber) can be compounded with all groups of polymers to produce polymer blends. The polymer market share is considerably enormous and a high amount of waste tires would be consumed with the addition of only 10 wt% of GTR into polymers [56]. GTR is added to each polymer group for a specific reason. GTR is added to thermoset polymers to improve their toughness, and the thermoplastic industry is interested in benefitting from the cheap filler of rubber in thermoplastic elastomers.

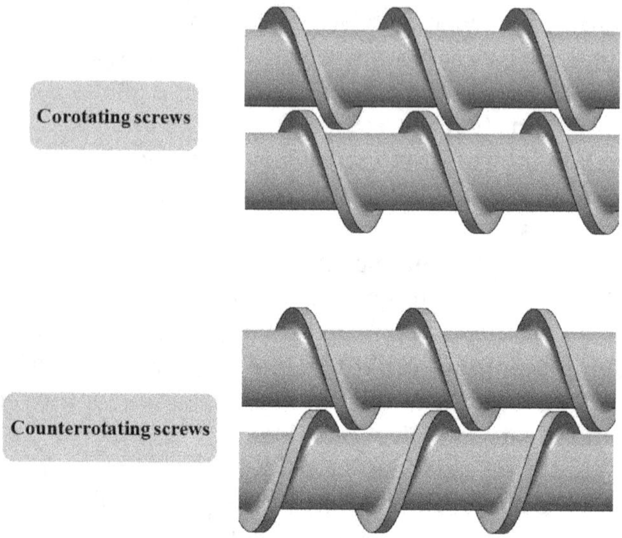

Figure 4.9 *Illustration of intermeshing corotating and counterrotating screws.*

In general, the final material properties are governed by the quality of the interface in all polymer groups. The blend of GTR or RTR with the polymeric matrix is incompatible in which GTR/RTR is dispersed in the polymer matrix. To obtain optimally balanced properties the weak interphase is required to be addressed. Compatibilization and crosslinking of the recycled rubber particles with the polymer matrix are two common practices. Both techniques will be discussed hereafter [57].

4.4.1 Compatibilization

Generally, the increase in the content of GTR/RTR leads to the reduction of properties in polymers because interfacial adhesion between the matrix and the dispersed phase is weak. In these incompatible blends, the components are physically mixed and the mechanical properties are mainly related to the continuous matrix phase. Compatibilizing the blend components is a key approach for achieving desirable properties. It is believed that the dynamic reaction of GTR/RTR and polymer in the presence of bridging role of a compatibilizer, improves the blend mechanical properties. Generally, compatibilization in necessary in these blends to: 1) decrease interfacial tension and enhance interfacial adhesion; 2) obtain finer dispersion of recycled rubber granulates in the matrix upon the blending process; 3) stabilize morphology upon the processing stage and service life [43, 58, 59].

Through compatibilization approach, special interaction sites are formed between the existing phases, or the interfaces between the phases become similar [60]. Compatibilization may be carried out through chemical processes, physico-mechanical processes, or their combination. To achieve an enhanced adhesion between the polymeric matrices and GTR/RTR, chemical modification is desired. The methods for chemical compatibilization may be reactive or non-reactive. Upon this approach, the interfacial wetting is enhanced and interfacial tension decreases which form good interphase for transferring stress.

The non-reactive method involves the incorporation of compatibilizers into the blend to enhance the blend compatibility. In this method, compatibilizers can be either specially prepared materials before mixing or commercially available graft and block copolymers such as Polybond™ of Uniroyal and Surlyn 1652™ of DuPont. There have been investigations on nonreactive compatibilization between various types of polymeric matrices and GTR/RTR. However, graft and block copolymers are costly materials [61].

In order for GTR/RTR to react with the polymer matrix and form chemically linked interphase, reactive additives are incorporated during mixing. This is called the reactive method. A reactive third component or the blend component might take part in the compatibilization process. EPDM-, PE-, and GTR-based thermoplastic elastomers were developed by reactive compatibilization. GTR was modified with acrylamide (AAm) or maleic anhydride (MA), while the modification of PE was done by functionalization with MA through grafting techniques. The results showed improved mechanical properties of the prepared thermoplastic elastomer [62]. Another practice for reactive compatibilization is in situ production of graft copolymer with one or both phases by adding reactive monomers and initiators. To compatibilize PP/GTR blends, Zhang et al. [63] utilized MA-grafted-styrene-ethylene-butylene-styrene and bitumen. Because of higher efficiency and low cost, it is preferred to use reactive compatibilization rather than non-reactive ones. Nevertheless, the reactive component should be chosen correctly so that the component fully reacts during compounding and produces no by-products that are hard to be removed.

At times, good properties of the blends can be obtained by taking advantage of both reactive compatibilization and non-reactive compatibilization. For instance, the modification of the GTR surface can be first carried out, subsequently, an additive may be incorporated to generate chemical bonds between the additive and modified GTR making it compatible with the matrix. This is commonly practiced in thermoplastic elastomer and thermoplastics blends. In the research by Shanmugharaj et al. [64], blends based on PP and GTR were compatibilized using allylamine grafted GTR and MA-grafted PP as compatibilizer. The results indicated that the compatibilized blends' mechanical properties improved in comparison with neat blends. In a similar study by Kim et al. [65], impact and tensile properties of the GTR/HDPE blend enhanced by compatibilizing of acrylamide modified GTR with MA-grafted PE. Although the modification of the surface is a cost-effective method, its efficacy is lower than that of reactive compatibilization.

4.4.2 Crosslinking

Crosslinking of recycled rubber waste particles with the polymer matrix in the blends is another approach for achieving good interphase. The crosslinking process is extensively used in rubber blends and is a preferred process in blends consisting of thermoplastic elastomers, whereas it is not widely used in thermoplastic blends. The crosslinking process includes peroxide crosslinking and sulfur crosslinking. Upon this process, the linkage between polymer chains is formed by chemical bonds. The formation of crosslinks (chemical bonds) between the polymers and recycled rubber waste leads to the improvement of their interfacial adhesion.

Vulcanization is the word used for sulfur crosslinking. In this process, in the presence of accelerators such as thiazole and thiuram and activators such as stearic acid and zinc oxide, sulfur acts as the vulcanizing agent. However, sulfur vulcanization is applicable in unsaturated materials like styrene butadiene and natural rubber. Sulfur mediates crosslinks between polymeric chains creating C–Sx–C links, which is widely utilized in rubber blends.

Peroxide crosslinking is a common practice in saturated polymers, e.g., silicone rubber, ethylene-propylene rubber (EPR), and PE. Benzoyl peroxide together with dicumyl peroxide are commonly used peroxides. In this process, peroxide is decomposed to radicals that subtract H atoms from the polymer chain resulting in the formation of macromolecular radicals. Subsequently, crosslinking between chains (carbon-carbon bonds) occurs by combinations of the macromolecular radicals. In thermoplastic blends, peroxide crosslinking binds GTR/RTR and the plastic phase to generate appreciable interphase [66].

Most thermoplastic elastomer and rubber blends show phase separation due to differences in their polarity. However, it has been revealed that phase boundary may be destructed if the blend is co-crosslinked above the order-disorder transition and in the range of lower and upper critical solution temperature of the blend. There are two main steps in the co-crosslinking process: (1) during vulcanization, the crosslinking agent forms graft or block co-polymer at the interphase; (2) the components are compatibilized by interphase thickening. Sonnier et al. [67] conducted research aimed at in situ compatibilization of the HDPE matrix and GTR particulates by peroxide catalyst for interphase dynamic co-crosslinking.

GTR and RTR are extensively compounded with rubber polymers including acrylonitrile butadiene rubber (NBR), natural rubber (NR), styrene butadiene rubber (SBR), and polybutadiene rubber (BR). It is believed that these recycled particles have good compatibility because they are composed of rubber components that are commonly used in tire formulations. However, RTR is favored in rubber blends because the soluble fraction (devulcanized section) in RTR can form a good interface, and also co-crosslinking between the rubber matrix and RTR can occur. The discussed blends are mostly vulcanized via sulfur vulcanization [57].

4.4.3 Morphology of reclaim blends

RTR and GTR are usually blended with NBR, SBR, BR, and NR. Since these rubbers constitute the major components in tire composition, the compatibility between GTR/RTR and these rubbers is believed to be good. However, RTR is more favored because the sol fraction (devulcanized section) of RTR allows for co-crosslinking between the rubber matrix and RTR as well as good interphase between them. Vulcanization of these blends is carried out by sulphur system [57, 68]. The vulcanization and co-crosslinking of the blends are schematically illustrated in Figure 4.10.

SEM photomicrographs of RTR or GTR blends with SBR and NR have indicated that RTR has better wetting with the matrix in comparison with GTR. Voids were observed around the GTR particles in the blends [69]. The cross-linked gel in GTR/RTR is not able to be dispersed in the virgin rubber as a continuous matrix. An increment in the GTR/RTR content in NR-BR blends led to less homogeneous fracture surface and different directions in crack paths resulting in lower mechanical properties [70].

Fracture surfaces of thermoplastic/GTR blends have shown that GTR tends to agglomerate when high content of GTR is incorporated in the blend. This is the most significant parameter in the fracture of materials. As a result of poor adhesion between thermoplastic polymers and GTR, the material failure occurs at the interface between the matrix and GTR. In general, by the GTR/RTR addition, impact and tensile properties deteriorate. This behavior is the result of weak adhesion at the interface of the plastic matrix and GTR/RTR. High interfacial tension is generated due to poor interphase leading to the formation of clusters/agglomerates of GTR particles and the void formation around GTR. With an increase in the GTR content, faults,

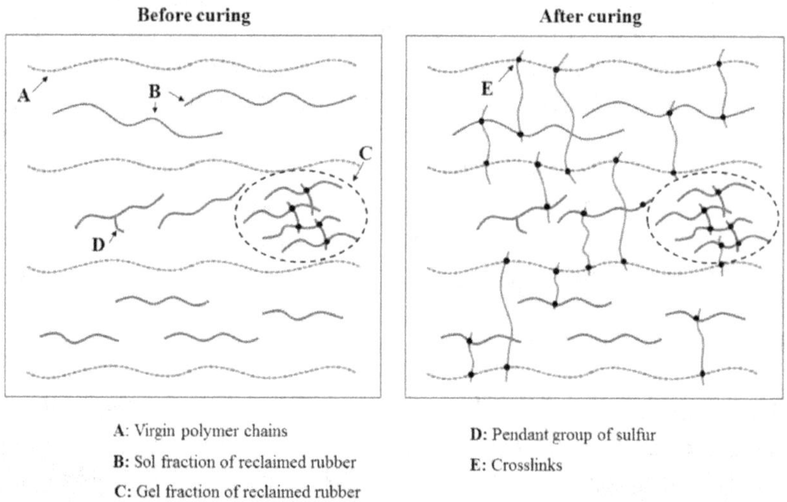

Figure 4.10 *Vulcanization and co-crosslinking of blends containing RTR.*

cracks, and defects have been indicated to increase in the matrix. The easy and clean removal of GTR particles from the matrix in fracture surfaces is the indicator of lack of interfacial adhesion [71].

In thermoplastic elastomer (TPE) blends, the rubber portions of TPE are partially substituted with RTR or GTR. Naskar et al. [72]

reported that in acrylated-HDPE/EPDM:GTR blends, EPDM encapsulated the GTR particles. It was indicated that when the content of GTR increased above 50% in the EPDM:GTR ratio, the interfacial adhesion, and compatibilization is reduced resulted from insufficient EPDM to encapsulate GTR [73].

4.5 Characterizing methods of devulcanized rubber

4.5.1 Sol and gel fraction

Principally, an increment in the rubber sol fraction and a decrease in crosslink density show the extent of network breakage. To measure the sol and gel proportions of reclaimed rubber, the crumb rubbers are initially extracted in acetone to remove low molecular weight substances such as accelerators, curatives, and processing oils. Then the sample is immersed in a solvent such as toluene for 72 h at room temperature to dissolve the non-crosslinked and devulcanized polymer according to ASTM D2765-11. The sample is vacuum-dried and weighed and the following equation can be used to calculate the sol fraction [74].

$$Sol\ fraction\ (\%) = \frac{m_1 - m_2}{m_1} \times 100 \tag{1}$$

where m_1 and m_2 are the weights of the samples before and after immersion in the solvent, respectively.

To achieve the crosslink density, the extracted sample is swelled in a solvent like toluene for 72 h at room temperature according to ASTM D6814-02. After taking out the sample the weight of the swelled sample is measured. To measure the crosslink density, the polymer volume fraction (V_r) is required to be determined. The following equation can be used to calculate this value:

$$V_r = \frac{\frac{W_r}{\rho_r}}{\frac{W_r}{\rho_r} + \frac{W_s}{\rho_s}} \tag{2}$$

In this equation, V_r is the volume of the swelled polymer; ρ_s and ρ_r, represent solvent and rubber densities, respectively; W_s and W_r are the weight of adsorbed toluene and dry rubber sample, respectively. With the obtained parameters, the Flory-Rehner equation can be used to calculate the crosslink density (Eq. (3)).

$$v_e = \frac{-\left[\ln(1-V_r) + V_r + \chi_1 V_r^{\,2}\right]}{\left[V_r\left(V_1^{\frac{1}{3}} - \frac{V_r}{2}\right)\right]} \tag{3}$$

In this equation, v_e and V_1 denote the crosslink density and the solvent molar volume, respectively [75, 76]. The interaction parameter, χ, is dependent on the solvent and polymer, for example, it is 0.56 for CIIR/toluene [77], 0.39 for NR/toluene [78], 0.34 for BR/toluene [79, 80], and 0.37 for SBR/toluene [81, 82].

4.5.2 Devulcanization percent

Another parameter for the evaluation of the devulcanization process is calculating the devulcanization percent. For calculating the reclaiming percent, the crosslink density reduction is considered as a measure of the reclaiming percent. This parameter can be calculated from the following equation [83]:

$$Reclaiming\ percent\ (\%) = \frac{\nu_{e1} - \nu_{e2}}{\nu_{e1}} \times 100 \tag{4}$$

where ν_{e1} denotes the crosslink density of rubber crumb before the reclaiming process and ν_{e2} denotes the crosslink density of rubber after the reclaiming process.

In an ideal reclaiming process, only the crosslinks are broken and the molecular weight of the main chains of the polymer is not changed [84, 85]. To study whether the crosslink scission or main chain scission occurred, Horikx theory can be used. The correlation between the crosslink density and the sol fraction was found by Horikx using statistical equations. The validity of these equations was confirmed by empirical values [86]. The Horikx equations are [22]:

$$1 - \frac{v_{e2}}{v_{e1}} = 1 - \frac{(1 - S_2^{\,1/2})^2}{(1 - S_1^{\,1/2})^2} \tag{5}$$

$$1 - \frac{v_{e2}}{v_{e1}} = 1 - \frac{\gamma_2(1 - S_2^{\,1/2})^2}{\gamma_1(1 - S_1^{\,1/2})^2} \tag{6}$$

In these equations, the indexes 1 and 2 denote before and after the reclaiming process, respectively. S is the sol fraction and γ denotes crosslink indexes or the average number of connected units along the chain. In the case of crosslink scission, the relation (5) exists between the relative reduction of crosslink density ($1-\upsilon_{e2}/\upsilon_{e1}$) and sol fraction. The equation (6) is true when both the main chain scission and crosslinks scission occur. The corresponding relationship between the sol fraction and the decrease in crosslink density is illustrated in Horik's theoretical diagram (Figure 4.11).

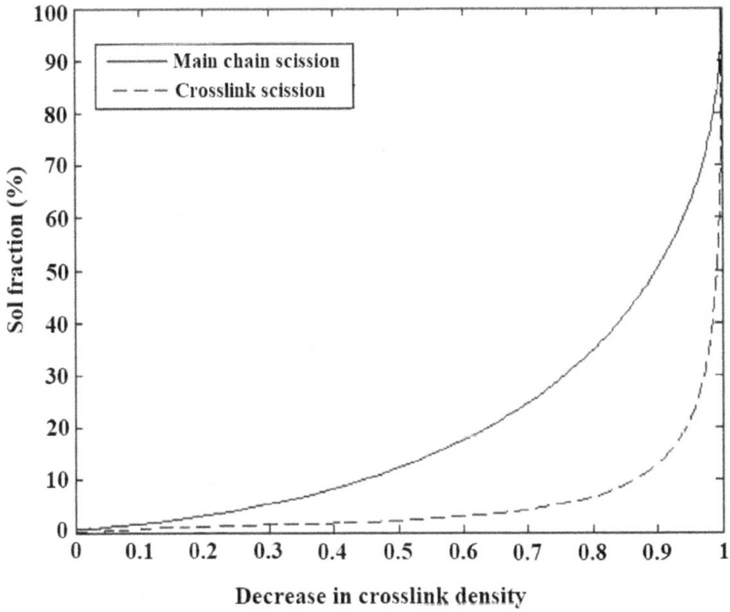

Figure 4.11 *Typical Horikx diagram of devulcanized samples.*

4.5.3 Mooney viscosity

An increase in the content of GTR/RTR leads to the rise of Mooney viscosity in the rubber blends that contain the recycled rubber wastes. The enhancement of the viscosity is caused by the existence of carbon black and cross-linked gel in GTR/RTR resulting in the restriction of chain mobility. An increase in the GTR/RTR content results in an increment in the concentration of carbon black and cross-linked gel leading to an increase in the viscosity of the blend. Additionally, the Mooney viscosity value of RTR is generally higher than that of the virgin rubber indicating higher viscosity value of blends containing RTR. A similar trend is observed for the shear viscosity as well as all GTR/RTR blends with virgin rubbers that exhibit pseudoplastic behavior. By increasing the content of GTR/RTR, a reduction of die swell occurs. The reason is that the mobility of the rubber chain and consequently, the compound elasticity are reduced by the increment of the carbon black content. Although the addition of GTR/RTR increases the viscosity of the blends, these ones can be still processed by conventional rubber processing equipment and techniques [70].

4.5.4 Curing behavior

Curing properties can determine the processing capacity of rubber compounds. To investigate the cure characteristics of rubber, scorch and cure times, as well as minimum and maximum torques, are measured by a rheometer [87, 88]. The curing behavior of rubber compounds changes with the addition of GTR/RTR to the rubber. Generally, an increase in the GTR/RTR content causes an increment in scorch time and maximum torque. On the other hand, an increase in minimum torque is observed, whereas a slight reduction in cure time occurs by adding GTR/RTR followed by stabilization upon further addition.

Stock viscosity is indicated by minimum torque. The rise in the minimum torque in the RTR-containing rubber blend is lower than that of the GTR-containing rubber blend. The reason is that RTR is partially devulcanized while GTR is vulcanized material. Therefore, the role of GTR is more like a filler without flowability which tends to form agglomerates in the rubber matrix leading to the increment of the blend viscosity. The rise of the minimum torque is the consequence of viscosity enhancement. As a result of higher mobility, only a slight increase in minimum torque is observed in RTR-containing compounds. In addition, the existence of processing oil (about 10–15%) in RTR facilitates the minimum torque reduction. Another constituent that causes the increase in the minimum torque is fillers like carbon black that is present in GTR/RTR. Overall, RTR shows better processability in comparison with GTR due to its lower minimum torque [89, 90].

Maximum torque is an indicator of elastic modulus associated with crosslink density. With the increase in RTR loading, a noticeable reduction in the maximum torque is observed. However, the introduction of GTR initially decreases the maximum torque and upon further increase of GTR, this value is stabilized. Because the reclaiming/devulcanization process generates smaller particles and shorter chains in RTR making them act as plasticizers. Moreover, the RTR matrix contains processing oil. RTR-containing blends exhibit lower maximum torque reading than do GTR-containing blends, which is the result of the cross-linked structure breakdown in RTR leading to lower crosslink density [91].

The point rubber compounds initiate crosslinking is indicated by scorch time. The scorch time of rubber compounds is reduced by the incorporation of GTR/RTR, which has been observed by many researchers. The reduction of scorch time is due to the existence of unreacted curatives and crosslink precursors or active sites in GTR/RTR for crosslinking leading to the acceleration of the crosslinking reaction by heating. Crosslink precursors (e.g. accelerator) could migrate to the virgin matrix from GTR/RTR resulting in the reduction of scorch time. This means that the crosslinking reaction is initiated in a shorter time which is undesirable in the industry. To increase the scorch time, De et al. [92] added N-cyclohexyl thiophthalimide (PVI) to RTR-containing NR. PVI has been reported to provide uniform curing within rubber products with a large thickness such as tires. In a study by Nelson and Kutty [93], reclaimed rubber powder was grafted with MA and added to NBR. The result showed that scorch and cure times were higher than those of the NBR/untreated RTR blend. Owing to the anhydride that is present in MA-grafted RTR, the retardation of curing process is observed, and the initiation of the curing process is delayed.

The time at which optimum physical properties are obtained for rubber compounds is the optimum cure time. With increasing the GTR/RTR content, optimum cure time is also reduced and stabilizes at higher loading. The same reasons that contribute to the reduction of scorch time, cause the decrease in optimum cure time. Further crosslink ability at higher loading is limited by cross-linked GTR/RTR leading to stabilization of the optimum cure time.

4.5.5 Mechanical properties

Mechanical properties of GTR/RTR-containing rubber blends are commonly measured by hardness, tear, tensile, compression set, resilience, abrasion resistance, and hysteresis. Table 4.1 represents some of the mechanical characteristics of rubber blends.

The type of virgin rubber in the blends was shown to determine tensile properties. The addition of GTR/RTR into NR has been shown to decrease elongation at break (EB) and tensile strength (TS), whereas an improvement in both properties has been observed in NBR blends. On the other hand, there are conflicting results about the addition of GTR/RTR into SBR. Some researchers have claimed deterioration while some researchers have reported an improvement in TS and EB. The decrease in TS and EB in the NR blend is the result of a less homogenous blend generation. The cross-linked gel in GTR/RTR is not able to be dispersed in the virgin rubber as a continuous matrix and acts as a stress concentrating point. The gel content in the matrix increases with the increase in the content of GTR/RTR leading to the reduction of TS and EB. Moreover, the molecular mobility of the continuous matrix and the molecular orientation are restricted by the existence of carbon black that creates stress concentration points, which leads to the failure of the blend at lower EB and TS. According to researchers, SBR and NBR have non-crystallizing nature which gives them poor gum strength and improves their TS and EB by the incorporation of GTR/RTR. Carbon black that is present in GTR/RTR renders a reinforcing property to these rubbers leading to the increase of TS. On the other hand, the processing oil in GTR/RTR offers plasticization effects in the blend causing the EB improvement [70, 89].

With the increment of the GTR/RTR content, tensile modulus increases because the crosslink density of the blends increases as a result of the gel part in the GTR/RTR and the resultant reduction in chain mobility. Therefore, it is necessary to apply a higher load to increase elongation. Since the density of crosslinks in GTR is higher than that of RTR, GTR blends have higher modulus [70].

By increasing the GTR/RTR loading, tear strength is either not influenced or improved. This is partly because of the existence of reinforcing filler in the recycled rubber. The filler enhances the intrinsic strength by the deviation of the tearing path as well as increasing energy dissipation. The reinforcing filler in GTR/RTR-containing blends has the role of deflecting or arresting the crack path, where the crack propagation is delayed or resisted, respectively. This results in stabilization or improvement of tear strength [90].

Similar to maximum torque, hardness depends on crosslink density as well as elastic modulus. With increasing GTR/RTR loading, the hardness of blends is enhanced owing to higher crosslink density along with the reinforcing filler [92].

Similar to TS, abrasion resistance is determined by the virgin rubber matrix. In general, the stability in the abrasion resistance is observed with the increase in the GTR content. In a study by Han and

Han [94], it was shown that abrasion resistance was inversely proportional to blend modulus and was directly affected by pattern spacing. On contrary, the abrasion resistance decreased in RTR-containing blends with an increase in the RTR loading. Lower abrasion resistance of RTR-containing blends is due to the weaker surface of RTR (smaller chain length and shorter fragments). Efficient crosslinking of the virgin matrix with weaker RTR surface could enhance abrasion resistance. As a result, shorter fragments in the RTR surface are linked to virgin matrix longer chains by crosslink bridges, which efficiently transfers the stress to the matrix [91].

With the increment of the content of GTR/RTR in the blends, heat build-up increases. As a result of an increase in GTR/RTR loading, the carbon black filler content and the gel content increase leading to the enhancement of the stiffness of the blends. Therefore, the rubber chains' molecular mobility is restricted and friction between chains is generated resulting in the increase in the heat build-up within the blends. Due to friction, energy loss occurs through filler breakdown or matrix-filler interphase. Hence, blend resilience is reduced which is more evident at higher loading of GTR/RTR. Loss of resilience and the increment of heat build-up affect the compression set. The synergistic influences of plasticizer, filler, and high temperature reduce the plasticity of the blend and cause loss of resilience (irreversible flowability of rubber chains under stress), which brings about unfavorable large compression set in the GTR/RTR-containing blends [70, 90].

Better tensile properties are obtained for the blends when smaller GTR particles are incorporated. RTR exhibits higher EB and TS than does GTR at an equivalent blend composition. Improved compatibility between the virgin matrix with RTR is provided owing to its partially devulcanized structure, which also makes co-crosslinking at the interphase possible, hence, improves the tensile properties of the blend [91, 94]. Generally, the variations of the mechanical properties of rubber blends containing GTR/RTR could be summarized as Figure 4.12.

Table 4.1 *Typical mechanical properties of rubber blends containing GTR/RTR*

Blends	GTR/RTR loading	Tensile strength (MPa)	Elongation at break (%)	Resilience (%)	Compression set (%)	Tear strength (N/mm)	Abrasion resistance (cc/h)
SBR/RTR [89]	20	17.5	350	-	-	-	-
	40	14.8	317	-	-	-	-
	60	8.7	260	-	-	-	-
SBR/RTR [90]	20	2.5	350	46	4.5	13	14.5
	40	3.8	525	45	5	17	19
	60	5.5	600	38	6.2	21	20
NBR/GTR [93]	20	4	325	30.5	15.5	22	4
	40	6	310	27	17.5	27	7
NBR/RTR [95]	20	2.5	320	30	18	27.9	2.5
	40	3	380	29	21	27.5	5
	60	5.5	500	28	26.5	27.3	8
NR/GTR [89]	20	15.9	500	-	-	-	-
	40	13.4	417	-	-	-	-
	60	11.6	260	-	-	-	-
NR/RTR [96]	20	15	610	58	18	27	10.6
	40	12.5	580	52	25	29	10.75
	60	12	60	50	25.2	30	10.7

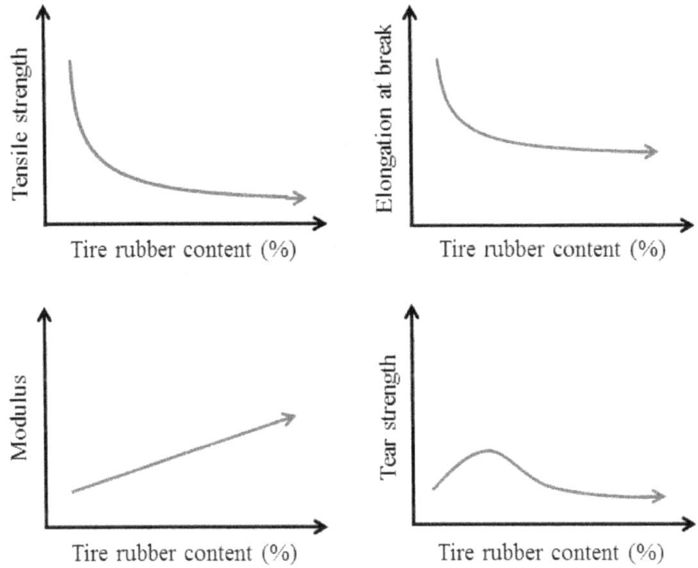

Figure 4.12 *Schematic variation of mechanical properties with the content of tire rubber content.*

4.5.6 Viscoelastic properties using DMTA

Damping characteristic and stiffness of GTR/RTR containing blends over a temperature range and also the variations of glass transition temperature (T_g) can be studied by dynamic mechanical analysis. With an increase in GTR/RTR loading in the rubber blends, storage modulus increases. The chain mobility is restricted by the increment of the carbon black content and crosslink density due to the increase in the content of GTR/RTR. The rigidity and thereby the storage modulus of the blend increase by these factors. Li et al. [97] reported lower storage modulus of RTR- containing SBR compared to GTR-containing SBR resulted from a lower density of crosslinks in RTR. Moreover, an increase in loss modulus, as an indicator of viscoelasticity, was also observed with the rise of GTR/RTR content. Contrarily, in some cases, the virgin rubber's loss modulus was higher than that of GTR/RTR-containing blends above the ambient temperature. In this case, heat build-up of blends would be lower than heat build-up of the virgin rubber vulcanizates [98].

With an increase in the content of GTR/RTR, the tan δ peak, which corresponds to T_g, has been found to increase. The chain mobility is restricted by the increase in carbon black content and crosslink density leading to the increment of glass transition temperature [92]. It has been indicated that the height of the tan δ peak is reduced with an increment of the GTR/RTR content revealing restriction of the chain movements and the increment of elastic behavior in the blends. In addition, the loss angle of carbon black is very low and thus, its presence in the blend can result in lower peak height. In certain cases, due to the presence of SBR, the width of the tan δ peak was observed to broaden with the increase in the GTR/RTR content. Greater mechanical loss is expected resulting from the presence of SBR which contains a hard styrene segment [69]. Lower height of the tan δ peak was reported for RTR/SBR compared to GTR/SBR by Li et al. [97] revealing enhanced elastic behavior originated from the RTR devulcanized structure. RTR/NR and GTR/NR blends also exhibited similar results.

4.6 Conclusion

Due to a great deal of waste tire produced every year, recycling of rubbers has attracted a significant consideration. In contrast to thermoplastics, elastomers cannot simply be recycled due to the crosslinking network. The aim of the devulcanization process is the selective breakage of the C-S bonds rather than C-C bonds. Higher selectivity leads to better mechanical properties of materials. Generally, physical, biotechnological, or chemical processes are employed for rubber waste devulcanization. Chlorinated hydrocarbons, hydroxide, thiol-amine reagents, disulfides mainly thiophenols, diphenyl disulfide, and their zinc salts are used as devulcanizing agents in the chemical approach. For acceleration of the devulcanization treatment, this method is generally accompanied by thermal and mechanical energy. SSSE is a new concept used in mechano-chemical devulcanization. It is believed that due to excessive shear stresses and high strain energy applied to the rubber granulate, fracture initiation and pulverization occurs. The processing conditions including screw configuration, screw rotation speed, and temperature profile influences the quality of rubber devulcanized by the SSSE method.

The addition of recycled waste rubber to various polymeric matrices has been investigated lately. In general, the final material properties are governed by the quality of the interface in all polymer groups. The blend of GTR or RTR with the polymeric matrix is incompatible in which GTR/RTR is dispersed in the polymer matrix. Compatibilization and crosslinking of the recycled rubber particles with the polymer matrix are two common practices to improve interfacial adhesion and interaction between the matrix and GTR/RTR. To characterize GTR- and RTR-containing blends, determination of sol and gel fraction, devulcanization percent, Mooney viscosity, curing behavior, mechanical properties, and viscoelastic properties are commonly used.

References

[1] Bilgili E, H Arastoopour, B Bernstein (2001) Pulverization of rubber granulates using the solid-state shear extrusion (SSSE) process: Part I. Process concepts and characteristics. Powder Technol 115:265-276. https://doi.org/10.1016/S0032-5910(00)00353-3

[2] Thaicharoen P, P Thamyongkit, S Poompradub (2010) Thiosalicylic acid as a devulcanizing agent for mechano-chemical devulcanization. Korean J Chem Eng 27:1177-1183. https://doi.org/10.1007/s11814-010-0168-9

[3] Isayev AI, T Liang, TM Lewis (2014) Effect of particle size on ultrasonic devulcanization of tire rubber in twin-screw extruder. Rubber Chem Technol 87:86-102. https://doi.org/10.5254/RCT.13.87926

[4] Simon DÁ, D Pirityi, P Tamás-Bényei, T Bárány (2020) Microwave devulcanization of ground tire rubber and applicability in SBR compounds. J Appl Polym Sci 137:48351. https://doi.org/10.1002/app.48351

[5] Formela K, JT Haponiuk (2014) Curing characteristics, mechanical properties and morphology of butyl rubber filled with ground tire rubber (GTR). Iran Polym J 23:185-194. https://doi.org/10.1007/s13726-013-0214-7

[6] Adhikari J, A Das, T Sinha, P Saha, JK Kim (2018) Grinding of Waste Rubber, Rubber Recycling: Challenges and Developments, Royal Society of Chemistry, Cambridge.

[7] Li X, X Xu, Z Liu (2020) Cryogenic grinding performance of scrap tire rubber by devulcanization treatment with $ScCO_2$. Powder Technol 374:609-617. https://doi.org/10.1016/j.powtec.2020.07.026

[8] Kim JK, P Saha, JT Haponiuk, MK Aswathi, S Thomas (2018) Rubber Recycling: Rubber Recycling. Royal Society of Chemistry, Cambridge.
[9] Böcker N, M Grahl, A Tota, P Häussinger, P Leitgeb, B Schmücker (2013) Nitrogen, Ullmann's Encyclopedia of Industrial Chemistry, Wiley-VCH, Weinheim.
[10] Liang S, Y Hao (2000) A novel cryogenic grinding system for recycling scrap tire peels. Adv Powder Technol 11:187-197. https://doi.org/10.1163/156855200750172303
[11] N.S. Enikolopov SAW, A.I. Nepomnjaschii, V.G. Nikolskii, in: G. Patent (Ed.) 1983.
[12] Enikolopian NS (1985) Some aspects of chemistry and physics of plastic flow. Pure Appl Chem 57:1707-1711. https://doi.org/10.1351/pac198557111707
[13] Enikolopow NS, SA Wolfson, AI Nepomnjaschtschie, WG Nikolskie, WA Teleschow, LA Filmakowa, H Brinkmann, E Pantzer, E Uhland, Method and apparatus for pulverizing polymers, Google Patents, 1986.
[14] Shutov F, G Ivanov, H Arastoopour, Solid state shear extrusion pulverization, Google Patents, 1995.
[15] Arastoopour H, Single-screw extruder for solid state shear extrusion pulverization and method, Google Patents, 1998.
[16] Dobrotă D, G Dobrotă (2018) An innovative method in the regeneration of waste rubber and the sustainable development. J Clean Prod 172:3591-3599. https://doi.org/10.1016/j.jclepro.2017.03.022
[17] Mangili I, E Collina, M Anzano, D Pitea, M Lasagni (2014) Characterization and supercritical CO_2 devulcanization of cryo-ground tire rubber: Influence of devulcanization process on reclaimed material. Polym Degrad Stab 102:15-24. https://doi.org/10.1016/j.polymdegradstab.2014.02.017
[18] Markl E, M Lackner (2020) Devulcanization technologies for recycling of tire-derived rubber: A review. Materials 13:1246. https://doi.org/10.3390/ma13051246
[19] Forrest M (2019) Recycling and re-use of waste rubber. De Gruyter, Berlin.
[20] Sun X, A Isayev (2008) Continuous ultrasonic devulcanization: Comparison of carbon black filled synthetic isoprene and natural rubbers. Rubber Chem Technol 81:19-46. https://doi.org/10.5254/1.3548195

[21] Seghar S, L Asaro, M Rolland-Monnet, NA Hocine (2019) Thermo-mechanical devulcanization and recycling of rubber industry waste. Resour Conserv Recycl 144:180-186. https://doi.org/10.1016/j.resconrec.2019.01.047
[22] Horikx M (1956) Chain scissions in a polymer network. J Polym Sci 19:445-454. https://doi.org/10.1002/pol.1956.120199305
[23] Edwards D, B Danon, P van der Gryp, J Görgens, A comparison of the selectivity for crosslink cleavage of three devulcanization processes, UNESCO/IUPAC Workshop & Conference on Macromolecules & Materials, 2015.
[24] Asaro L, M Gratton, S Seghar, N Aït Hocine (2018) Recycling of rubber wastes by devulcanization. Resour Conserv Recycl 133:250-262. https://doi.org/10.1016/j.resconrec.2018.02.016
[25] Akca E, A Gursel, N Sen (2018) A review on devulcanization of waste tire rubber. Period Eng Nat Sci 6:154-160. http://dx.doi.org/10.21533/pen.v6i1.167
[26] de Sousa FD, CH Scuracchio, G-H Hu, S Hoppe (2017) Devulcanization of waste tire rubber by microwaves. Polym Degrad Stab 138:169-181. https://doi.org/10.1016/j.polymdegradstab.2017.03.008
[27] Ghavipanjeh F, ZZ Rad, M Pazouki (2018) Devulcanization of ground tires by different strains of bacteria: optimization of culture condition by taguchi method. J Polym Environ 26:3168-3175. https://doi.org/10.1007/s10924-017-1169-0
[28] Garcia P, F De Sousa, J De Lima, S Cruz, C Scuracchio (2015) Devulcanization of ground tire rubber: Physical and chemical changes after different microwave exposure times. EXPRESS Polym Lett 9. https://10.3144/expresspolymlett.2015.91
[29] Isayev AI, SP Yushanov, S-H Kim, VY Levin (1996) Ultrasonic devulcanization of waste rubbers: Experimentation and modeling. Rheol Acta 35:616-630. https://doi.org/10.1007/BF00396511
[30] Temram C, K Wattanakul, Investigation the degradation and devulcanization reaction of thermoplastic vulcanizate using peroxide compound, AIP Conf Proc, American Institute of Physics, 2012, pp. 246-251.
[31] Halle LL, A Palmqvist, K Kampmann, FR Khan (2020) Ecotoxicology of micronized tire rubber: Past, present and future considerations. Sci Total Environ 706:135694. https://doi.org/10.1016/j.scitotenv.2019.135694

[32] Molanorouzi M, SO Mohaved (2016) Reclaiming waste tire rubber by an irradiation technique. Polym Degrad Stab 128:115-125. https://doi.org/10.1016/j.polymdegradstab.2016.03.009

[33] Mark JE, B Erman (2011) Science and Technology of Rubber. Elsevier Science

[34] Sabzekar M, MP Chenar, SM Mortazavi, M Kariminejad, S Asadi, G Zohuri (2015) Influence of process variables on chemical devulcanization of sulfur-cured natural rubber. Polym Degrad Stab 118:88-95. https://doi.org/10.1016/j.polymdegradstab.2015.04.013

[35] Yamashita S, N Kawabata, S Sagan, K Hayashi (1977) Reclamation of vulcanized rubbers by chemical degradation. V. Degradation of vulcanized synthetic isoprene rubber by the phenylhydrazine–ferrous chloride system. J Appl Polym Sci 21:2201-2209. https://doi.org/10.1002/app.1977.070210817

[36] Jalilvand A, I Ghasemi, M Karrabi, H Azizi (2008) An investigation on the EPDM devulcanization in co-rotating twin screw extruder by response surface methodology. Prog Rubber Plast Recycl Technol 24:33-46. https://doi.org/10.1177%2F147776060802400103

[37] Tsuchii A, T Suzuki, K Takeda (1985) Microbial degradation of natural rubber vulcanizates. Appl Environ Microbiol 50:965-970. https://doi.org/10.1128/aem.50.4.965-970.1985

[38] Jana G, R Mahaling, C Das (2006) A novel devulcanization technology for vulcanized natural rubber. J Appl Polym Sci 99:2831-2840. https://doi.org/10.1002/app.22984

[39] Dorigato A, Rigotti D, Fredi G (2022). Recent advances in the devulcanization technologies of industrially relevant sulfur-vulcanized elastomers. Adv Ind Eng Polym Res 6:288-309. https://doi.org/10.1016/j.aiepr.2022.11.003

[40] Warner WC (1994) Methods of devulcanization. Rubber Chem Technol 67:559-566. https://doi.org/10.5254/1.3538692

[41] Liu Z, X Li, X Xu, X Wang, C Dong, F Liu, W Wei (2015) Devulcanizaiton of waste tread rubber in supercritical carbon dioxide: Operating parameters and product characterization. Polym Degrad Stab 119:198-207. https://doi.org/10.1016/j.polymdegradstab.2015.05.017

[42] Joseph AM, B George, K Madhusoodanan, R Alex (2016) The current status of sulphur vulcanization and devulcanization chemistry: devulcanization. Rubber Sci 29:62-100.

[43] De SK, A Isayev, K Khait (2005) Rubber Recycling. CRC Press, Florida.

[44] Chaubey T, H Arastoopour (2011) Studying the pulverization mechanism of rubber with a modified design of the solid-state shear extrusion process. J Appl Polym Sci 119:1075-1083. https://doi.org/10.1002/app.32643

[45] Bilgili E, H Arastoopour, B Bernstein (2000) Analysis of rubber particles produced by the solid state shear extrusion pulverization process. Rubber Chem Technol 73:340-355. https://doi.org/10.5254/1.3547595

[46] Jalilvand AR, I Ghasemi, M Karrabi, H Azizi (2007) A study of EPDM devulcanization in a co-rotating twin-screw extruder. Iran Polym J 16:327-335.

[47] Yazdani H, I Ghasemi, M Karrabi, H Azizi, GR Bakhshandeh (2013) Continuous devulcanization of waste tires by using a Co-rotating twin screw extruder: Effects of screw configuration, temperature profile, and devulcanization agent concentration. J Vinyl Addit Technol 19:65-72. https://doi.org/10.1002/vnl.20322

[48] Yazdani H, M Karrabi, I Ghasmi, H Azizi, GR Bakhshandeh (2011) Devulcanization of waste tires using a twin-screw extruder: The effects of processing conditions. J Vinyl Addit Technol 17:64-69. https://doi.org/10.1002/vnl.20257

[49] Rauwendaal C (1991) Mixing in polymer processing. CRC Press, Florida.

[50] Gogos CC, Z Tadmor, MH Kim (1998) Melting phenomena and mechanisms in polymer processing equipment. Adv Polym Tech 17:285-305. https://doi.org/10.1002/(SICI)1098-2329(199824)17:4%3C285::AID-ADV1%3E3.0.CO;2-N

[51] Giles HF, JR Wagner, EM Mount (2013) Extrusion: The Definitive Processing Guide and Handbook. Elsevier Science

[52] Kutz M (2011) Applied plastics engineering handbook: processing and materials. William Andrew, Norwich.

[53] König T (2008) Geometry of the Co-Rotating Extruders: Conveying, and Kneading Elements, in: K. Kohlgrüber (Ed.), Co-rotating Twin-screw Extruder, Hanser, Liberty Township.

[54] Shrivastava A (2018) 5 - Plastics Processing, in: A. Shrivastava (Ed.), Introduction to Plastics Engineering, William Andrew Norwich.

[55] Chang S-C, Counter-rotating twin screw extruder, Google Patents, 2013.

[56] Karger-Kocsis J, L Mészáros, T Bárány (2013) Ground tyre rubber (GTR) in thermoplastics, thermosets, and rubbers. J Mater Sci 48:1-38. https://doi.org/10.1007/s10853-012-6564-2

[57] Ramarad S, M Khalid, C Ratnam, AL Chuah, W Rashmi (2015) Waste tire rubber in polymer blends: A review on the evolution, properties and future. Prog Mater Sci 72:100-140. https://doi.org/10.1016/j.pmatsci.2015.02.004
[58] Sonnier R, E Leroy, L Clerc, A Bergeret, J Lopez-Cuesta (2007) Polyethylene/ground tyre rubber blends: Influence of particle morphology and oxidation on mechanical properties. Polym Test 26:274-281. https://doi.org/10.1016/j.polymertesting.2006.10.011
[59] Bazli L, A Khavandi, MA Boutorabi, M Karrabi (2017) Correlation between viscoelastic behavior and morphology of nanocomposites based on SR/EPDM blends compatibilized by maleic anhydride. Polymer 113:156-166. https://doi.org/10.1016/j.polymer.2017.02.057
[60] Bazli L, MH Bagherian, M Karrabi, F Abbassi-Sourki, H Azizi (2020) Effect of starch ratio and compatibilization on the viscoelastic behavior of POE/starch blends. J Appl Polym Sci 137:48877. https://doi.org/10.1002/app.48877
[61] Mészáros L, T Bárány, T Czvikovszky (2012) EB-promoted recycling of waste tire rubber with polyolefins. Radiat Phys Chem 81:1357-1360.
https://doi.org/10.1016/j.radphyschem.2011.11.058
[62] Grigoryeva O, A Fainleib, J Grenet, J Saiter (2008) Reactive compatibilization of recycled polyethylenes and scrap rubber in thermoplastic elastomers: chemical and radiation-chemical approach. Rubber Chem Technol 81:737-752. https://doi.org/10.5254/1.3548229
[63] Zhang SL, ZX Zhang, ZX Xin, K Pal, JK Kim (2010) Prediction of mechanical properties of polypropylene/waste ground rubber tire powder treated by bitumen composites via uniform design and artificial neural networks. Mater Des 31:1900-1905. https://doi.org/10.1016/j.matdes.2009.10.057
[64] Shanmugharaj A, JK Kim, SH Ryu (2005) UV surface modification of waste tire powder: characterization and its influence on the properties of polypropylene/waste powder composites. Polym Test 24:739-745. https://doi.org/10.1016/j.polymertesting.2005.04.006
[65] Kim J, S Ryu, Y Chang (2000) Mechanical and dynamic mechanical properties of waste rubber powder/HDPE composite. J Appl Polym Sci 77:2595-2602. https://doi.org/10.1002/1097-4628(20000919)77:12%3C2595::AID-APP60%3E3.0.CO;2-C
[66] Akiba Ma, A Hashim (1997) Vulcanization and crosslinking in elastomers. Prog Polym Sci 22:475-521. https://doi.org/10.1016/S0079-6700(96)00015-9

[67] Sonnier R, E Leroy, L Clerc, A Bergeret, JM Lopez-Cuesta, AS Bretelle, P Ienny (2008) Compatibilizing thermoplastic/ground tyre rubber powder blends: efficiency and limits. Polym Test 27:901-907. https://doi.org/10.1016/j.polymertesting.2008.07.003

[68] Jalilvand AR, I Ghasemi, M Karrabi, H Azizi (2007) Using devulcanized EPDM in PP/HDPE/EPDM ternary blend: mechanical properties and morphology. Iran Polym J 16:637-644.

[69] Kumnuantip C, N Sombatsompop (2003) Dynamic mechanical properties and swelling behaviour of NR/reclaimed rubber blends. Mater Lett 57:3167-3174. https://doi.org/10.1016/S0167-577X(03)00019-3

[70] Sombatsompop N, C Kumnuantip (2003) Rheology, cure characteristics, physical and mechanical properties of tire tread reclaimed rubber/natural rubber compounds. J Appl Polym Sci 87:1723-1731. https://doi.org/10.1002/app.11698

[71] Datta J, M Włoch (2015) Morphology and properties of recycled polyethylene/ground tyre rubber/thermoplastic poly(ester-urethane) blends. Macromol Res 23:1117-1125. https://doi.org/10.1007/s13233-015-3155-5

[72] Naskar AK, AK Bhowmick, S De (2001) Thermoplastic elastomeric composition based on ground rubber tire. Polym Eng Sci 41:1087-1098. https://doi.org/10.1002/pen.10809

[73] Cañavate J, P Casas, X Colom, F Nogues (2011) Formulations for thermoplastic vulcanizates based on high density polyethylene, ethylene-propylene-diene monomer, and ground tyre rubber. J Compos Mater 45:1189-1200. https://doi.org/10.1177%2F0021998310369596

[74] Saiwari S, WK Dierkes, JW Noordermeer (2014) Comparative investigation of the devulcanization parameters of tire rubbers. Rubber Chem Technol 87:31-42. https://doi.org/10.5254/rct.13.87933

[75] Hergenrother WL, AS Hilton (2003) Use of χ as a function of volume fraction of rubber to determine crosslink density by swelling. Rubber Chem Technol 76:832-845. https://doi.org/10.5254/1.3547775

[76] Hrnjak-Murgić Z, J Jelenčić, M Bravar, M Marović (1997) Influence of the network on the interaction parameter in system EPDM vulcanizate–solvent. J Appl Polym Sci 65:991-999. https://doi.org/10.1002/(SICI)1097-4628(19970801)65:5%3C991::AID-APP17%3E3.0.CO;2-V

[77] Jose T S, J Rani (2007) EPDM/CIIR Blends: Effect of EPDM Grade on Mechanical Properties. Int J Polym Mater Polym Biomater 56:743-758. https://doi.org/10.1080/00914030601100771

[78] Kojima M, M Tosaka, Y Ikeda (2004) Chemical recycling of sulfur-cured natural rubber using supercritical carbon dioxide. Green Chem 6:84-89. https://doi.org/10.1039/B314137C

[79] Hergenrother W (1972) Characterization of networks from the peroxide cure of polybutadiene. J Appl Polym Sci 16:2611-2622. https://doi.org/10.1002/app.1972.070161014

[80] Parker WO, A Ferrando, D Ferri, V Canepari (2007) Cross-link density of a dispersed rubber measured by 129Xe chemical shift. Macromolecules 40:5787-5790. https://doi.org/10.1021/ma070793a

[81] George SC, Ka Ninan, S Thomas (1999) Effect of degree of cross-linking on swelling and mechanical behaviour of conventionally vulcanised styrene-butadiene rubber membranes. Polym Polym Compos 7:343-353.

[82] Abd-El-Messieh S, K Abd-El-Nour (2003) Effect of curing time and sulfur content on the dielectric relaxation of styrene butadiene rubber. J Appl Polym Sci 88:1613-1621. https://doi.org/10.1002/app.11686

[83] Shi J, K Jiang, D Ren, H Zou, Y Wang, X Lv, L Zhang (2013) Structure and performance of reclaimed rubber obtained by different methods. J Appl Polym Sci 129:999-1007. https://doi.org/10.1002/app.38727

[84] Bani A, G Polacco, G Gallone (2011) Microwave-induced devulcanization for poly (ethylene–propylene–diene) recycling. J Appl Polym Sci 120:2904-2911. https://doi.org/10.1002/app.33359

[85] Jalilvand A, E Ghasemi, H Azizi, M Karrabi (2007) A study of EPDM devulcanization in a co-rotating twin-screw extruder. Iran Polym J 16:327-335.

[86] Seghar S, L Asaro, N Aït Hocine (2019) Experimental Validation of the Horikx Theory to be Used in the Rubber Devulcanization Analysis. J Polym Environ 27:2318-2323. https://doi.org/10.1007/s10924-019-01513-z

[87] Bazli L, A Khavandi, MA Boutorabi, M Karrabi (2016) Morphology and viscoelastic behavior of silicone rubber/EPDM/Cloisite 15A nanocomposites based on Maxwell model. Iran Polym J 25:907-918. https://doi.org/10.1007/s13726-016-0477-x

[88] Shafiee S, L Bazli, M Karrabi, MHR Ghoreishy, M Bazli (2022) Effect of Organoclay Addition on Rheological, Thermal, and Mechanical

Properties of Nitrile Rubber/Phenolic Resin Blend. Polymers 14:1463. https://doi.org/10.3390/polym14071463

[89] Yehia A, M Ismail, Y Hefny, E Abdel-Bary, M Mull (2004) Mechano-Chemical Reclamationof Waste Rubber Powder andIts Effect on the performanceof NR and SBR Vulcanizates. J Elastomers Plast 36:109-123. https://doi.org/10.1177%2F0095244304039896

[90] Sreeja T, S Kutty (2003) Styrene butadiene rubber/reclaimed rubber blends. Int J Polymer Mater 52:599-609. https://doi.org/10.1080/00914030304902

[91] Li S, J Lamminmäki, K Hanhi (2005) Effect of ground rubber powder and devulcanizates on the properties of natural rubber compounds. J Appl Polym Sci 97:208-217. https://doi.org/10.1002/app.21748

[92] De D, S Maiti, B Adhikari (2000) Reclaiming of rubber by a renewable resource material (RRM). III. Evaluation of properties of NR reclaim. J Appl Polym Sci 75:1493-1502. https://doi.org/10.1002/(SICI)1097-4628(20000321)75:12%3C1493::AID-APP8%3E3.0.CO;2-U

[93] Nelson P, S Kutty (2003) Studies on Maleic anhydride grafted reclaimed rubber/acrylonitrile butadiene rubber blends. Prog Rubber Plast Recycl Technol 19:171-188. https://doi.org/10.1177%2F147776060301900303

[94] Han SC, MH Han (2002) Fracture behavior of NR and SBR vulcanizates filled with ground rubber having uniform particle size. J Appl Polym Sci 85:2491-2500. https://doi.org/10.1002/app.10575

[95] Sreeja T, S Kutty (2002) Studies on acrylonitrile butadiene rubber/reclaimed rubber blends. J Elastomers Plast 34:145-155. https://doi.org/10.1106%2F009524402024278

[96] Sreeja T, S Kutty (2000) Cure characteristics and mechanical properties of natural rubber/reclaimed rubber blends. Polym Plast Technol Eng 39:501-512. https://doi.org/10.1081/PPT-100100043

[97] Li Y, S Zhao, Y Wang (2012) Microbial desulfurization of ground tire rubber by Sphingomonas sp.: a novel technology for crumb rubber composites. J Polym Environ 20:372-380. https://doi.org/10.1007/s10924-011-0386-1

[98] De D, PK Panda, M Roy, S Bhunia (2013) Reinforcing effect of reclaim rubber on natural rubber/polybutadiene rubber blends. Mater Des 46:142-150. https://doi.org/10.1016/j.matdes.2012.10.014

5

Waste Rubber Recycling by Mechano-chemical and Thermo-mechanical Devulcanization

Lucia Asaro[a], Michel Gratton[b], Said Seghar[c] and Nourredine Aït Hocine[d,*]

[a]*Institute of Materials Science and Technology (INTEMA), University of Mar del Plata and National Research Council (CONICET), Av. Colón 10850, 7600 Mar del Plata, Argentina.*
[b,d]*INSA CVL, Univ. Tours, Univ. Orléans, LaMé, 3 rue de la Chocolaterie, CS 23410, 41034 Blois Cedex, France.*
[c]*RubberGreen Industrie, Route de Bavay 44, Frameries 7080, Belgium.*
**nourredine.aithocine@insa-cvl.fr*

5.1 Introduction

Rubber parts and tires manufacturing industries are constantly growing. Nonetheless, the inability to recycle their wastes represents a serious environmental problem [1, 2]. The disposal of rubber wastes and end-of-life tires is one of the biggest challenges of the 21st century in waste management. Putting this in numbers, it is appropriate to mention that nearly 70% of the rubber produced worldwide is used in tires. The amount of waste tires discarded in the world each year is approximately 800 million (~10 million tons) and taking into account that the amount of synthetic and natural rubber in tires is around 60%, 6 million tons of tire wastes are produced each year [3-5]. Despite the efforts made to recycle the waste rubber, an important fraction is still deposited in landfills, creating a stock of scraps [6] which represent a real environmental and health hazard [7]. In fact, the stacking of rubber wastes outdoors promotes the growth of rodents and mosquitoes, and also they can produce spontaneous heating resulting in dangerous fires [8-10].

For the aforementioned, end-of-life tires' stacking was banned by the European Commission in 1999 [11]. Since that, the reutilization and recycling of rubber wastes has become a crucial subject of study in many countries.

When considering reutilization, some applications of the waste rubber are road and rail equipment, sports and playing surfaces, floor and walkways tiles, acoustic isolation, rubber-concrete, thermal isolation, footwear, activated carbon production, etc. [6, 12, 13]. However, due to the large amount of rubber wastes produced in the world each year, recycling is the best way to carry out its disposal, including all methods where rubber scrap is converted to make new commodities [12, 14-17].

Because of the three-dimensional network of rubbers caused by the vulcanization process which produces strong bonds, and the particular composition that includes numerous additives, the recycling of this kind of polymers is still a current technological challenge. As an example, a tire is a composite made up of different types of synthetic and natural rubbers, chemical additives (oils, sulfur, etc.), particulate fillers (carbon black, silica, etc.), and textile and/or metal reinforcements [18-20]. The composition, in percentage by weight, of a passenger's vehicle tire is presented in Figure 5.1 highlighting the complex formulation of tires which make them difficult to be recycled.

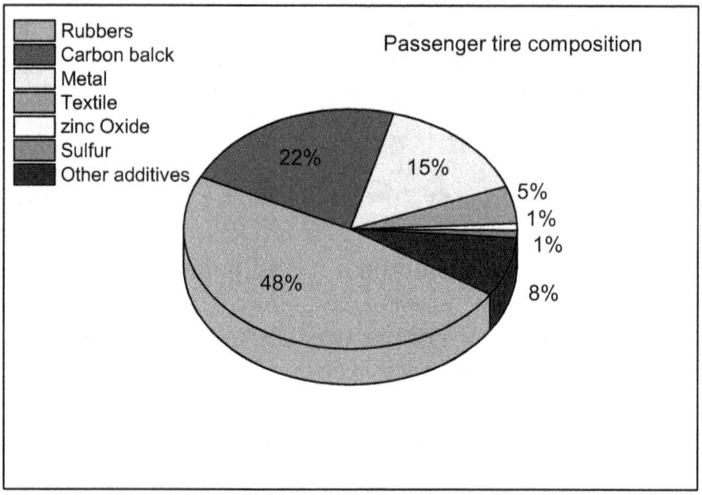

Figure 5.1 *Passenger tire composition in percentage by weight.*

With the aim of reducing the above-mentioned problematic, it is crucial to find a safety way to perform rubber recycling, i.e., a way to cleave their crosslinked structure. The concept of devulcanization has appeared in the recent years. Rubber devulcanization can be defined as a process where poly-, di-, and mono-sulfidic bonds, formed during vulcanization, are totally or partially broken. This means that devulcanization can be defined as a process that causes the selective breakup of sulfidic bonds without main chain scissions or degradation of the polymer. More generally, rubber devulcanization is a process in which the vulcanized waste rubber is transformed into the state in which it can be mixed, reprocessed, and revulcanized, resulting in a new material [21, 22]. Performing a successful devulcanization is a great challenge, since the energies required to achieve the breaking of the sulfur-sulfur (S-S) and carbon-sulfur (C-S) bonds (227 and 273 kJ/mol, respectively) are close to the energy needed to break the main chains of the rubber i.e., the carbon-carbon (C-C) bonds (348 kJ/mol) [20, 23, 24].

No matter of the selected devulcanization method, some of the main polymer chains are broken during the process and, thus, the properties of the treated rubber are different compared to the original one. Depending on the desired application, the devulcanized rubber can be reused as obtained or mixed in different quantities with virgin rubber or other polymers. In economic terms, it is possible to ensure that the use of devulcanized rubber in new formulations can reduce the cost of the final products.

The chapter aims to address the rubber recycling by mechano-chemical and thermo-mechanical devulcanization. It will be focused on the rubber devulcanization processes, the characterization of the devulcanized materials and in the analysis of the obtained results. The implemented devulcanization methods are respectful with the environment.

5.2 Bibliography summary

Several devulcanization processes have been developed and presented in the literature in the last decade, including chemical [25, 26], biological [23, 24, 28], ultrasound [29–31], microwaves [32–34], mechano-chemical [35], and thermo-mechanical [32, 33].

In the thermo-mechanical devulcanization, the material is submitted to high shear strain, at a given temperature, stretching the network of the rubber. This type of devulcanization can be divided into two large groups, the one that is carried out in a reactor and sometimes in the presence of chemicals, and the one that is carried out in an extruder, that is a continuous process. Some additives, such as reclaiming agents or oils, could be added during the devulcanization. In both cases, the supercritical carbon dioxide ($scCO_2$) is sometimes used as atmosphere during the treatment. That is because in the supercritical state the CO_2 has high diffusivity and acts as an excellent swelling agent in the rubber, accelerating the permeation of devulcanizing reagents into it, facilitating the sulfidic bonds breaking and, therefore, the devulcanization [38]. In addition, the use of $scCO_2$ allows the devulcanization reaction to took place homogeneously all over the rubber and make the process more efficient.

In the mechano-chemical devulcanization the waste rubber is mixed in a mixer or a reactor with the chemical agents, at controlled temperature and pressure. This technique is generally combined with thermal and mechanical energy to speed up the devulcanization treatment. Disulfides (such as diphenyl disulfide, thiophenols and their zinc salts), thiol-amine reagents, hydroxide or chlorinated hydrocarbons are examples of devulcanizing chemical agents. Devulcanization reagents are typically used in concentrations in the range of 0.5–15 wt%.

During the last years, many authors have been working on the thermo-mechanical devulcanization of different type of rubbers. On the one hand, there are the works related to devulcanization carried out in autoclaves or reactors. M Kojima *et al.* [39] studied the devulcanization of unfilled polyisoprene rubber with devulcanizing reagents and $scCO_2$, in an autoclave. Between the different used chemicals, thiophenol (PhSH)-n–butylamine (n–BuNH2) combination and diphenyl disulfide (DD) gave as a result 100% soluble materials and were the most appropriate. When analyzing the results, it was also pointed that the CO_2 made the devulcanization more efficient when it was in the supercritical state. In 2004, the same authors [40] analyzed the devulcanization of sulfur-cured unfilled natural rubber (NR) in an autoclave using $scCO_2$ and DD. 60 % of sol fraction was obtained in the samples treated during 60 min at 10 MPa of pressure and 180 °C, with 10% of DD. Even if the reaction or mechanism between the rubber, DD and $scCO_2$ was not fully understood, they proposed a cleavage reaction mechanism during devulcanization which is explained as follows. Through the action of temperature, DD is dissociated and dissolved in the $scCO_2$, then it can easily penetrate the rubber structure and attack the sulfidic bonds to break them, resulting in a selective bond breaking, and therefore in a devulcanized material. Kojima *et al.* [41] also studied the main role of the $scCO_2$ during isoprene rubber devulcanization. They conclude that the $scCO_2$ helps in the homogeneous penetration of the devulcanizing reagent into the rubber; basically, it facilitates the dispersion of DD into the elastomer. The mentioned group also investigated the devulcanization of NR-based truck tire rubber, and NR filled with different contents of carbon black (CB), in the presence of DD and $scCO_2$, in an autoclave [42]. One of the main conclusions was that the devulcanization process was not affected by the CB, which is an important finding. Another conclusion claims that it was possible to obtain blends based on devulcanized tire rubber and virgin rubber with reasonable physical properties. As an example, a reduction of about 10% was obtained in the tensile strength when 40 parts per hundred rubber (phr) of devulcanized tire rubber were added to the virgin matrix. Jiang *et al.* [43] carried out the devulcanization of sulfur-cured butyl rubber in a high-pressure reactor, with $scCO_2$ atmosphere and DD as devulcanizing reagent. The analysis of the devulcanized material was based on the soluble fraction content, the highest value for the treated butyl rubber was found to be 98.5%. When studying the revulcanization of the recycled rubber to be used as a new material, it was highlighted that large amounts of DD could affect such process, and in fact, are not recommended. Mangili *et al.* [44, 45] worked on the GTR devulcanization with supercritical CO_2, using DD as devulcanizing reagent. A full factorial experimental design was used to determine the tests conditions. After correlating the results, it was pointed that the DD fraction, the temperature, and their interaction were the parameters that most affect the devulcanization process. The effect of DD in the mechanical properties of the blends based on devulcanized GTR/virgin GTR was also analyzed. It was found that the use of DD favors the increase of the strain at break and the decrease of the Young's modulus i.e. decrease of the material rigidity. Summarizing, they proposed a devulcanization process that takes few times, and is done at a reasonable pressure, which means energy saving and promotes the environmental protection. The devulcanization of waste tire rubber in an autoclave, using $scCO_2$ and DD as devulcanizing reagent, was also done by Liu *et al.* [38]. The changes in the crosslinked network of the rubber caused by the devulcanization with and without $scCO_2$ were analyzed. It was found that the $scCO_2$ brings the DD into the structure of the rubber and promotes the devulcanization reaction to occur on the surface and inside the material at the same time, which results in a high molecular weight and a low polydispersity index of the soluble fraction. This finding is represented in Figure 5.2, where it can be seen that after the devulcanization with $scCO_2$ the polymer chains of the sol remained intact, which suggests that mainly crosslink scission occurs. The concentration of devulcanizing reagent, in this case DD, was pointed as one of the main features of the process.

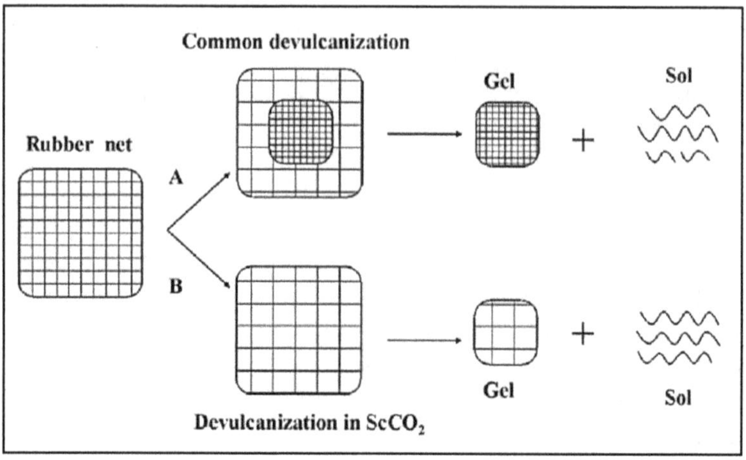

Figure 5.2 *Representation of the devulcanization process with and without scCO$_2$ [38].*

On the other hand, there are the works dealing with thermo-mechanical and mechano-chemical devulcanization in extruders or mixers. The thermo-mechanical devulcanization of EPDM rubber in an extruder was investigated by several authors [46–50] a mechanism for the mentioned devulcanization was proposed (Figure 5.3).

It was considered that the energy required to break the mono-sulfidic bonds is greater than that needed to break the poly sulfidic ones, and so, the poly-sulfidic bonds are broken in the first step. As a consequence, the rubber structure changes to be composed by mono-sulfidic bonds, that then are broken by the shear stress, as shown in Figure 5.3. During the process various gaseous species are released, such as hydrogen sulfide (H_2S), sulfur dioxide (SO_2), and carbon disulfide (CS_2), among others, also a few main chains of the polymer are broken.

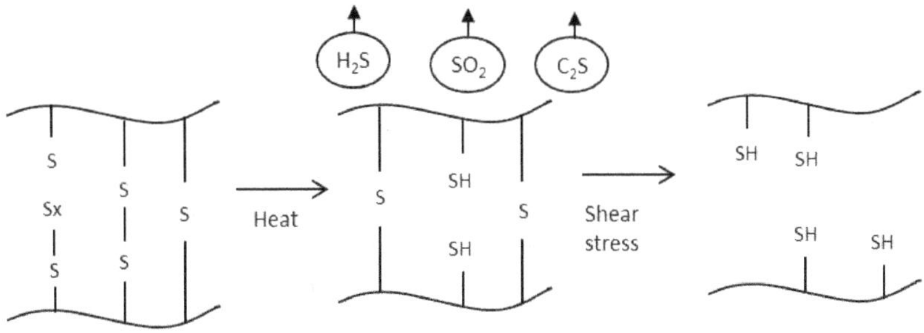

Figure 5.3 *Representation of crosslink scission during thermo-mechanical devulcanization [48].*

Li *et al.* [51] performed ground rubber powder (GRP) devulcanization by continuous shear flow in an extruder. The use of devulcanized rubber (dev.R) and GRP as fillers for NR was studied. A moderately satisfactory devulcanization was obtained, with crosslink scission but also with main chain scission. It was pointed that the addition of 10 phr of dev.R to the virgin NR did not reduce remarkably its mechanical performance. The material lost just 13% of its original strength and 9% of its strain at break. It was mainly concluded that the compounds filled with dev.R showed better properties than those filled with the virgin GRP. Lamminmaki *et al.* [52] studied the devulcanization of end of life tires by continuous shear flow, and the effect of compounding the obtained material with virgin rubber. It was found that the devulcanized material can be compounded, up to 50 phr with virgin rubber without decline remarkably its mechanical properties. The elongation at break, tensile strength, abrasion resistance and hardness were enhanced by the addition of the treated rubber into the compounds formulation. The mechano-chemical devulcanization of waste natural rubber in a two-roll cracker-cum mixing mill, with and without diphenyl disulfide (DD) was studied by Jana *et al.* [53]. It was pointed that DD plays a key role in the devulcanization treatment and that the revulcanized rubber produced by the mechano-chemical process could replace virgin NR in many applications. Devulcanization of waste ground rubber (WGR) in a twin screw extruder was performed by Maridass *et al.* [54]. The optimal tests conditions in their study were 170 °C and 30 rpm. Temperature and screw speed rotation were pointed as the most important parameters during the process. It was concluded that the treated rubber can be used to replace about 65 wt.% of virgin NR, in many applications, which is a significant finding. Zhang *et al.* [55] worked on the devulcanization of GTR in a pan-mill mechano-chemical reactor, they also analyzed the mixture of the devulcanized rubber with virgin NR, at different ratios. Results showed that the mechanical properties of the blends NR/devulcanized GTR were higher than those of the blends NR/no devulcanized GTR, fact that was attributed to the better interfacial interaction between NR and the devulcanized GTR. It was concluded that, in the thermo-mechanical process, the quality of the devulcanized material is mostly related to the operation conditions, especially to the screw rotational speed and the temperature. The devulcanization of NR by chemical and mechano-chemical processes, using benzoyl peroxide as devulcanizing reagent, was studied by Rooj *et al.* [23]. As in another studies, it was found that the mechanical properties of the revulcanized material were better than those of the original rubber. It was pointed that the reaction time and the concentration of benzoyl peroxide were the most important devulcanization parameters, and that the mechano-chemical process was more efficient than the chemical one. Rubber devulcanization in a twin screw extruder equipped with a special screw design was performed by Formela *et al.* [56]. The effects of feed rate and screw speed on the quality of the devulcanized material were evaluated. It was found that the use of a plasticizing system that induces high shearing stress in the rubber, allows the use of lower devulcanization temperature, and therefore, energy saving. The mechanical properties of the obtained devulcanized materials were similar to those of commercial reclaimed rubbers. The devulcanization of SBR rubber blended with EPDM in a twin screw extruder, with subcritical water as reaction medium, was carried out by Wang *et al.* [57]. It was concluded that the subcritical water assists the devulcanization process, protects the material from oxidative degradation, reduces the size of the gel particles in the devulcanized product, favors the crosslink breakage, and considerably improves the mechanical properties of the revulcanizates. Hu Si *et al.* [58] analyzed the devulcanization of different GTR materials in a twin screw extruder. Results showed that increasing the temperature and the screw speed favor the devulcanization. It was revealed that the crosslinking network was easily broken when the NR was the major component of the GTR. On the other hand, better mechanical properties were obtained in the revulcanized material when the SBR was the major component of the rubber. The devulcanization behavior of SBR/NR in a twin-screw extruder was studied by Yazdani *et al.* [59]. It was shown that the screw configuration determines the residence time of the rubber in each step of the process, and that it also controls the sol fraction and the crosslink density of the obtained material. Jinwei Shi *et al.* [60] analyzed the thermo-mechanical devulcanization of rubber, in a twin screw extruder. They produced liquid reclaimed rubber (LRR) from GTR, in a continuous process. A maximum soluble fraction of 73.5% was obtained. The LRR was then used as plasticizer in NR to replace the conventional oils. It was concluded that the plasticizing efficiency of LRR was lower than that of environmental aromatic oil (EAO). Nevertheless, it appeared that the LRR-plasticized NR has higher extension strength, Young modulus at 100 % and 300% elongations, hardness and thermo stability than EAO-plasticized NR. Mechano-chemical reclaiming of waste EPDM rubber using disulfide oil (DSO) as devulcanizing reagent was studied by Sabzekar *et al.* [2]. It was pointed that the use of DSO increases the success of the performed devulcanization. Results revealed that it was possible to mix up to 40 wt% of reclaimed rubber with virgin EPDM without influencing the material curing time. Also, that the mechanical properties of the recycled blends were not very deteriorated. The devulcanization of natural rubber in an open roll mixing mill, by using bis (3-triethoxysilyl propyl) tetrasulfide (TESPT) as a devulcanization reagent was performed by Ghorai *et al.* [17]. A successful devulcanization was obtained. When analyzing the scanning electron microscopy observations, it was revealed that the devulcanized material was more homogenous and without granular structure, as the devulcanization time increases. The maximum reaction time was set as 60 min. In the work done by Zedler *et al.* [61], ground tire rubber (GTR) was reclaimed by the mechano-chemical process, in the presence of bitumen and additives (peptizer P300, tetramethylthiuram 21 disulfide (TMTD), and di(2-tert-butyl-peroxyisopropyl)benzene 22 (BIB1)), at low temperature. The obtained treated rubber was then blended with virgin nitrile butadiene rubber (NBR). It was concluded that the additives used during the mechano-chemical reclaiming of GTR could affect the performance of the devulcanized rubber and the properties of its blends with other polymers. The devulcanization of EPDM rubber by high shear mixing was analyzed by Diaz *et al.* [62]. For that, a high shear mixing machine for small batches was built and instrumented. Results showed that the smaller particles seemed to have lower crosslink density compared to the larger ones, at equivalent treatment conditions. It was highlighted that the lower the average rubber particle size, the smaller the energy needed for the devulcanization treatment.

There are numerous publications regarding thermo-mechanical and mechano-chemical devulcanization. In most of them it was possible to carry out a partial or quite-complete devulcanization and the key parameters of the process were identified. In some works, the behavior of the treated material mixed with virgin rubber was also analyzed. One of the advantages of the devulcanized rubber is that they could be mixed with virgin components and be revulcanized to form a new material.

5.3 Mechano-chemical and thermo-mechanical devulcanization

Mechano-chemical and thermo-mechanical devulcanization of rubber were performed.

On one hand, a novel device allowing performing mechano-chemical rubber devulcanization in $scCO_2$ atmosphere was developed [63]. For that, a triaxial compression reactor was manufactured. It was integrated into a dynamic hydraulic universal testing machine equipped with a heating chamber. In Figure 5.4, the devulcanization equipment is presented. It mainly consists of a cylinder and a piston that allows applying a triaxial compression stress on the rubber. The reactor is placed into a heating chamber in a dynamic hydraulic universal testing machine (100 kN). It has a cavity to contain the rubber particles in a controlled atmosphere and, if necessary, a devulcanizing reagent. In the case of the CO_2, the compression loading generated by the testing machine results in an increase of the gas pressure surrounding the rubber particles, turning it from the gaseous state to the supercritical state. The main idea of the reactor is to devulcanize the rubber by submitting it to an equi-triaxial mechanical stress (hydrostatic pressure loading) and by using small quantities of CO_2 (less than 5 wt%). It is possible to see in Figure 5.4 that the pressure inside the reactor is controlled by the movement of a piston (38 mm diameter), using the software of the testing machine (WaveMatrix). The movement of the piston has to be gradual, during unloading, to avoid the rapid gas decompression that could damage the rubber o'rings, and therefore, the sealing system. To obtain a desired pressure by moving the piston, the following procedure is required. First two compression loading tests must be carried out, with and without CO_2 in the reactor chamber. From the difference between the obtained force-displacement curves, the experimental gas pressure-full load curve is deduced. Then, the force to be imposed to reach a desired gas pressure is obtained from the mentioned curve. Finally, the piston has to be moved till reached the mentioned force, and the gas pressure can be satisfactorily achieved. During a devulcanization test in the reactor, the parameters that can be controlled are: amount of rubber, temperature (up to 160 °C), CO_2 content, CO_2 pressure, amount of devulcanizing reagent, time, and compression loading and unloading steps.

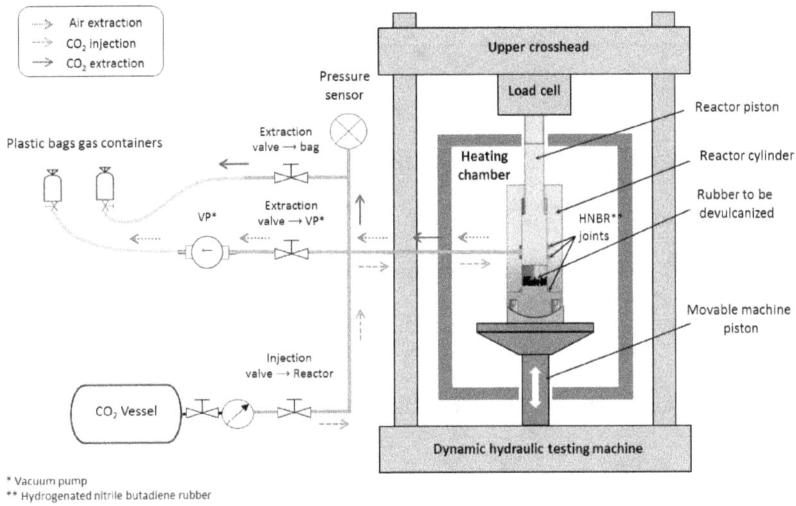

Figure 5.4 *Scheme of the devulcanization reactor.*

In the other hand, thermo-mechanical devulcanization of waste rubber was conducted in a large-scale twin-screw extruder Leistritz ZSE 27 MAXX 48D, schematically presented in Figure 5.5a [64]. The maximum flow rate of the extruder was 30 kg/h. The screw (28 mm) was specifically designed to performed rubber devulcanization, its configuration is presented in Figure 5.5b. It is divided in three zones; the first one is composed by conveying and kneading elements that prepare the material to be devulcanized. The second one is the central devulcanization zone that is followed by a vacuum extraction, and finally the discharge.

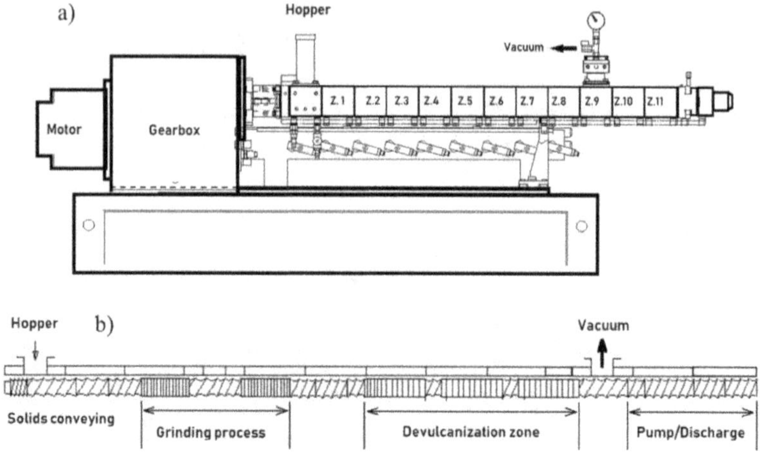

Figure 5.5 a) *Twin screw extruder and* ***b)*** *screw configuration used to perform the devulcanization tests.*

5.3.1 Materials and devulcanization processes

Two kinds of waste natural rubber were used in the present study. For the mechano-chemical devulcanization the starting material was waste sulfur-cured NR, provided by an industry. Its formulation is presented in Table 5.1. The average size of the rubber particles, measured by sieving was less than 4 mm. The rubber was used as received, and no further details were provided by the industry. Diphenyl disulfide (DD) from Sigma Aldrich was used as devulcanizing reagent. The amount of DD was 15 wt% in all cases.

Table 5.1 *Waste sulfur-cured NR formulation*

Component	Sulfur-cured NR (phr)
NR	100
Carbon black	35
Processing oil	3
Stearic acid	1
CBS*	1
Sulfur	1.5

*CBS: N-Cyclohexyl-2-benzothiazole sulfonamide

The atmosphere during the devulcanization tests was CO_2 because it is non-toxic, cheap, its supercritical point is easy to reach (31.1 °C and 7.38 MPa), is non-flammable, and easily removable from the rubber [35, 60]. In the supercritical state the CO_2 has high diffusivity and acts as an exceptional swelling agent in the rubber, which helps in the bonds breaking and thus, in the devulcanization process [38]. The purity of the used CO_2 was ≥ 99.7%. To perform each devulcanization test, 3 g of DD mixed with 20 g of NR were placed in the cavity of the reactor, at room temperature. Then the device was closed and heated to the specified temperature. Tests were performed at 120, 140 and 160 °C, during 1 hour. Before injecting the CO_2, the air inside the reactor was removed using a vacuum pump. Afterwards, the compression was executed, producing an increase in the pressure and allowing the CO_2 to become supercritical. The imposed force was equal to -18 kN, inducing a pressure of 12 MPa inside the reactor. After each test, the system was slowly cooled to room temperature. Then, the device was opened and the devulcanized rubber was extracted in the form of one agglomerate piece. Finally, the treated material was cut and reserved for its later characterization. The degree and quality of the performed devulcanization were evaluated by the Horikx theory, thermogravimetry and scanning electron microscopy.

For the thermo-mechanical devulcanization, which us a continuous process, the starting material was also NR industry waste rubber, but this time coming from injected parts. The average particles size was between 0.8 and 4 mm. No further details were given by the company, rubber formulation was not available in this case. This material was used for the continuous devulcanization because it was available in large quantities that were in accordance with the extruder requirements. In each devulcanization test, the temperature was constant along the 11 zones of the barrel (Figure 5.5a). Tests were performed at 80, 100, 120, 140, 160, 180, 200 and 220 °C. The remaining extrusion parameters were constant for all the tests: the material feed rate was 5 kg/h, the screw speed was 240 rpm, and the die temperature was 80 °C. For simplicity, the devulcanized samples were named as dNR-X, where X corresponds to the barrel temperature. At the end of each test, the temperature of the devulcanized material was measured at the output of the extruder with a laser thermometer. Eight materials were obtained after the devulcanization tests. The degree and quality of the performed devulcanization were evaluated by the swelling test combined with the Horikx theory. Three of those materials, dNR-100, dNR-160 and dNR-220, were revulcanized according to a common industrial formulation, detailed in Table 5.2 [19, 61]. During revulcanization, the exact amounts of components (Table 5.2) were mixed in a roll mixing mill and then revulcanized during 5 minutes, at 160 °C and 150 MPa. The obtained materials were named as dNRv-X, where X corresponds to the devulcanization temperature.

Table 5.2 *Revulcanization formulation*

Material	Devulcanized NR (phr)			Components (phr)			
	dNR-100	dNR-160	dNR-220	Stearic Acid	ZnO	CBS*	S
dNRv-100	100	0	0	2	4	1.5	1.5
dNRv-160	0	100	0	2	4	1.5	1.5
dNRv-220	0	0	100	2	4	1.5	1.5

*CBS: N-Cyclohexyl-2-benzothiazole sulfonamide

5.3.2. Material characterization

The mechano-chemical devulcanization was performed on a laboratory scale, and the thermo-mechanical devulcanization was done on a pilot plant scale in a company, that is why the starting materials and the characterization techniques for each case are different, according to the capabilities of each place. Anyway, results are comparable and will be analyzed in the next section.

5.3.2.1 Evaluation of the quality and degree of devulcanization: application of the Horikx theory

As direct access at molecular level is non-trivial for rubber materials, an indirect method is required to analyze the changes generated in the structure by the devulcanization process. The most widely used indirect technique is based on the swelling test (ASTM D 6814-02). Through this test, it is possible to obtain many characteristics of the rubber such as the crosslink density, the degree of swelling, the soluble and gel fractions, etc. The degree of devulcanization is estimated as the decrease in the material crosslink density, after comparing that of the devulcanized rubber with that of the virgin vulcanized rubber. It is worth underlining that the swelling test results are also used to perform the Horikx chart [67]. The Horikx chart [67] is a tool that allows to perform an evaluation of the devulcanization. In this diagram, the devulcanization degree and the soluble fraction of the rubber are related, so that both the degree and quality of devulcanization, in terms of crosslink scission versus main chain scission, can be qualitatively deducted. The theoretical Horikx chart for one of the studied NR (%S = 1.17 %) is presented in Figure 5.6. The solid line corresponds to only selective crosslink scission, and the dashed line corresponds to only main chain scission. It is considered that the experimental points located near the solid curve correspond to the rupture of the sulfidic bonds, denoting a successful devulcanization, and those positioned near the dashed curve correspond to the carbon-carbon chains scission i.e., the polymer degradation. If there are points located between both curves, they are considered to undergo regeneration i.e., a combination between devulcanization and degradation of the rubber. To perform this analysis, the soluble fraction and the devulcanization degree of each sample has to be measured by the swelling test, as explained hereafter.

The first step for the devulcanization degree measurement was washing the rubber with acetone, in a Soxhlet extractor, during 16 h (Standard ASTM D297-02). Later, the devulcanization degree was evaluated according to the Standard ASTM D 6814-02, by the swelling test in toluene. For the swelling test the following procedure was followed: after washing, the specimen was dried and weighted (m_i), and then immersed in a flask with toluene at ambient temperature, during 72 h. The solvent was refreshed every 24 h. The sample impregnated with toluene was weighed (m_t) and then dried in an oven at 70 °C overnight. The dried sample was finally weighed (m_d). In all cases, three samples of each material were tested, and the results were averaged.

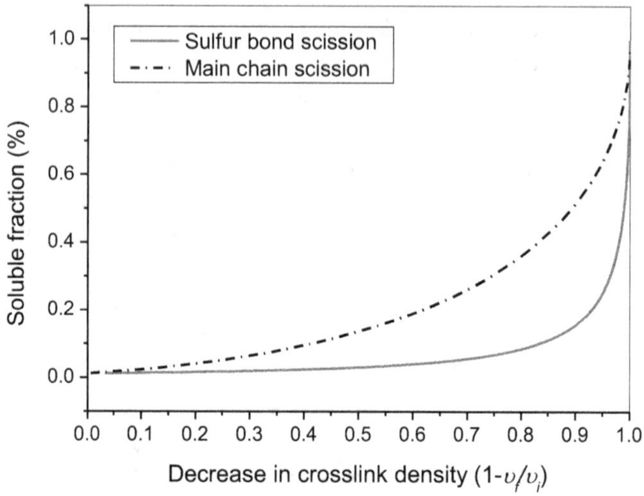

Figure 5.6 *Theoretical Horikx chart for the studied NR.*

The soluble fraction, %S, and the crosslink density, v_e, were evaluated by using the Flory-Rehner equations:

$$\%S = \frac{m_i - m_d}{m_i} \times 100 \quad (1)$$

$$v_e = \frac{-[ln(1-V_r)+V_r+\chi V_r^{2}]}{[V_1\,(V_r^{1/3}-V_r/2)]} \quad (2)$$

where V_r is the rubber volume fraction in the swollen sample, χ is the rubber-solvent interaction parameter (χ=0.391 for NR-toluene interaction), and V_1 is the molar volume of toluene. V_1=106.3 mL/mol. V_r was calculated using the following equation:

$$V_r = \frac{m_d/\rho_d}{m_d/\rho_d + m_s/\rho_s} \quad (3)$$

where m_d and ρ_d is respectively the mass and the density of the dried rubber, m_s is the mass of the solvent absorbed by the sample, in this case toluene, and ρ_s its density at room temperature. As explained in the Standard ASTM D 6814-02, the density of the dried rubber (ρ_d) was evaluated by hydrostatic weighing.

The devulcanization degree, *%Devulc.*, was calculated as follows:

$$\%Devulc. = \left[1 - \frac{\nu_f}{\nu_i}\right] \quad (4)$$

where ν_i and ν_f are the crosslink densities of the rubber before and after devulcanization, respectively.

Once the values corresponding to soluble fraction and devulcanization degree were obtained, they were transferred to the Horikx diagram with the corresponding theoretical curves.

5.3.2.2 Thermogravimetry

Thermogravimetric analyses (TGA) were performed in a Perkin Elmer Diamond TG/DTA. Samples of about 10 mg were heated from room temperature up to 800 °C, at a heating rate of 10 °C/min, under nitrogen atmosphere. The mass variation of the samples was gradually measured during the thermal degradation of the material. Tests were performed in the vulcanized NR and the mechano-chemical devulcanized rubber.

5.3.2.3 Scanning electron microscopy

Morphological analysis of the vulcanized NR and the mechano-chemical devulcanized rubbers was achieved by scanning electron microscopy (SEM). Images were taken in a TESCAN MIRA3 FEG-SEM. The working accelerating voltage was 20 kV.

5.3.2.4 Mooney viscosity

The viscosity of the thermo-mechanical devulcanized NR was measured with a Mooney Viscometer MV2000 from Alpha Technologies, according to the ASTM D1646 Standard. The test consists of a continuous rotation of a disk in a heated chamber filled with rubber, for a specific time. The resistance to this rotation induced by the rubber is measured, in arbitrary torque units, as the Mooney viscosity. Samples were preheated during 1 minute, and the viscosity was measured after 4 minutes of testing, at 100 °C. The results are presented as ML (1+4) 100 °C.

5.3.2.5 Mechanical properties

Tensile tests were carried out on the revulcanized rubbers (Table 5.2), according to the Standard ISO-37, using a Shimadzu AGS-X traction machine equipped with a 5 kN load cell, at ambient temperature. H2 dumbbell specimens of 2 mm thickness were used. The crosshead speed was 500 mm/min and the strain was measured with an extensometer.

5.4 Results

After carrying out the devulcanization and the swelling test, the obtained results were plotted in the Horikx diagram. For the mechano-chemical and the thermo-mechanical processes, the theoretical curves and the obtained experimental points corresponding to the studied materials are reported in Figure 5.7 and 5.8, respectively. Figure 5.7 highlights that all experimental points are near the solid curve, which corresponds to sulfur bond scission. This result confirms that the mechano-chemical devulcanization was effective, with a selective bonds scission and no significant rubber degradation. Figure 5.7 shows also that, as the temperature increases, the experimental points are moved to the right side of the diagram i.e., to higher devulcanization levels. A maximum devulcanization degree of ~90 % was reached, which is a great achievement. According to the devulcanization mechanisms proposed in the literature (Section 2), it is not possible to obtain a successful devulcanization if the temperature is not enough to dissociate the DD molecule, which was the case of the test carried out at 120 °C (Figure 5.7). The reactivity of DD seems to increase with increasing temperature, allowing to obtain a greater devulcanization degree at the higher temperature (160 °C). Based on the literature [21, 44, 68] it is expected that the mechanical properties of the revulcanized NR are similar or better than that of the virgin rubber.

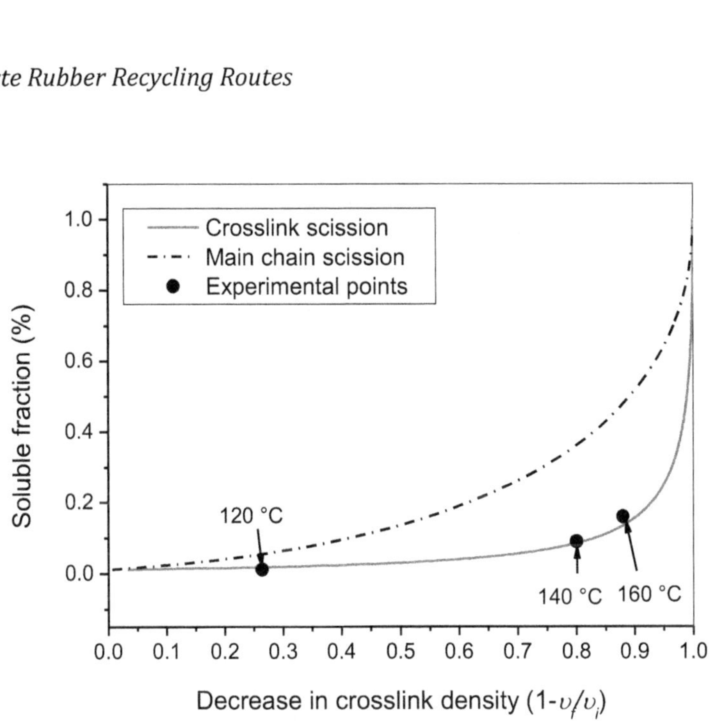

Figure 5.7 *Theoretical Horikx curves and experimental points for the mechano-chemical treated NR.*

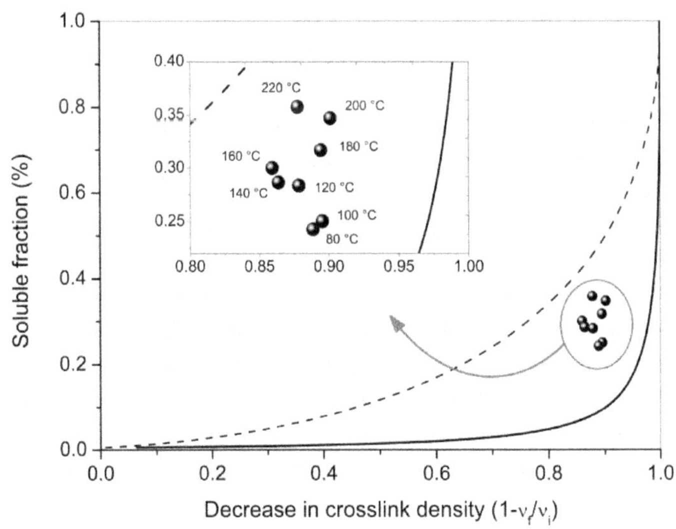

Figure 5.8 *Theoretical curves and experimental points obtained for the thermo-mechanical treated NR.*

Figure 5.8 shows that, in the case of thermo-mechanical devulcanization in the extruder, all experimental points are positioned between both curves, which mean that the rubber underwent a regeneration phenomenon, i.e., a part of the material was devulcanized and the other part was degraded. Furthermore, it can be seen that a high degree of reclaiming, near 90%, was obtained for all the samples, which is an important finding, as the proposed procedure could be directly applied on an industrial scale to recycle greater quantities of waste rubber. It can be pointed that, as the temperature increases the experimental points are shifted near the upper curve. Thus, in this study and contrary to what it was obtained by the other devulcanization method (Figure 5.7), increasing temperature just induce rubber degradation. The most devulcanized sample was that treated at 100 °C, that is why it was selected to perform the revulcanization.

It is evident that each devulcanization process is different. In one hand, the mechano-chemical method carried out with a moderate amount of sample gives as a result a successfully devulcanized material when using high treatment temperature. On the other hand, the thermo-mechanical method performed with a large amount of rubber, gives a highly regenerated material at a moderate working temperature. This means that depending on the needs and availability, one method or another is recommended.

Complementary tests, as TGA and SEM, were performed to further check the conclusions drawn by the Horikx analysis. Figure 5.9 presents the TGA results for the vulcanized NR and the NR devulcanized at 160 °C in the reactor. It can be seen that the devulcanized material starts the thermal degradation process before the vulcanized NR. The average degradation temperature (375 °C) was the same for both rubbers, which confirms that the polymer chains have not been deteriorated throughout the process. It was also found that the thermal degradation rate was slower for the devulcanized material, result that was associated to the lower sulfur content of this rubber. Finally, the residual mass corresponding to the virgin rubber (20%) is less than that of the devulcanized rubber (34%). This result was related to the carbon black content (Table 5.1). In fact, as the devulcanized rubber was treated by the swelling in toluene and lost its soluble fraction, the relative carbon black content of this rubber is higher and is observed as a higher residual mass.

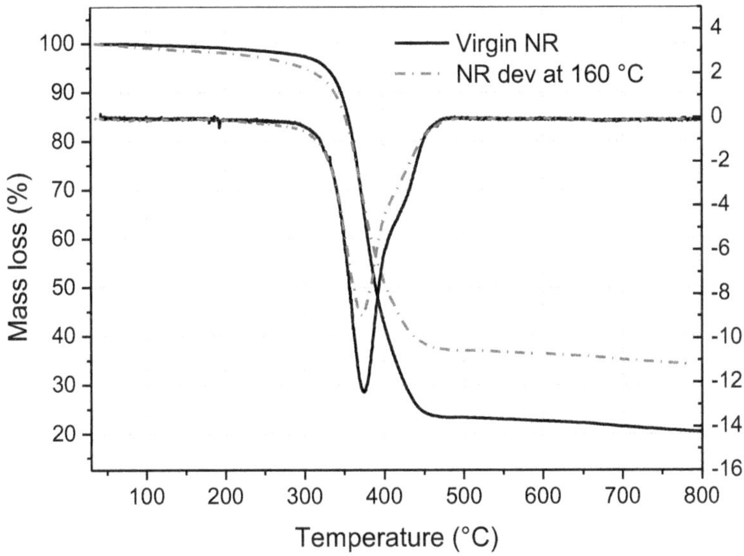

Figure 5.9 *Thermogravimetric results for the vulcanized NR and the NR devulcanized at 160 °C in the reactor.*

The analysis regarding the rubber morphology was carried out by SEM. Figure 5.10a shows that the starting material treated in the reactor was homogeneous, its morphology was continuous and smooth. The morphology of the NR devulcanized at 120 °C (Figure 5.10b) was similar to that of the virgin NR, suggesting that the rubber underwent slight changes during the devulcanization process. The NR treated at 140 °C (Figure 5.10c) was partially devulcanized and presents two phases, a homogeneous cohesive phase and a discontinuous non-cohesive phase, corresponding respectively to the devulcanized and non-devulcanized areas of the rubber. Figure 5.10d displays that the microstructure of the material devulcanized at 160 °C is homogeneous, smooth, and continuous, confirming that the rubber was effectively devulcanized, as concluded from the Horikx analysis (Figure 5.7).

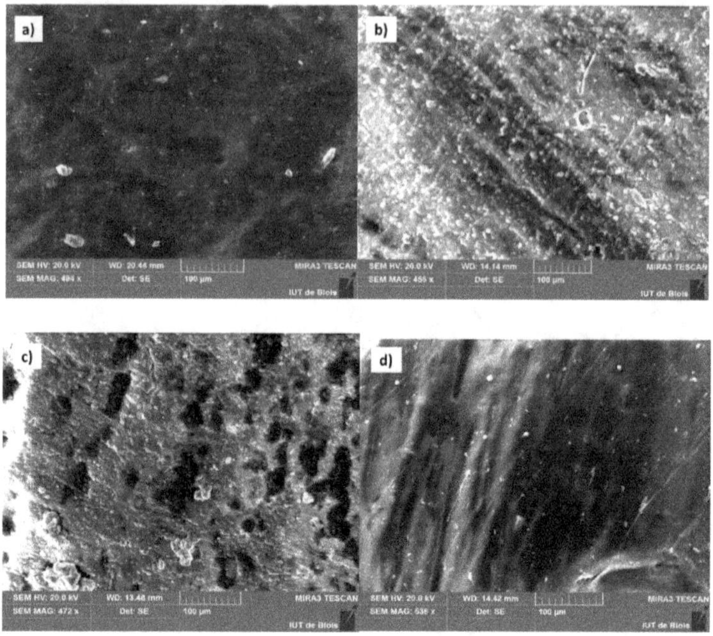

Figure 5.10 *SEM images of a) vulcanized NR, b) NR treated at 120 °C, c) NR treated at 140 °C, and d) NR treated at 160 °C in the reactor.*

For the thermo-mechanical devulcanized material, Mooney viscosity measurements and mechanical test were performed as complementary characterization. The Mooney viscosity is related to the network configuration of the material, it can be quickly measured, and it gives information related the ability of the material to be reprocessed. High viscosity values limit the rubber to be manufactured in a particular shape, while a very low viscosity obstruct processing [69]. It is worth noting that a low Mooney viscosity value of the treated rubber, denotes a higher number of broken bonds during the devulcanization process, as reported by Si et al. [58] during the devulcanization of ground tire rubber in a twin-screw extruder. Nevertheless, this parameter cannot give an idea about the quality of the performed devulcanization treatment, i.e., if was done by crosslink scission or main chain scission. Therefore, results are complemented with those of the Horikx theory. Figure 5.11 shows that the Mooney viscosity of the devulcanized NR decreases strongly as the barrel temperature increases up to 180 °C and remains almost constant for higher temperatures. The obtained values are in concordance with those reported in bibliography [43], and also with the results obtained in the Horikx analysis.

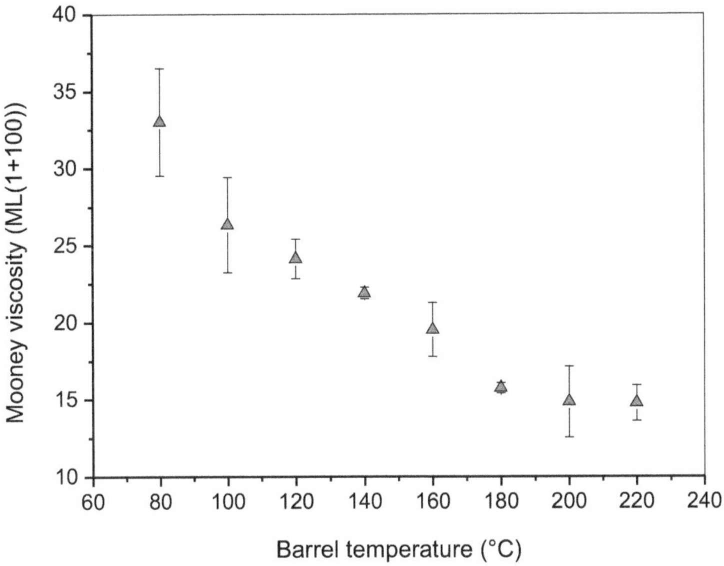

Figure 5.11 *Mooney viscosity as a function of the barrel temperature.*

The last characterization was performed in the revulcanized rubber. Three devulcanized materials were revulcanizated, dNRv-100, dNRv-160 and dNRv-220. After revulcanization, the swelling test was performed to analyze the crosslink densities and soluble fractions of these materials. The obtained results are presented in Table 5.3. dNRv-100 has the higher crosslink density, fact that implies the better ability of this material to be revulcanized and a probable preservation of long polymer chains after the devulcanization treatment. It can be seen that the soluble fraction of the revulcanized materials increases with increasing devulcanization temperatures. This could be related to an increase in the quantity of low molecular weight compounds, and therefore, to more degraded materials. Results corresponding to the mechanical properties of the revulcanized materials are also presented in Table 5.3. In accordance with the changes in the crosslink density, the mechanical properties of these materials decrease as the barrel temperature increases. The highest mechanical properties were obtained for the revulcanizate produced with the rubber devulcanized at 100 °C. Therefore, the revulcanized materials mechanical properties suggest that dNR-100 is the best devulcanized material to be blend with virgin rubber.

Table 5.3 *Crosslink density, soluble fraction and mechanical properties of the revulcanized materials*

Material/ Property	Soluble fraction [%]	Crosslink density [x 10^{-4}mol/cm^3]	M 100% [MPa]	Stress at break [MPa]	Elon-gation at break [%]
dNRv-100	4.57 ± 0.31	2.325 ± 0.026	1.25 ± 0.03	11.00 ± 0.29	518 ± 8
dNRv-160	5.73 ± 0.06	2.059 ± 0.040	1.11 ± 0.01	10.20 ± 0.49	488 ± 11
dNRv-220	7.55 ± 0.58	1.812 ± 0.029	1.06 ± 0.04	8.87 ± 0.08	481 ± 13

5.5 Concluding remarks

Throughout this chapter, natural rubber devulcanization was analyzed using two techniques, mechano-chemical devulcanization and thermo-mechanical devulcanization. In the first case, tests were performed on a laboratory scale with a moderate amount of sample; and in the second one, tests were performed on a pilot plant scale in a company using great quantities of rubber. In both cases the starting rubber was an industrial waste. Tests were performed in static and dynamic conditions, under different temperatures. In both cases the Horikx chart was used to analyze the quality of the performed devulcanization.

It was found that, through both processes, high reclaiming levels ~90 % could be achieved. The selection of the accurate devulcanization parameters plays an important role on the processes, and allows obtaining materials of a good quality, with little or no degradation. It was pointed that each devulcanization process is different. In one hand, the mechano-chemical method carried out with a moderate amount of sample gives as a result a successfully devulcanized material when using high treatment temperature. On the other hand, the thermo-mechanical method performed with a large amount of rubber, gives a highly regenerated material at a moderate working temperature. This means that depending on the needs and availability, one method or another is recommended. Results suggest that the NR industry waste could be almost entirely recycled with low energy consuming, which represents a real contribution to the industrial recycling and thus to the environmental protection.

References

1. Liu HL, Wang XP, Jia DM (2020) Recycling of waste rubber powder by mechano-chemical modification. J Clean Prod 245:118716. https://doi.org/10.1016/j.jclepro.2019.118716
2. Sabzekar M, Zohuri G, Chenar MP, et al (2016) A new approach for reclaiming of waste automotive EPDM rubber using waste oil. Polym Degrad Stab 129:56–62. https://doi.org/10.1016/j.polymdegradstab.2016.04.002
3. Rozada F, Otero M, Morán A, García AI (2005) Activated carbons from sewage sludge and discarded tyres: Production and optimization. J Hazard Mater 124:181–191. https://doi.org/10.1016/j.jhazmat.2005.05.002
4. Mui ELK, Lee VKC, Cheung WH, McKay G (2008) Kinetic modeling of waste tire carbonization. Energy and Fuels 22:1650–1657. https://doi.org/10.1021/ef700601g
5. Pehlken A, Müller DH (2009) Using information of the separation process of recycling scrap tires for process modelling. Resour Conserv Recycl 54:140–148. https://doi.org/10.1016/j.resconrec.2009.07.008
6. Molino A, Donatelli A, Marino T, et al (2018) Waste tire recycling process for production of steam activated carbon in a pilot plant. Resour Conserv Recycl 129:102–111. https://doi.org/10.1016/j.resconrec.2017.10.023
7. Burlakovs J, Kriipsalu M, Klavins M, et al (2017) Paradigms on landfill mining: From dump site scavenging to ecosystem services revitalization. Resour Conserv Recycl 123:73–84. https://doi.org/10.1016/j.resconrec.2016.07.007
8. Colom X, Faliq A, Formela K, Cañavate J (2016) FTIR spectroscopic and thermogravimetric characterization of ground tyre rubber devulcanized by microwave treatment. Polym Test 52:200–208. https://doi.org/10.1016/j.polymertesting.2016.04.020

9. Adhikari B, De D, Maiti S (2000) Reclamation and recycling of waste rubber. Progress in Polymer Science (Oxford) 25:909–948. https://doi.org/10.1016/S0079-6700(00)00020-4
10. Yun J, Isayev AI (2003) Superior mechanical properties of ultrasonically recycled EPDM rubber. Rubber chemistry and technology 76:253–270
11. (1999) Landfill of Waste Directive
12. Myhre M, MacKillop DA (2002) Rubber Recycling. Rubber Chemistry and Technology 75:429–474. https://doi.org/10.5254/1.3547678
13. Smith FG, Daniels EJ, Teotia APS (1995) Testing and evaluating commercial applications of new surface-treated rubber technology utilizing waste tires. Resour Conserv Recycl 15:133–144. https://doi.org/10.1016/0921-3449(95)00036-I
14. Nicholas P.P. (1982) The scission of polysulfide crosslinks in scrap rubber particles through phase transfer catalysis. Rubber Chem. Technol. 55, 1499–1515. http://dx.doi.org/10.5254/1.3535945
15. Accetta A, VJM, LPP (1982) Kinetic parameters of the overall reaction of scrap rubber vulcanization by 2% sulfur. Thermochim. Acta 59, 14. Thermochim Acta 149–156
16. (1994) Methods of devulcanization. Rubber Chem Technol 67, 559–566 http://dx.doi.org/105254/13538692 559–566
17. Ghorai S, Bhunia S, Roy M, De D (2016) Mechanochemical devulcanization of natural rubber vulcanizate by dual function disulfide chemicals. Polym Degrad Stab 129:34–46. https://doi.org/10.1016/j.polymdegradstab.2016.03.024
18. Fang Y, Zhan M, Wang Y (2001) The status of recycling of waste rubber. Mater Des 22:123–128. https://doi.org/10.1016/S0261-3069(00)00052-2
19. Sienkiewicz M, Kucinska-Lipka J, Janik H, Balas A (2012) Progress in used tyres management in the European Union: A review. Waste Management 32:1742–1751. https://doi.org/10.1016/j.wasman.2012.05.010
20. Karger-Kocsis J, Mészáros L, Bárány T (2013) Ground tyre rubber (GTR) in thermoplastics, thermosets, and rubbers. J Mater Sci

21. Isayev AI (2013) Recycling of Rubbers. The Science and Technology of Rubber 697–764. https://doi.org/10.1016/B978-0-12-394584-6.00020-0
22. Feng W, Isayev AI (2006) Recycling of tire-curing bladder by ultrasonic devulcanization. Polym Eng Sci 46:8–18
23. Rooj S, Basak GC, Maji PK, Bhowmick AK (2011) New Route for Devulcanization of Natural Rubber and the Properties of Devulcanized Rubber. J Polym Environ 19:382–390. https://doi.org/10.1007/s10924-011-0293-5
24. Ramarad S, Khalid M, Ratnam CT, et al (2015) Waste tire rubber in polymer blends: A review on the evolution, properties and future. Prog Mater Sci 72:100–140. https://doi.org/10.1016/j.pmatsci.2015.02.004
25. Anu Mary J, Benny G, Madhusoodanan KN, Rosamma A (2016) The current status of sulphur vulcanization and devulcanization chemistry: devulcanization. Rubber Science 29:62–100
26. Zhang X, Saha P, Cao L, et al (2018) Devulcanization of waste rubber powder using thiobisphenols as novel reclaiming agent. Waste Management 78:980–991. https://doi.org/10.1016/j.wasman.2018.07.016
27. Li Y, Zhao S, Wang Y (2012) Microbial Desulfurization of Ground Tire Rubber by Sphingomonas sp.: A Novel Technology for Crumb Rubber Composites. J Polym Environ 20:372–380. https://doi.org/10.1007/s10924-011-0386-1
28. Yao C, Zhao S, Wang Y, et al (2013) Microbial desulfurization of waste latex rubber with Alicyclobacillus sp. Polym Degrad Stab 98:1724–1730. https://doi.org/10.1016/j.polymdegradstab.2013.06.002
29. Isayev AI, Liang T, Lewis TM (2014) Effect of Particle Size on Ultrasonic Devulcanization of Tire Rubber in Twin-Screw Extruder. Rubber Chemistry and Technology 87:86–102. https://doi.org/10.5254/RCT.13.87926
30. Mangili I, Lasagni M, Huang K, Isayev AI (2015) Modeling and optimization of ultrasonic devulcanization using the response surface methodology based on central composite face-centered design. Chemometrics and Intelligent

Laboratory Systems 144:1–10. https://doi.org/10.1016/j.chemolab.2015.03.003
31. Hon YJ, Jeong KM, Saha P, et al (2015) Processing and Characterization of Microwave and Ultrasonically Treated Waste-EPDM/LDPE Polymer Composites. Polym Eng Sci. https://doi.org/DOI 10.1002/pen.23916
32. Novotny DS, Richard L M, FranK C M, David N T (1978) Microwave Devulcanization of Rubber.
33. Aoudia K, Azem S, Aït Hocine N, et al (2017) Recycling of waste tire rubber: Microwave devulcanization and incorporation in a thermoset resin. Waste Management 60:471–481. https://doi.org/10.1016/j.wasman.2016.10.051
34. Simon DÁ, Pirityi D, Tamás-Bényei P, Bárány T (2020) Microwave devulcanization of ground tire rubber and applicability in SBR compounds. J Appl Polym Sci 137:1–8. https://doi.org/10.1002/app.48351
35. Jana GK, Mahaling RN, Das CK (2006) A novel devulcanization technology for vulcanized natural rubber. J Appl Polym Sci 99:2831–2840. https://doi.org/10.1002/app.22984
36. Seghar S, Asaro L, Rolland-Monnet M, Aït Hocine N (2019) Thermo-mechanical devulcanization and recycling of rubber industry waste. Resour Conserv Recycl 144:180–186. https://doi.org/10.1016/j.resconrec.2019.01.047
37. Diaz R, Colomines G, Peuvrel-Disdier E, Deterre R (2018) Thermo-mechanical recycling of rubber: Relationship between material properties and specific mechanical energy. J Mater Process Technol 252:454–468. https://doi.org/10.1016/j.jmatprotec.2017.10.014
38. Liu Z, Li X, Xu X, et al (2015) Devulcanizaiton of waste tread rubber in supercritical carbon dioxide: Operating parameters and product characterization. Polym Degrad Stab 119:198–207. https://doi.org/10.1016/j.polymdegradstab.2015.05.017
39. Kojima M, Ogawa K, Mizoshima H, et al (2003) Devulcanization of Sulfur-Cured Isoprene Rubber in Supercritical Carbon Dioxide. Rubber Chemistry and Technology 76:957–968. https://doi.org/10.5254/1.3547784

40. Kojima M, Tosaka M, Ikeda Y (2004) Chemical recycling of sulfur-cured natural rubber using supercritical carbon dioxide
41. Kojima M, Kohjiya S, Ikeda Y (2005) Role of supercritical carbon dioxide for selective impregnation of decrosslinking reagent into isoprene rubber vulcanizate. Polymer (Guildf) 46:2016–2019. https://doi.org/10.1016/j.polymer.2004.12.053
42. Kojima M, Tosaka M, Ikeda Y, Kohjiya S (2005) Devulcanization of carbon black filled natural rubber using supercritical carbon dioxide. J Appl Polym Sci 95:137–143. https://doi.org/10.1002/app.20806
43. Jiang K, Shi J, Ge Y, et al (2013) Complete devulcanization of sulfur-cured butyl rubber by using supercritical carbon dioxide. J Appl Polym Sci 127:2397–2406. https://doi.org/10.1002/app.37542
44. Mangili I, Collina E, Anzano M, et al (2014) Characterization and supercritical CO2 devulcanization of cryo-ground tire rubber: Influence of devulcanization process on reclaimed material. Polym Degrad Stab 102:15–24. https://doi.org/10.1016/j.polymdegradstab.2014.02.017
45. Mangili I, Oliveri M, Anzano M, et al (2014) Full factorial experimental design to study the devulcanization of ground tire rubber in supercritical carbon dioxide. Journal of Supercritical Fluids 92:249–256. https://doi.org/10.1016/j.supflu.2014.06.001
46. Fukumori K, Matsushita M, Okamoto H, et al (2002) Recycling technology of tire rubber. JSAE Review 23:259–264. https://doi.org/10.1016/S0389-4304(02)00173-X
47. Suzuki Y. OM, MM, et al. (1998) in: Toyota Technical Review 48, 1, 53-57.
48. Mouri M, SN, OH, et al. (2000) A new devulcanization process. Continuous reclamation of rubber by shear flow reaction control technology (Part I). International Polymer Science and Technology 27 17–22.
49. Mouri M, OH, MM, et al. (2000) Devulcanization conditions and mechanical properties of re-vulcanized rubber for EPDM. Continuous reclaimation of rubber by shear flow reaction control technology (Part II). International Polymer and Technology 27 23–28.

50. Mouri M, OH, MM, et al. (2000) Continuous devulcanization by shear flow stage reaction control technology for rubber recycling. Part 3. Study of the devulcanization process for EPDM. International Polymer Science and Technology 27 12–16.
51. Li S, Lamminmäki J, Hanhi K (2005) Effect of ground rubber powder and devulcanizates on the properties of natural rubber compounds. J Appl Polym Sci 97:208–217. https://doi.org/10.1002/app.21748
52. Lamminmäki J, Li S, Hanhi K (2006) Feasible incorporation of devulcanized rubber waste in virgin natural rubber. J Mater Sci 41:8301–8307. https://doi.org/10.1007/s10853-006-1010-y
53. Jana GK, Mahaling RN, Rath T, et al (2007) Mechanochemical recycling of sulfur cured natural rubber. Polimery/Polymers 52:131–136
54. Maridass B, Gupta BR (2007) Effect of extruder parameters on mechanical properties of revulcanized ground rubber tire powder. Polimery/Polymers 52:456–460
55. Zhang X, Lu C, Liang M (2009) Properties of natural rubber vulcanizates containing mechanochemically devulcanized ground tire rubber. Journal of Polymer Research 16:411–419. https://doi.org/10.1007/s10965-008-9243-x
56. Formela K, Cysewska M, Haponiuk J, Stasiek A (2013) The influence of feed rate and shear forces on the devulcanization process of ground tire rubber (GTR) conducted in a co-rotating twin screw extruder. Polimery/Polymers 58:906–912. https://doi.org/10.14314/polimery.2013.906
57. Wang X, Shi C, Zhang L, Zhang Y (2013) Effects of shear stress and subcritical water on devulcanization of styrene-butadiene rubber based ground tire rubber in a twin-screw extruder. J Appl Polym Sci 130:1845–1854. https://doi.org/10.1002/app.39253
58. Si H, Chen T, Zhang Y (2013) Effects of high shear stress on the devulcanization of ground tire rubber in a twin-screw extruder. J Appl Polym Sci 128:2307–2318. https://doi.org/10.1002/app.38170
59. Yazdani H, Karrabi M, Ghasmi I, et al (2011) Devulcanization of waste tires using a twin-screw extruder: The

effects of processing conditions. Journal of Vinyl and Additive Technology 17:64–69

60. Shi J, Zou H, Ding L, et al (2014) Continuous production of liquid reclaimed rubber from ground tire rubber and its application as reactive polymeric plasticizer. Polym Degrad Stab 99:166–175. https://doi.org/10.1016/j.polymdegradstab.2013.11.010
61. Zedler Ł, Przybysz M, Klein M, et al (2017) Processing, physico-mechanical and thermal properties of reclaimed GTR and NBR/reclaimed GTR blends as function of various additives. Polym Degrad Stab 143:186–195. https://doi.org/10.1016/j.polymdegradstab.2017.07.004
62. Diaz R, Colomines G, Peuvrel-Disdier E, Deterre R (2018) Thermo-mechanical recycling of rubber: Relationship between material properties and specific mechanical energy. J Mater Process Technol 252:454–468. https://doi.org/10.1016/j.jmatprotec.2017.10.014
63. Asaro L, Gratton M, Poirot N, et al (2020) Devulcanization of natural rubber industry waste in supercritical carbon dioxide combined with diphenyl disulfide. Waste Management 118:647–654. https://doi.org/10.1016/j.wasman.2020.09.026
64. Seghar S, Asaro L, Rolland-Monnet M, Aït Hocine N (2019) Thermo-mechanical devulcanization and recycling of rubber industry waste. Resour Conserv Recycl 144:180–186. https://doi.org/10.1016/j.resconrec.2019.01.047
65. Ikeda Y (2014) Recycling of sulfur cross-linked natural rubber (NR) using supercritical carbon dioxide. Chemistry, Manufacture and Applications of Natural Rubber 436–451. https://doi.org/10.1533/9780857096913.3.436
66. Ghorai S, Bhunia S, Roy M, De D (2016) Mechanochemical devulcanization of natural rubber vulcanizate by dual function disulfide chemicals. Polym Degrad Stab 129:34–46. https://doi.org/10.1016/j.polymdegradstab.2016.03.024
67. Horikx M (1956) Chain scissions in a polymer network. J Polym Sci B Polym Phys 19:445–454.

https://doi.org/https://doi.org/10.1002/pol.1956.120199305

68. Li S, LJ, HK (2005) Effect of ground rubber powder and devulcanizates on the properties of natural rubber compounds. J Appl Polym Sci 97 208–217
69. Zhang X, Saha P, Cao L, et al (2018) Devulcanization of waste rubber powder using thiobisphenols as novel reclaiming agent. Waste Management 78:980–991. https://doi.org/10.1016/j.wasman.2018.07.016

6

Thermo-mechanical Devulcanization of Waste Rubber through Supercritical CO_2-assisted Reactive Extrusion

Costas Tzoganakis* and **Maria Tzoganakis**

Department of Chemical Engineering, University of Waterloo, 200 University Avenue West, Waterloo, Ontario N2L3G, Canada.
*costas.tzoganakis@uwaterloo.ca

6.1 Introduction

Unvulcanized natural rubber is a sticky material with poor mechanical properties which cannot be used as is in the manufacturing of industrially useful rubber products. The discovery of rubber vulcanization by Charles Goodyear [1] was an extremely important milestone. Vulcanization increases rubber elasticity and has contributed immensely in the development of a plethora of high performance rubber products which can withstand high temperatures and harsh chemical environments. This high performance of vulcanized rubber is due to its three-dimensional structure which is imparted by the formation of covalent bonds between linear rubber macromolecules during chemical crosslinking. Although originally the term vulcanization was used to denote sulfur-initiated chemical crosslinking, today it is used widely for both sulfur and non-sulfur crosslinking of natural as well as synthetic rubbers. The complexity of sulfur crosslinking mechanisms has been studied extensively over the years and several reviews have been presented [see for example references 2-6]. The types of crosslinks formed during sulfur crosslinking have been identified [7] and the relative amount of these crosslinks depends on the curing recipe. Accelerated vulcanization recipes can be generally classified in three categories, conventional (CV), semi-efficient (semi-EV) and efficient (EV) vulcanization, depending on the sulfur-to-accelerator ratio. In high sulfur systems (CV), more poly-sulfidic links are formed while in lower sulfur systems (semi-EV and EV) more di-sulfidic and mono-sulfidic links are formed [6].

The discovery of vulcanization contributed immensely to the development of rubber manufacturing processes and thus created a growing demand for natural and synthetic rubber. The annual production and consumption of natural and synthetic rubber has increased significantly over the last half century. A recent study published by the Malaysian Rubber Council [8] indicates a tremendous growth during the last 20 years as shown in Figure 6.1. According to this study, global consumption of rubber in 2019 was about 29 million metric tons comprising of 13.6 million metric tons of natural rubber (NR) and 15.5 million metric tons of synthetic rubber (SR).

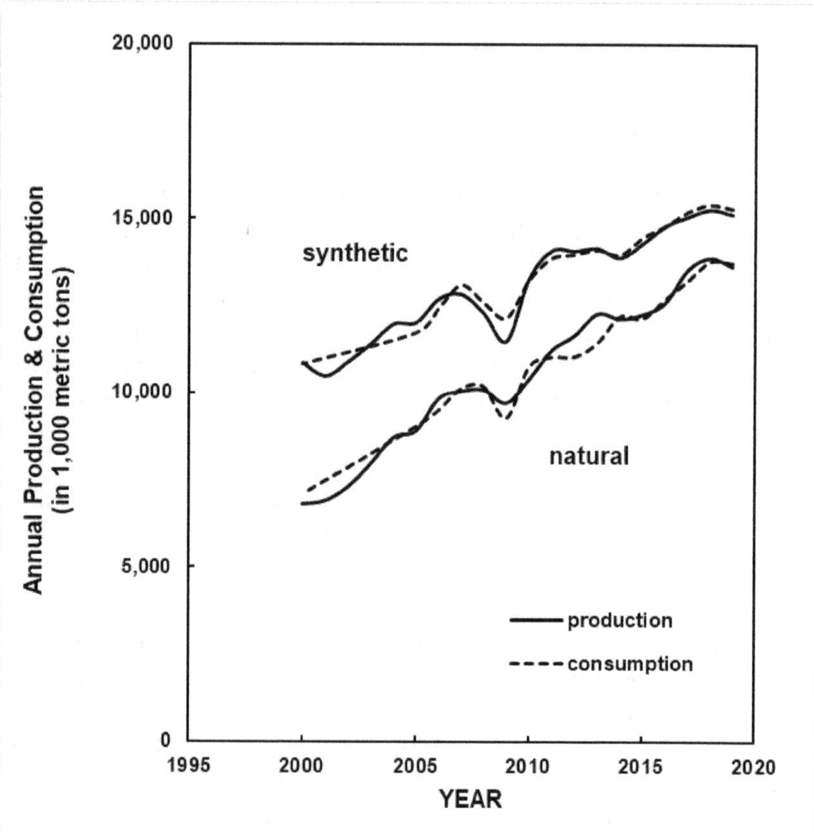

Figure 6.1 *Annual production and consumption of natural and synthetic rubber (data extracted from ref [7])*

Recent statistics [9] indicate that the majority of rubber (about 70%) is used in tire manufacturing applications as clearly shown in

Figure 6.2 for data during the period 2016-2020. It is estimated [10] that the global rubber tires market is set to grow at an approximately 5% CAGR steady pace in the foreseeable future (2019-2027). Annually, the world collectively uses and eventually discards over one billion End-of-Life Tires (ELTs). This is a growing environmental problem as global production of tires is projected to exceed 2.5 billion units by 2022. The rubber tire industry, as many others, operates in a linear economy fashion (take-make-waste) in which "Raw Material" from the resource sector is used for "Production" of tires which are used and eventually end up as "Waste" (Figure 6.3).

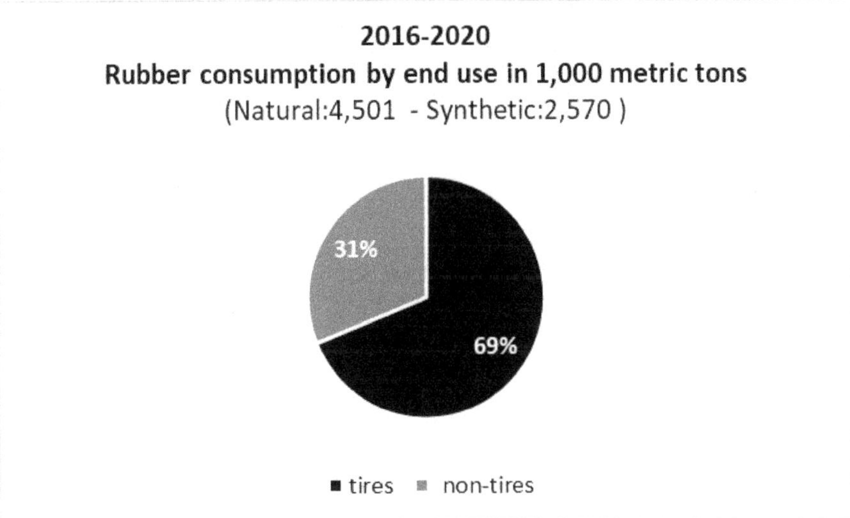

Figure 6.2 *Rubber consumption in tire and non-tire markets (data extracted from ref [8])*

ELTs are extremely difficult to recycle due to the chemically crosslinked nature of the rubber they contain. The same crosslinks that improve performance of vulcanized rubber make it impossible to reprocess tire rubber by dissolution or melting. Generally, recycling of ELTs can be achieved by techniques such as size reduction for crumb production, energy recovery, pyrolysis, and reclamation and devulcanization [11]. To prevent their massive stockpiling, ELTs are often burned for their energy content as tire-derived fuel (TDF) because they contain nearly 80% energy-intense rubber. Currently in the EU, nearly 30% of ELTs are used as TDF and in North America the amount is approximately 50%. Energy recovery is the lowest form of recycling and is harmful to the environment. In this new era of social responsibility and sustainability, better solutions are required. Recent data on global ELT management [12] indicate that the majority of the 30.9 million metric tons of ELTs generated in 2019, were used for energy and material recovery (Figure 6.4). However, even in this recent study, the importance of ELT management through devulcanization routes is largely not addressed.

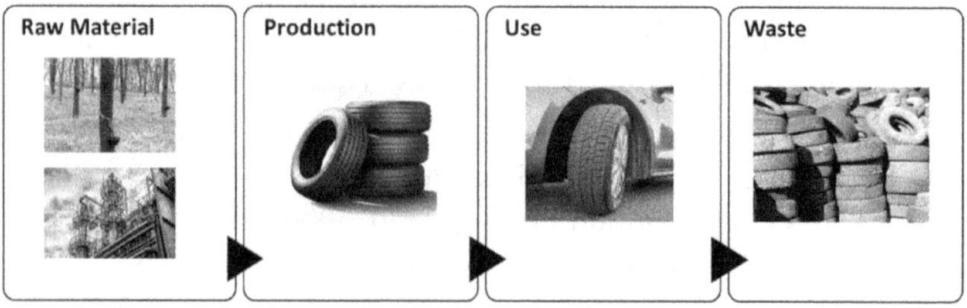

Figure 6.3 *Linear economy in rubber tire manufacturing.*

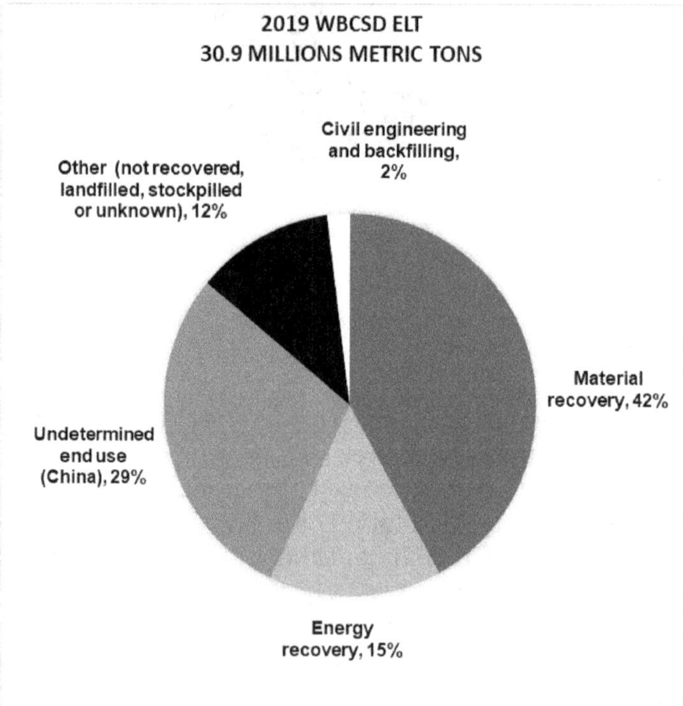

Figure 6.4 *End-of-life tire recycling statistics (2019 data extracted from ref [12]).*

Devulcanization allows the regeneration of valuable rubber in discarded ELTs and its reuse in high value products. Devulcanization involves the selective scission of sulfur-sulfur (S-S) bonds while leaving intact the backbone polymer carbon-carbon (C-C) covalent bonds. This differentiates devulcanization from reclamation processes in which all types of bonds (S-S and C-C) are broken randomly as reviewed recently in the literature [13-14]. Devulcanization of rubber involves the destruction of sulfur chemical crosslinks in a vulcanized rubber [15]. According to ASTM STP183A devulcanization is defined as "a combination of depolymerization, oxidation and increased plasticity" [16]. However, devulcanization is understood to be the reversal of the vulcanization process, that is, the cleavage of chemical bonds formed during vulcanization. Bonds formed during sulfur vulcanization are C-S and S-S and only such bonds should be cleaved during devulcanization. Therefore, devulcanization is the process of cleaving the mono-sulfidic (C–S–C), di-sulfidic (C–S–S–C), and poly-sulfidic (C–Sx–C) cross-links of vulcanized rubber while leaving the backbone of the rubber polymer molecules intact. However, in real applications, partial cleavage of backbone polymer chains is unavoidable regardless of the devulcanization method employed.

In view of these definitions, the majority of physical and chemical methods converting waste rubber into reusable rubber should be classified as reclaiming rather than devulcanization. These methods involve random scission of chemical bonds as opposed to targeted scission of sulfur crosslinks. Table 6.1 shows a classification of such processes into two categories, physical and chemical, which have been previously reviewed [17].

In terms of devulcanization, there are several methods that have been developed over the years and they have been reviewed widely [for example see 11, 17-27]. The methods most extensively studied can be categorized as: (i) chemical [28-52], (ii) ultrasonic [53-59], (iii) thermo-mechanical/chemical [60-87], (iv) microwave [88-92], (v) biological [93-98], (vi) chemical in supercritical-CO_2 ($scCO_2$) [99-106], and (vii) $scCO_2$-assisted thermo-mechanical [107-118].

Achieving devulcanization through these various methods in research labs at low throughput has been successfully demonstrated, but the true test of economic viability is scaling to commercial output levels that match the needs of large manufacturers. In 2004, the California Integrated Waste Management Board [119] conducted an evaluation study on approximately twenty-five potential devulcanization technologies which were identified throughout the world. It was found that only a very small number of small-capacity devulcanization systems were operating. Overall, it was concluded that these devulcanization methods were not economically viable given the low price of virgin rubber.

Among all devulcanization methods currently explored, the continuous thermomechanical method in an extruder seems to be the most promising for industrial applications [22, 120]. However, existing studies still indicate a significant degree of random backbone scission and there is room for much required process development and optimization. In addition, production rates in existing studies are far from what is necessary for industrial economic viability.

An improvement of this thermo-mechanical method is one in which devulcanization takes place in the presence of supercritical CO_2 ($scCO_2$) in a screw extruder. This method is one of the few currently being practiced industrially [www.tyromer.com] and it will be briefly reviewed here.

Table 6.1 *Rubber Reclaiming/Devulcanization Processes*

Physical Processes	Chemical Processes
Mechanical	Radical scavengers
Thermo-mechanical	Nucleophilic additives
Microwave	Catalyst systems
Ultrasonic	Chemical probes

6.2 Tyromer® process overview

The process of thermo-mechanical devulcanization through a CO_2-assisted reactive extrusion (REX) process was originally developed at the University of Waterloo in Canada [112-115]. This was part of a larger effort for development of REX operations for plastics using $scCO_2$. Subsequently, this technology was commercialized through licensing and joint venture operations by Tyromer Inc., a University of Waterloo start-up company.

The technology was based on the well-known differences of the various types of bonds in vulcanized rubber as well as the performance of $scCO_2$ as a plasticizer and foaming agent in plastics and rubber extrusion operations. Table 6.2 shows the energies of C-C, C-S and S-S bonds involved in vulcanized rubbers as well as their corresponding elastic force constants [121-122]. It is evident that the difference in bond energies is not that large. Therefore, relying on just energy differences to affect cleavage will result in scission of all types of bonds. That is why devulcanization methods employing mainly thermal methods result in relatively lower quality reclaimed rubber. On the other hand, considering the elastic constants values of these bonds, there is two orders of magnitude difference between C-C and S-S bonds. Therefore, under deformation S-S bonds will extend substantially more than C-C bonds and lead to preferential cleavage of sulfur crosslinks in vulcanized rubber. Supercritical CO_2 can be easily dissolved into rubber and upon expansion can generate significant localized deformation to break S-S bonds. Over a decade of research and development has allowed Tyromer Inc. to commercialize this invention and scale up production equipment and throughput. The company has demonstrated that it is possible to selectively break the sulfur links without the use of chemical solvents and devulcanization chemical reagents.

Table 6.2 *Bond Energies and force constants*

Bond type	Energy (KJ/mol) [121]	Elastic constant (relative value) [122]
C-C	346	100
C-S	272	-
S-S	226	3

The basis of this technology is graphically illustrated in Figure 6.5. In this REX process, rubber crumb is metered into a co-rotating twin-screw extruder and CO_2 is injected under supercritical conditions along the extruder barrel (Figure 6.5a). Supercritical conditions can be easily maintained in a pressurized section of the extruder. scCO2 is dissolved into the rubber and upon expansion leads to stretching of S-S bonds (Figure 6.5b) which can easily be broken under the deformation imparted onto the rubber by the rotating action of the screws in fully filled sections. As a result, a high degree of devulcanization is achieved and the devulcanized rubber flows readily as a thermoplastic material through a die in the form of a sheet or strands for pelletization. In addition to screw design, the main processing variables are screw speed, rubber flow rate and temperature profile. To demonstrate the fact that preferential sulfur crosslink scission is taking place, a theory previously developed to study crosslinking and scission reactions during polymer irradiation [123] has been used to differentiate between random or crosslink scissions. This theoretical model is based on the determination of the amount of soluble material generated from such reactions and it was further adopted by Horikx to model random and crosslink scissions in a vulcanized rubber [124]. This theory considers the relationship between the degree of devulcanization (change in crosslink density) and the soluble fraction content of the rubber before and after the devulcanization process. The effect of selective cleavage of sulfur crosslinks on degree of devulcanization is expressed by equation (1), while that of random chain scission is expressed by equation (2). Equation (3) is used to calculate the crosslinking index γ.

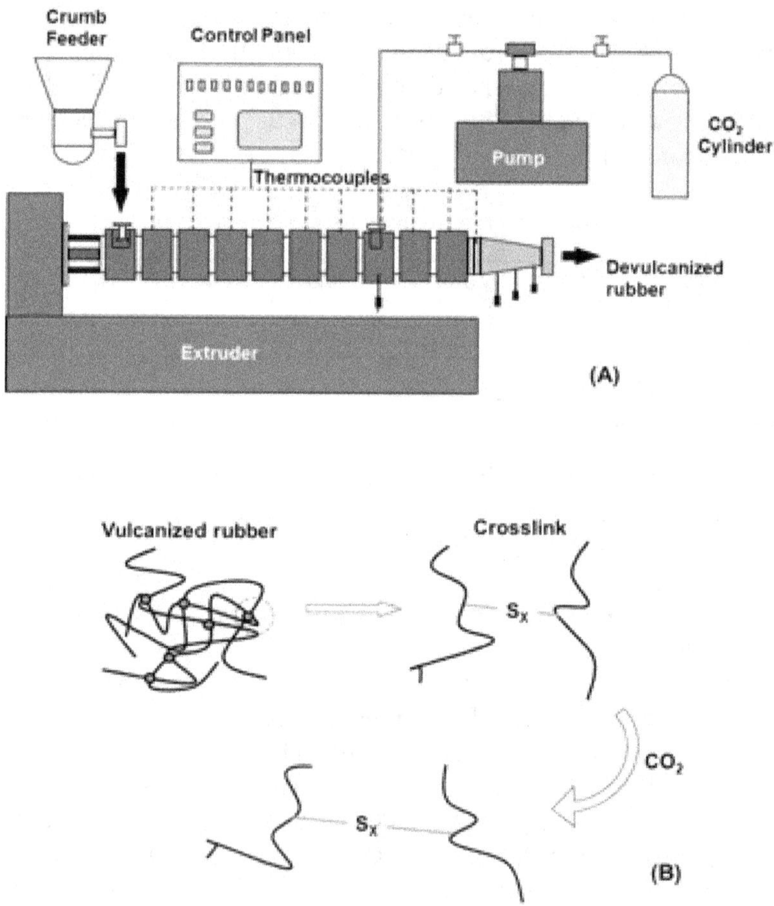

Figure 6.5 *Schematic illustration of the scCO₂-assisted thermo-mechanical REX devulcanization process.*

Selective crosslink scission

$$\text{degree of devulcanization} = 1 - \frac{v_{e2}}{v_{e1}} = 1 - \frac{\gamma_2\left(1-s_2^{1/2}\right)^2}{\gamma_1\left(1-s_1^{1/2}\right)^2} \tag{1}$$

Random scission

$$\text{degree of devulcanization} = 1 - \frac{v_{e2}}{v_{e1}} = 1 - \frac{\left(1-s_2^{1/2}\right)^2}{\left(1-s_1^{1/2}\right)^2} \tag{2}$$

where

$$s = 1/(1+\gamma-\gamma s)^2 \tag{3}$$

and

v_e	= crosslink density
s	= soluble fraction content
γ	= crosslinking index
Subscript 1	= index used for values before devulcanization
Subscript 2	= index used for values after devulcanization

The predictions from these model equations can be used to assess experimental data of crosslink densities and soluble content before and after devulcanization. This is done in the form of a Horikx plot, comparing soluble content to the degree of devulcanization. Figure 6.6 shows the Horikx plot for devulcanized rubber samples obtained from the Tyromer® process under varying processing conditions using tire tread crumb. It is evident from this figure that these samples demonstrate primarily selective sulfur scission. The Horikx theory has been employed in several studies and it has been recently validated experimentally for use in rubber devulcanization analysis by employing an elegant approach using virgin samples having varying degrees of vulcanization [125].

Using special chemical probes [126-128], detailed characterization of the crosslink types was performed [116] before and after devulcanization. It was found that primarily poly-sulfidic bonds undergo scission during the Tyromer® devulcanization process. Figure 6.7 shows partial results of the effect of screw speed on crosslink density by type of links. It can be clearly observed that mainly polysulfidic bonds break and that higher screw speeds do not have a significant effect on crosslink density while they may be detrimental to resulting rubber quality.

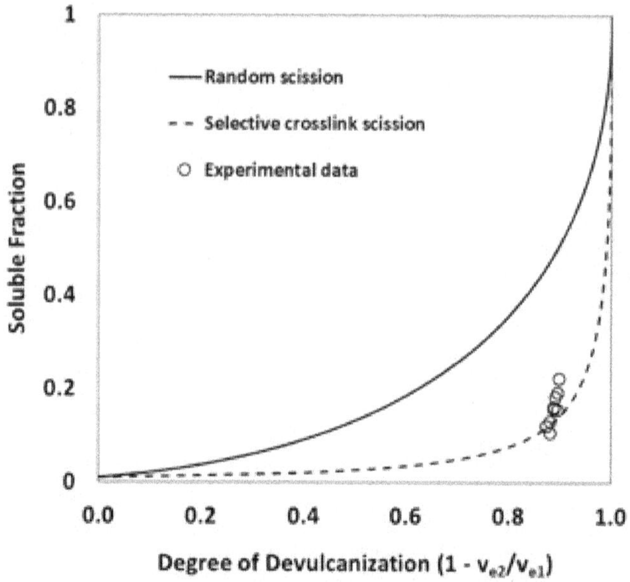

Figure 6.6 *Horikx plot for devulcanized samples obtained from the Tyromer® devulcanization process using tire tread crumb.*

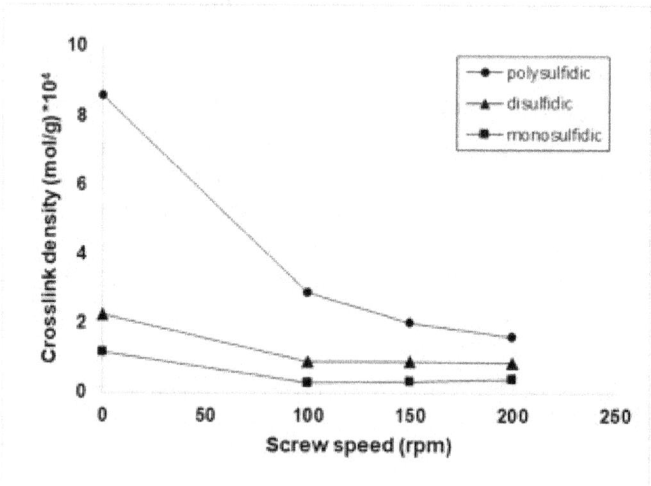

Figure 6.7 *Effect of screw speed on crosslink density by type of sulfidic links for devulcanized samples obtained from the Tyromer® devulcanization process (data extracted from ref [116]).*

Detailed product characterization has been performed in terms of product crosslink density, Mooney viscosity, and mechanical properties, and the effects of processing conditions on properties has been studied through statistical design of experiments [107]. Statistical analysis of the results revealed that, for a given screw configuration, the crumb feed rate was found to be the most significant variable affecting the degree of devulcanization. While the effects of processing conditions on degree of devulcanization were mostly non-linear, Mooney viscosity correlated linearly with crosslink density and soluble content of the devulcanized rubber as seen in Figure 6.8.

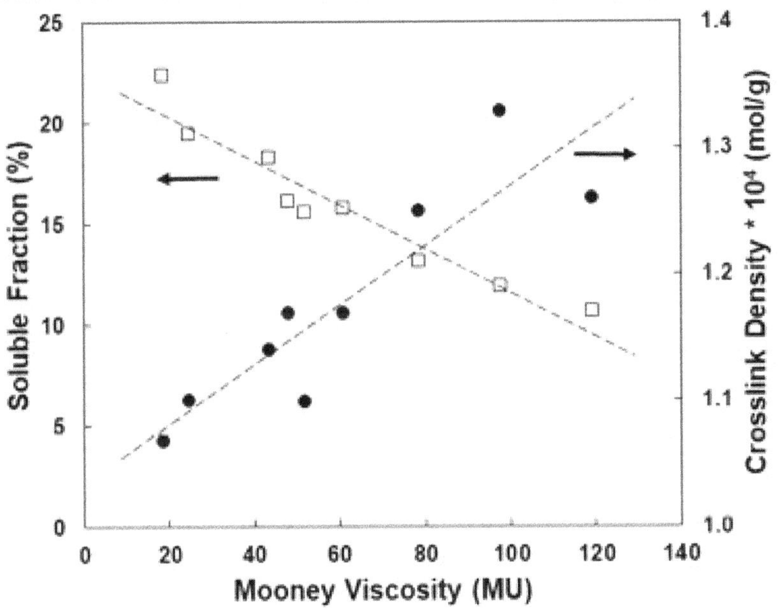

Figure 6.8 *Correlation between Mooney viscosity and soluble fraction/crosslink density for devulcanized samples obtained from the Tyromer® devulcanization process (data extracted from ref [116]).*

Tyromer Inc. has worked primarily with Leistritz to develop and scale-up the production process (https://extruders.leistritz.com/en/industries/special-application) using co-rotating twin-screw extruders from a 34 mm LSM30.34 up to a ZSE 110 MAXX machine. The evolution of production capacity for different size machines is shown in Figure 6.9. Scale-up between these machines was based on specific energy consumption criteria and product temperature while ensuring similar degrees of devucanization. Tyromer Inc. has successfully scaled up a lab invention into an industrially reliable economic process whereby ELT rubber crumb is converted continuously into a tire-derived polymer (TDP) product, without using chemical reagents or solvents. TDP production is: (i) rapid (<2 minutes from crumb to TDP), (ii) energy efficient (<400 kWh/MT) and (iii) complete (no rubber waste by-product).

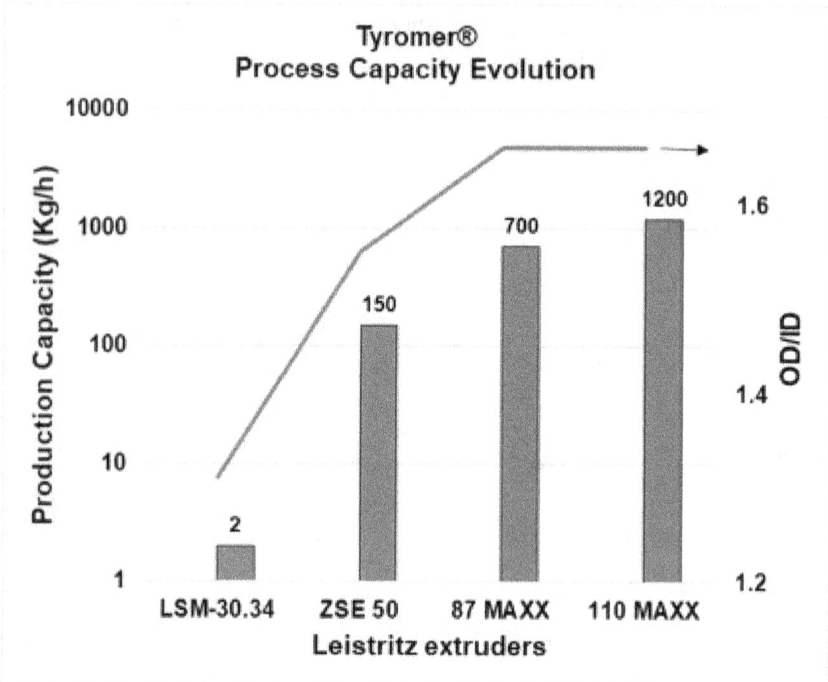

Figure 6.9 *Tyromer® devulcanization process production rate evolution for Leistritz co-rotating twin-screw extruders having various ratios of outside to inside diameters (OD/ID).*

6.3 Applications

The devulcanized rubber product from the Tyromer® process, tire-derived polymer (TDP), can be produced by using crumb from various tire sources. Typical property values are shown in Table 6.3 for TDP-B produced from truck tire tread crumb (TDP-B is a product code). Those properties can be adjusted by varying devulcanization processing conditions and curing packages depending on individual application and customer requirements. Although the stand-alone properties of TDP-B are very good, they do not always provide a clear indication of final compound quality and performance. Devulcanized materials are almost always blended in to a master batch rubber compound and not used at 100% content. As such, TDP devulcanized material must first be mixed into a master batch according to certain guidelines and then compared against the control compound. TDP-B

Table 6.3 *Tyromer® TDP-B rubber master batch properties*

TEST	ASTM	TYPICAL VALUES*	NOTES
Mooney Viscosity (1+4 @ 100C)	D1646		
MI		64	
ML		45	+/- 15**
Relax (Tx80)		12	
Mooney Scorch (@ 138C)	D1646		
ML		30	
T5		3.3	
T35		4.75	
Rheometer - MDR (30' @ 140C)	D5289		
ML		0.7	
Ts2		3.5	
Tc50		6	
Tc80		10	
Tc90		13	
MH		9.50	
Tensile Strength (20' @ 140C)	D412		
Tensile (MPa)		10	+/- 1
Elongation (%)		270	+/- 30
Modulus (MPa) @ 100%		2.3	
Tear (N/mm)	D624	35	
Durometer	D2240	58.0	+/- 3
Density	D297	1.14	+/- 0.03
Flexometer	D623		
Hysteresis (°C)		25	
Blow out Minutes		3.5	
DIN Abrasion	D5963	160	
Rebound	D7121	35	
**Typical results for TDP-B produced from truck tire tread with IS6306-1971 standard cure* ***Mooney viscosity is adjustable based on process conditions*			

is not a filler and can actually be used to replace rubber master batch in diverse applications such as rubber mats, conveyor belts, and tire compounds. Of course, the best use of recycled material is for its originally intended purpose, and TDP from tire rubber crumb is mainly used in tire compound formulations.

Typical applications of TDP-B are in truck and off-the-road (OTR) tire retreading compounds. Its use not only leads to improved economics and compound performance but also contributes to carbon footprint reduction. Figure 6.10 shows data on two OTR retreading compounds at 20% and 30% levels of incorporation. It can be seen that, compared to the control formulation, incorporation of TDP-B leads to improvement in key properties such as tear, abrasion resistance, mass loss, hysteresis, and blow-out time. During tire rotation, the rubber compound undergoes a cyclic compression and relaxation process where energy dissipation leads to heat generation due to hysteresis. Excess heat can lead to accelerated aging of rubber, compromise of tire life, and potential premature tread failure (e.g. "alligator skins" on highways). Data in Figure 6.10 indicates that TDP-B incorporation in an OTR compound not only improves tear and abrasion resistance and mass loss, but it additionally improves safety through reduction in hysteresis temperature and increased blow-out time. Additionally, reduction in hysteresis improves rolling resistance and savings in green-house gas (GHG) emissions.

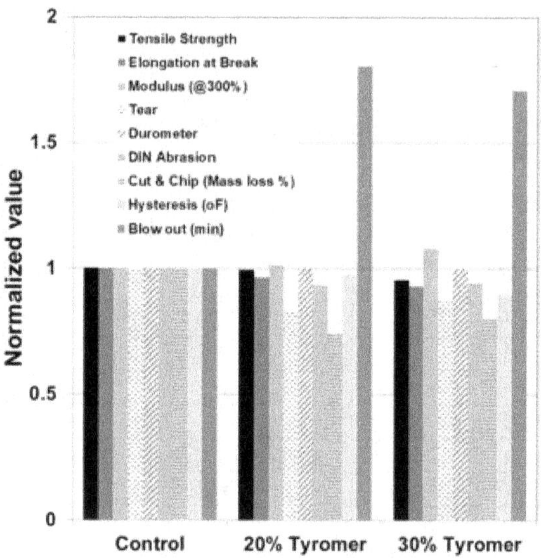

Figure 6.10 *Effect of TDP content on tire compound properties.*

Table 6.4 compares typical energy values required to produce tire compounds with the energy required to produce TDP through devulcanization. The results show that use of TDP in tire compounds can lead to energy savings as high as 94% compared to virgin tire compound manufacturing. In terms of GHG emissions saving potential, the use of 1 MT of TDP will provide tire manufacturers 17 MT of GHG reduction, based upon typical methods of electricity generation in North America.

Table 6.4 *Energy Savings from using Tyromer® TDP in tire compounds*

Energy content of tire compound (kWh/kg) [129]	26
Energy required to produce crumb (kWh/kg) [129]	1.2
Energy required to produce TDP (kWh/kg)	0.4
Net energy value of TDP (kWh/kg)	24.4
Energy conserved by using TDP	94%

In terms of performance, consistent results have been obtained for incorporation of TDP-B in higher quality OTR recipes. Figure 6.11 summarizes results obtained at an incorporation level of 23% for three different TDP-B grades having varying Mooney viscosities. It can be clearly seen that tear and abrasion resistance, cut-and-chip and hysteresis improve for all grades.

Tyromer Inc. has collaborated with Airboss of America Corp., one of the largest rubber compounders in North America, to implement a first-of-its-kind commercial scrap tire rubber crumb devulcanization system. The TDP is used in creation of new AirBoss compounds that include recycled content and perform like new rubber, rather than simply diverting rubber crumb to lower value applications such as filler for asphalt. For those tire remanufacturers and their customers that are looking to further "green" their operations, AirBoss now offers EcoBoss™, a line of rubber compounds for off-the-road retreading applications that incorporate up to 20% recycled material [130]. In addition to the collaboration with AirBoss, Tyromer Inc. is expected to start up two new joint venture manufacturing facilities in 2021: one in Windsor, Ontario, Canada and one in the Netherlands, in order to supply a leading tire OEM.

Experience and customer feedback on OTR retreading compounds have been very positive. For over four years, TDP has been used to replace approximately 20% of virgin tire rubber master batch compound in OTR tire retreading by KALTIRE [131], one of the largest mining tire distribution and service companies in the world. This collaborative effort between KALTIRE and Tyromer Inc. has demonstrated that TDP use for tires is the best ELT rubber sustainable reuse option which promotes a circular economy in tire manufacturing.

Through process optimization and continuous improvement, Tyromer has demonstrated that TDP meets the three key adoption drivers in the tire and rubber industry: (i) TDP performs well as a tire rubber compound replacement material, (ii) TDP offers tire producers material cost savings, and (iii) TDP reduces GHG emission and reduces manufacturing carbon footprint.

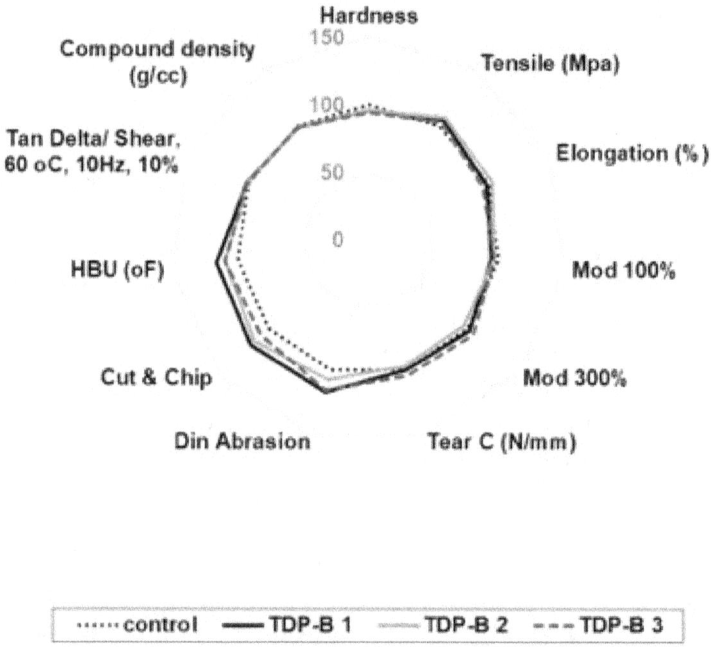

Figure 6.11 *Effect of TDP type on OTR tire compound properties.*

6.5 Concluding Remarks

Management of end-of-life tires (ELTs) is and will continue to be a global environmental problem. The tire industry faces major challenges and its sustainability initiatives do not fully address the products they produce at their end-of-life. Climate change and global sustainability awareness and requirements are driving businesses toward a new circular economy. The tire industry will be no different and it will adopt this new economic model. Several devulcanization technologies which have been developed must be seriously considered for adoption as they can contribute towards the tire manufacturing circular economy. The Tyromer® non-chemical REX devulcanization technology is leading the way towards this goal as it has already been proven at an industrial scale. Figure 6.12 illustrates how TDP brings ELT recycling into the value chain of the tire retreading and new tire manufacturing industry. This devulcanization process continuously converts ELT rubber crumb into TDP while significantly preserving the integrity of the rubber. As a result, TDP exhibits excellent dynamic properties and is suitable for use as replacement for virgin tire rubber compounds to reduce the use of "Raw Material" from the resource sector. Using ELT rubber crumb as a feedstock for conversion to TDP reduces the "Waste". Therefore, converting ELT rubber to TDP for reuse in new tires will bring the Circular Economy to the tire industry and "Best Practice" to the global management of ELT. TDP production is environmentally sustainable and TDP can meet the demanding requirements of the tire industry as well as those of the General Rubber Goods (GRG) industry. Most importantly, TDP production is economically viable as cost reduction is the key driver for innovation adoption.

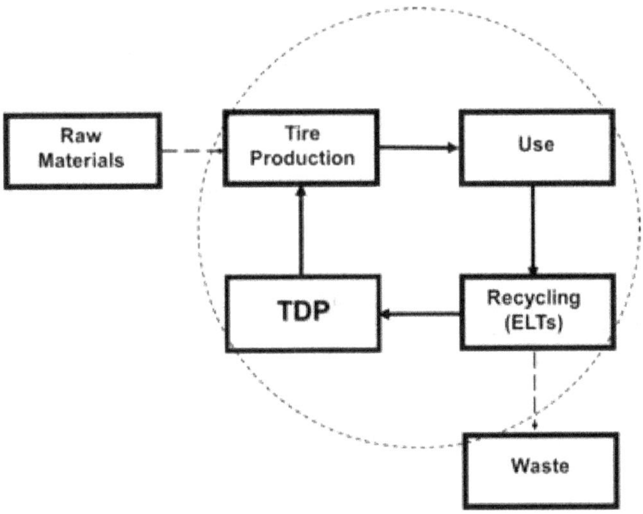

Figure 6.12 *Tyromer® TDP use in circular economy for tire compound manufacturing.*

References

1. Goodyear, C. (1844) Improvement in India-Rubber Fabrics. US Patent 3633.
2. Bateman, L., Moore, C.G., Porter, M., and Saville, B. (1963) Chapter 19 In: The Chemistry and Physics of Rubber-Like Substances, Bateman, L. (ed.), John Wiley & Sons, New York.
3. Alliger, G. and Sjothun, I.J. (1964) Vulcanization of Elastomers, Reinhold Publishing Co., New York
4. Hofmann, W. (1967) Vulcanization and Vulcanizing Agents, Maclaren and Sons Ltd., London
5. Coran, A.Y. (2005) Chapter 7 In: Science and Technology of Rubber, F.R. Eirich (ed.), 3rd ed., Academic Press, New York, pp. 321-366
6. M. Akiba and A.S. Hashim, (1997) Vulcanization and crosslinking in elastomers. Prog. Polym. Sci. 22:475-521
7. Y. Tanaka, (1991) Recent Advances in Structural Characterization of Elastomers. Rubber Chem. Technol. 64:325-385
8. Malaysian Rubber Council, http://www.rubberstudy.com/welcome, http://www.myrubbercouncil.com/industry/world_production.ph
9. https://www.statista.com/statistics/275390/world-usage-distribution-of-natural-rubber/
10. https://www.transparencymarketresearch.com/rubber-tires-market.html
11. De, S.K., Isayev, A.I., and Khait, K. (2005) Rubber Recycling, CRC Press, Boca Raton, FL, USA
12. World Business Council for Sustainable Development, "Global ELT Management – A global state of knowledge on regulation, management systems, impacts of recovery and technologies" (2019). https://www.wbcsd.org/Sector-Projects/Tire-Industry-Project/End-of-Life-Tires-ELTs
13. Tzoganakis, C., and Visaisouk, J. (2019) Understanding devulcanization, the path to a circular economy and potential market impact. Rubber World. 259(5): 18-23
14. Noordermeer, J.W.M., Dierkes, W., Blume, A., van Hoek, H., Reuvekamp, L., Dijkhuis, K., and Saiwari, S. (2020) Cradle-to-cradle devulcanization options for various elastomers Rubber World. 262(5): 20-28
15. ASTM D1566-15, "Standard Terminology Related to Rubber".

16. ASTM Special Technical Publication N184A, version 29 (1987) Recent Developments in Polymer Recycling
17. Myhre, M., Saiwari, S., Dierkes. W., and Noordermeer, J. (2012) Rubber recycling: chemistry, processing and applications. Rubber Chem. Technol., 85: 408-449
18. Myhre, M. and MacKillop, D.A. (2002) Rubber recycling. Rubber Chem. Technol. 75: 429-474
19. Warner, W.C. (1994) Methods of devulcanization. Rubber Chem. Technol. 67: 559-566
20. Adhikari, B., De, D., and Maiti, S. (2000) Reclamation and recycling of waste rubber. Prog. Polym. Sci. 25: 909-948
21. Le Beau, D.S. (1967) Science and technology of reclaimed rubber. Rubber Chem. Technol. 40: 217-237
22. Asaro, L., Gratton, M., Seghar, S., and Aït Hocine, N. (2018) Recycling of rubber wastes by devulcanization. Resources, Conservation & Recycling 133: 250-262
23. Markl, E., and Lackner, M. (2020) Devulcanization Technologies for Recycling of Tire-Derived Rubber: A Review. Materials 13: 1246
24. Isayev, A.I. (2001) Recycling of Elastomers, In: Encyclopedia of Materials: Science and Technology, K. H. J. Buschow (ed.), Elsevier, Amsterdam
25. Isayev, A.I. (2013) Recycling of Rubbers. Chapter 15 In: The Science and Technology of Rubber, J. E. Mark, B. Erman and F. R. Eirich, (eds), 4th Ed., Academic Press, New York
26. Forrest, M. (2014) Recycling and Re-use of Waste Rubber. Smithers Rapra, London.
27. Fang, Y., Zhan, M., and Wang, Y. (2001) The status of recycling of waste rubber. Mater. Des. 22: 123–128
28. Nicholas, P.P. (1979) Devulcanized rubber composition and process for preparing same. US Patent 4,161,464
29. Nicholas, P. P. (1982) The Scission of Polysulfide Crosslinks in Scrap Rubber Particles through Phase Transfer Catalysis. Rubber Chem. Technol. 55: 1499-1515
30. Myhre, M., and MacKillop, D.A. (1995) paper #21, presented at a meeting of the *Rubber Division, ACS*, Oct. 17-20, Cleveland, OH, USA; abstract in Rubber Chem. Technol. 69: 142 (1996).
31. Yamashita, S., (1976) Japan-USSR Polym. Symp. (Proc. Eng.) 355-364; Chem. Abstracts (1976) 85: 34333f

32. Chapman, A.V., and Porter, M. (1998) Natural Rubber Science & Technology. D. Roberts (ed.), Oxford University Press, New York.
33. Novotny, D.S., Marsh, R.L., Masters, F.C., and Tally, D.N. (1978) Microwave devulcanization of rubber. US Patent 4,104, 205
34. Oliveira Da Cunha Lima, L.C. (1997) Process for reclaiming cured or semi-cured rubber. US Patent 5,677,354
35. Baranwal, K., Rogers, J. W., and Standley, P. M. (1998) paper #39 presented at a meeting of the *Rubber Division, ACS*, Sept. 29-Oct. 2, Nashville TN, USA; abstract in Rubber Chem. Technol. 72: 235 (1999)
36. Sekhar, B.C., Kormer, V.A., Sotnikova, E.N., Mironyuk, V.P., Trunova, L.N., and Nikitina, N.A. (1998) Reclaiming of elastomeric materials. US Patent 5,770,632
37. Pillai, C. R. (1998) Process for the manufacture of high grade devulcanized rubbers from scrap and reject rubber items. US Patent 5,731,358
38. Dufton, P.W. (2001) End-of-Life Tyres – Exploiting their Value, Rapra Technology Ltd, UK
39. Hon, K. K., Siesseger, F., Sekhar, B.C., and Sin, S.W. (1995) paper #55, presented at a meeting of the *Rubber Division, ACS*, Oct. 17-20, Cleveland, OH, USA; abstract in Rubber Chem. Technol. 69: 151 (1996)
40. Kohler, R., and O'Neill, J. (1995) paper #58, presented at a meting of the *Rubber Division, ACS*, Oct. 8-11, Louisville, KY, USA; abstract in Rubber Chem. Technol. 70: 162 (1997)
41. R. Newell (1996) The De Link Process: A Unique Opportunity to Recycle a Variety of Polymer Types. Int. Rubber Conf., Manchester UK
42. Onouti, Y., Inagaki, S., Okamoto, H., and Furukawa J. (1980) Reclamation of disposal rubber vulcanizates (II) Reclamation of crushed tire scrap with thiol and amine derivatives. Rubber Chem Technol. 53: 756–762
43. De, D., Maiti, S., and Adhikari, B. (1999) Reclaiming of rubber by a renewable resource material (RRM). II. Comparative evaluation of reclaiming process of NR vulcanizate by RRM and diallyl disulfide. J. Appl. Polym. Sci., 73: 2951-2958
44. Watabe, Y., Fuji, Y., Anzai, S., and Furukawa, J. (1980) Process for reclaiming scrap vulcanized rubber. US Patent 4,211,676

45. Mouri, M., Usuki, A., and Sato, N. (1977) Process for producing reclaimed rubber or unvulcanized reclaimed rubber. US Patent 5,672,630
46. De, D., Maiti, S., and Adhikari, B. (2000) Reclaiming of rubber by a renewable resource material (RRM). III. evaluation of properties of NR reclaim. J. Appl. Polym. Sci. 75: 1493-1502
47. Adhikari, B., De, D., Maiti S. (2000) Reclamation and recycling of waste rubber. Prog Polym Sci. 25: 909-948.
48. Maiti, S., and Adhikari, B. (2000) Reclaiming of rubber by a renewable resource material -Assessment of vulcanized SBR reclaiming process. Kautsch. Gummi Kunstst., 53: 346
49. Mangili, I., Collina, E., Anzano, M., Pitea, D., Lasagni, M. (2014) Characterization and supercritical CO2 devulcanization of cryo-ground tire rubber: Influence of devulcanization process on reclaimed material. Polym. Degrad. Stab. 102: 15–24
50. Anu Mary, J., Benny, G., Madhusoodanan, K.N., Rosamma, A. (2016) The current status of sulphur vulcanization and devulcanization chemistry: devulcanization. Rubber Sci. 29: 62–100
51. Rajan, V.V. (2005) Devulcanization of NR Based Latex Products for Tyre Applications. Ph.D Thesis, University of Twente, Enschede, The Netherlands
52. Sabzekar, M., Chenar, M.P., Mortazavi, S.M., Kariminejad, M., Asadi, S., Zohuri, G. (2015) Influence of process variables on chemical devulcanization of sulfur-cured natural rubber. Polym. Degrad. Stab. 118: 88–95
53. Pelofsky, A.H., (1973) Rubber reclamation using ultrasonic energy. US Patent 3,725,314
54. Isayev, A.I., Chen, J., Tukachinsky, A. (1995) Novel Ultrasonic Technology for Devulcanization of Waste Rubbers. Rubber Chem. Technol. 68: 267–280
55. Isayev, A.I., Yushanov, S.P., Kim, S.-H.H., Levin, V.Y. (1996) Ultrasonic devulcanization of waste rubbers: Experimentation and modeling. Rheol. Acta 35: 616–630
56. Isayev, A.I., Liang, T., Lewis, T.M. (2014) Effect of Particle Size on Ultrasonic Devulcanization of Tire Rubber In Twin-Screw Extruder. Rubber Chem. Technol. 87: 86–102
57. Yu Levin, V., Kim, S.H., Isayev, A.I. (1997) Vulcanization of Ultrasonically Devulcanized SBR Elastomers. Rubber Chem. Technol. 70: 120–128

58. Yun, J., Yashin, V.V., Isayev, A.I. (2004) Ultrasonic devulcanization of carbon black–filled ethylene propylene diene monomer rubber. J. Appl. Polym. Sci. 91: 1646–1656
59. Yushanov, S.P., Isayev, A.I., Levin, V.Y. (1996) Percolation simulation of the network degradation during ultrasonic devulcanization. J. Polym. Sci. B Polym. Phys. 34: 2409–2418
60. W. Klingensmith, K. Kelly, I. Morris, V. Nilolski (1998). paper #54 presented at a meeting of the Rubber Division, ACS, Sept. 29-Oct. 2, Nashville; abstract in Rubber Chem. Technol. 72: 241 (1999)
61. M. Mouri, et. al. (1999). paper #84 presented at a meting of the Rubber Division, ACS, April 13-16, Chicago; abstract in Rubber Chem. Technol. 72: 805 (1999)
62. L. K. Kostanski et. al (1998). paper #58, presented at a meeting of the Rubber Division, ACS, Sept. 29-Oct. 2, Nashville; abstract in Rubber Chem. Technol. 72: 242 (1999)
63. L. K. Kostanski et. al (1999). paper #34 presented at a meting of the Rubber Division, ACS, Sept. 21-24, Orlando; abstract in Rubber Chem. Technol. 73: 148 (2000)
64. Suzuki Y., Owaki, M., Mouri, M. et al. (1998). In: Toyota Technical Review 48(1): 53–57
65. Mouri, M., Sato, N., Okamoto, H., Matsushita, M., Honda, H., Nakashima, K. Takeushi, K., Suzuki, Y., Owaki, M. (2000). Continuous devulcanization by shear flow stage reaction control technology for rubber recycling. Part 3. Study of the devulcanization process for EPDM. Int. Polym. Sci. Technol. 27: 12–16
66. Mouri, M., Sato, N., Okamoto, H., Matsushita, M., Honda, H., Nakashima, K. Takeushi, K., Suzuki, Y., Owaki, M. (2000). Devulcanization conditions and mechanical properties of revulcanized rubber for EPDM. Continuous reclaimation of rubber by shear flow reaction control technology (Part II). Int. Polym. Technol. 27: 23–28
67. Mouri, M., Sato, N., Okamoto, H., Matsushita, M., Honda, H., Nakashima, K. Takeushi, K., Suzuki, Y., Owaki, M. (2000). A new devulcanization process: continuous reclamation of rubber by shear flow reaction control technology (Part I), Int. Polym. Sci. Technol. 27: 17–22
68. Mouri, M., Usuki, A., Sato, N. (1977) Process for producing reclaimed rubber or unvulcanized reclaimed rubber US Patent 5,672,630

69. Maridass, B., Gupta, B. R. (2003) Recycling of Waste Tire Rubber Powder. KGK Kautschuk Gummi Kunststoffe. 56 Jahrgang, Nr. 5.

70. Maridass, B., Gupta, B.R. (2007) Effect of extruder parameters on mechanical properties of revulcanized ground rubber tire powder. Polimery 52: 456–460

71. Li, S., Lamminmäki, J., Hanhi, K. (2005) Effect of ground rubber powder and devulcanizates on the properties of natural rubber compounds. J. Appl. Polym. Sci. 97: 208–217

72. Scaffaro, R., Dintcheva, N.T., Nocilla, M.A., La Mantia, F.P. (2005) Formulation, characterization and optimization of the processing condition of blends of recycled polyethylene and ground tyre rubber: mechanical and rheological analysis. Polym.Degrad. Stab. 90: 281–287

73. Sutanto, P., Laksmana, F.L., Picchioni, F., and Janssen, L.P.B.M. (2006) Modeling on the kinetics of an EPDM devulcanization in an internal batch mixer using an amine as the devulcanizing agent. Chem. Eng. Sci. 61: 6442–6453

74. Jana, G.K., Mahaling, R.N., Rath, T., Kozlowska, A., Kozlowski, M., Das, C.K. (2007) Mechano-chemical recycling of sulfur cured natural rubber. Polimery/Polymers 52: 131–136

75. Jana, K. and Das, C.K. (2005) Devulcanization of Natural Rubber Vulcanizates by Mechanochemical Process. Polymer-Plastics Technology and Engineering 44: 1399-1412

76. Rooj, S., Basak, G.C., Maji, P.K., Bhowmick, A.K. (2011) New route for devulcanization of natural rubber and the properties of devulcanized rubber. J. Polym. Environ. 19: 382–390

77. Tao, G., He, Q., Xia, Y. Jia, G., Yang, H., Ma, W. (2013) The effect of devulcanization level on mechanical properties of reclaimed rubber by thermal-mechanical shearing devulcanization. J. Appl. Polym. Sci. 129: 2598–2605

78. Yazdani, H., Ghasemi, I., Karrabi, M., Azizi, H. (2013) Continuous devulcanization of waste tires by using a co-rotating twin screw extruder: effects of screw configuration, temperature profile, and devulcanization agent concentration. J. Vinyl. Addit. Technol. 19: 65–72

79. Sabzekar, M., Zohuri, G., Chenar, M.P., Mortazavi, S.M., Karimi-nejad, M., Asadi, S. (2016) A new approach for reclaiming of waste automotive EPDM rubber using waste oil. Polym. Degrad. Stab. 129: 56–62

80. Formela, K., Cysewska, M., Haponiuk, J. (2014) The influence of screw configuration and screw speed of co-rotating twin screw extruder on the properties of products obtained by thermomechanical reclaiming of ground tire rubber. Polimery 59: 170–177
81. Formela, K. and Cysewska, M. (2014) Efficiency of thermomechanical reclaiming of ground tire rubber conducted in counter-rotating and co-rotating twin screw extruder. Polimery 59: 231–238
82. Formela, K., Cysewska, M. and Haponiuk, J. T. (2016) Thermomechanical reclaiming of ground tire rubber via extrusion at low temperature: Efficiency and limits. J. Vinyl Addit. Technol. 22: 213–221
83. Simon, D.Á., Pirityi, D.Z. and Bárány, T. (2020) Devulcanization of ground tire rubber: microwave and thermomechanical approaches. Sci. Rep. 10: 16587
84. Shi, J., Zou, H., Ding, L., Li, X., Jiang, K., Chen, T., Zhang, X., Zhang, L., Ren, D. (2014) Polym. Degrad. Stab. 99: 166-175
85. Ghorai, S., Bhunia, S., Roy, M., De, D. (2016) Mechanochemical devulcanization of natural rubber vulcanizate by dual function disulfide chemicals. Polym. Degrad. Stab. 129: 34–46
86. Zedler, Ł., Przybysz, M., Klein, M., et al. (2017) Processing, physico-mechanical and thermal properties of reclaimed GTR and NBR/reclaimed GTR blends as function of various additives. Polym. Degrad. Stab. 143: 186–195
87. Diaz, R., Colomines, G., Peuvrel-Disdier, E., et al. (2018) Thermo-mechanical recycling of rubber: relationship between material properties and specific mechanical energy. J. Mater. Process. Technol. 252: 454–468
88. Garcia, P.S., de Sousa, F.D.B., de Lima, J.A., Cruz, S.A., Scuracchio, C.H. (2015) Devulcanization of ground tire rubber: physical and chemical changes after different microwave exposure times. Express Polym. Lett. 9: 1015–1026
89. Molanorouzi, M., Mohaved, S.O. (2016) Reclaiming waste tire rubber by an irradiation technique. Polym. Degrad. Stab. 128: 115–125
90. Seghar, S., Aït Hocine, N., Mittal, V., Azem, S., Al-Zohbi, F., Schmaltz, B., Poirot, N. (2015) Devulcanization of styrene butadiene rubber by microwave energy: effect of the presence of ionic liquid. Express Polym. Lett. 9: 1076–1086

91. Colom, X., Faliq, A., Formela, K., Canavate, J. (2016) FTIR spectroscopic and thermogravimetric characterization of ground tyre rubber devulcanized by microwave treatment. Polym. Test. 52: 200–208
92. de Sousa, F.D.B., Scuracchio, C.H., Hu, G.H., Hoppe, S. (2017) Devulcanization of waste tire rubber by microwaves. Polym. Degrad. Stab. 138: 169–181
93. Tatangelo, V., Mangili, I., Caracino, P., Anzano, M., Najmi, Z., Bestetti, G., Collina, E., Franzetti, A., Lasagni, M. (2016) Biological devulcanization of ground natural rubber by Gordonia desulfuricans DSM 44462T strain. Appl. Microbiol. Biotechnol. 100: 8931–8942
94. Tsuchii, A., Takeda, K. (1990) Rubber-Degrading enzyme from a bacterial culture. Appl. Environ. Microbiol. 56: 26–274
95. Tsuchii, A., Tokiwa, Y. (2001) Microbial degradation of tyre rubber particles. Biotechnol. Lett. 23: 963–969
96. Tsuchii, A., Suzuki, T., Takeda, K. (1985) Microbial degradation of natural rubber vulcanizates. Appl. Environ. Microbiol. 50: 965–970
97. Tsuchii, A., Takeda, K., Tokiwa, Y. (1996) Degradation of the rubber in truck tires by a strain of Nocardia. Biodegradation 7: 405–413
98. Yao, C., Zhao, S., Wang, Y., Wang, B., Wei, M., Hu, M. (2013) Microbial desulfurization of waste latex rubber with Alicyclobacillus sp. Polym. Degrad. Stab. 98: 1724–1730
99. Kojima, M., Ogawa, K., Mizoshima, H., Tosaka, M., Kohjiya, S., Ikeda, Y. (2003) Devulcanization of sulfur-cured isoprene rubber in supercritical carbon dioxide. Rubber Chem. Technol. 76: 957–968
100. Kojima, M., Tosaka, M., Ikeda, Y. (2004) Chemical recycling of sulfur-cured natural rubber using supercritical carbon dioxide. Green Chem., 6: 84-89
101. Kojima, M., Kohjiya, S., Ikeda, Y. (2005) Role of supercritical carbon dioxide for selective impregnation of decrosslinking reagent into isoprene rubber vulcanizate. Polymer 46: 2016–2019
102. Kojima, M., Tosaka, M., Ikeda, Y., Kohjiya, S. (2005) Devulcanization of carbon black filled natural rubber using supercritical carbon dioxide. J. Appl. Polym. Sci. 95: 137–143
103. Mangili, I., Oliveri, M., Anzano, M., Collina, E., Pitea, D., Lasagni, M. (2014) Full factorial experimental design to study the

devulcanization of ground tire rubber in supercritical carbon dioxide. J. Supercrit. Fluids 92: 249–256

104. Oliveri, M. (2013) Optimization of Supercritical CO_2 Devulcanization Process of Cryo Ground Tires and Characterization of the Treated Material, PhD Thesis. University of Milano–Bicocca
105. Jiang, K., Shi, J., Ge, Y., Zou, R., Yao, P., Li, X., Zhang, L. (2013) Complete devulcanization of sulfur-cured butyl rubber by using supercritical carbon dioxide. J. Appl. Polym. Sci. 127: 2397–2406
106. Liu, Z., Li, X., Xu, X., Wang, X., Dong, C., Liu, F., Wei, W., (2015) Devulcanizaiton of waste tread rubber in supercritical carbon dioxide: operating parameters and product characterization. Polym. Degrad. Stab. 119: 198–207
107. Meysami, M., Tzoganakis, C., Mutyala, P., Zhu, S., and Bulsari, M. (2017) Devulcanization of scrap tire rubber with supercritical CO2: A study of the effects of process parameters on the properties of devulcanized rubber. Intern. Polymer Processing, XXXII: 183-193
108. Meysami, M., Mutyala, P., Zhu, S., and Tzoganakis, C. (2014) Continuous Devulcanization of Scrap EPDM Rubber with Supercritical CO_2: Effect of Process Parameters on Devulcanized Rubber Properties. Proceedings of the 72nd Annual Technical Conference of the Society of Plastics Engineers, Las Vegas, NV, USA
109. Meysami, M. Mutyala, P., Zhu, S., and Tzoganakis, CV. (2013) Effect of process parameters on properties of devulcanized rubber obtained from a supercritical CO_2 assisted devulcanization process. Proceedings of the 71st Annual Technical Conference of the Society of Plastics Engineers, Cincinnati, OH, USA
110. Meysami, M., Zhu, S., and Tzoganakis, C. (2010) Effect of Incorporation of Devulcanized Rubber on the Properties of a Virgin Tire Compound. Proceedings of the 68th Annual Technical Conference of the Society of Plastics Engineers, Orlando, FL, USA
111. Zhang, Q., and Tzoganakis, C. (2004) Devulcanization of Recycled Tire Rubber Using Supercritical Carbon Dioxide. 62nd Annual Technical Conference of the Society of Plastics Engineers, Chicago, IL, USA

112. Tzoganakis, C., Zhang, Q., and Zhu, S. (2004) Devulcanization of Recycled Tire Rubber Using Supercritical Carbon Dioxide and Use in Thermoplastic Vulcanizate Preparation. Polymer Processing Society, Asia/Australia Regional Meeting, Gyeongju, Korea
113. Zhang, Q., and Tzoganakis, C. (2004) Devulcanization of Recycled Tire Rubber Using Supercritical Carbon Dioxide. Global Plastics Environmental Conference Proceedings, paper#49, Society of Plastics Engineers, Detroit, MI, USA
114. Tzoganakis, C. (2011) Method of modifying crosslinked rubber. CA Patent 2,500,372
115. Tzoganakis, C. (2007) Method of modifying crosslinked rubber. US Patent 7,189,762
116. Meysami, M. (2012) A Study of Scrap Rubber Devulcanization and Incorporation of Devulcanized Rubber into Virgin Rubber Compounds, PhD Thesis, University of Waterloo, Canada
117. Asaro, L., Gratton, M., Seghar, S., Poirot, N., and Aït Hocine, N. (2018) Devulcanization of waste rubber using thermomechanical method combined with supercritical CO2. Int. J. Environ. Ecological Eng. 12(3): 1
118. L. Asaro, M. Gratton, S. Seghar, N. Aït Hocine (2017) Thermomechanical devulcanization of waste rubber with supercritical CO2. Workshop on the Multiphysics Behavior of Rubber, Marseille, France, https://wmbr2017.sciences-conf.org/169584/document
119. California Integrated Waste Management Board, CalRecovery Inc (2004) Evaluation of Waste Tire Devulcanization Technologies, USA
120. Seghar, S., Asaro, L., Rolland-Monnet, M., and Aït Hocine, N. (2019) Thermo-mechanical devulcanization and recycling of rubber industry waste. Resources, Conservation & Recycling. 144: 180–186
121. https://www.wiredchemist.com/chemistry/data/bond_energies_lengths.html
122. Fukumori, K. and Matsushita, M. (2004) Materials recycling technology of crosslinked rubber waste. R&D Review of Toyota CRDL. 38(1): 39-47
123. Charlesby, A. (1953) Solubility and molecular size distribution of crosslinked polystyrene. J. Polym. Sci. 11: 513-520
124. Horikx, M.M. (1956) Chain scissions in a polymer network. J. Polym. Sci. 19: 445-454

125. Seghar, S., Asaro, L., and Aït Hocine, N. (2019) Experimental Validation of the Horikx Theory to be Used in the Rubber Devulcanization Analysis. J. Polymers and the Environment. 27: 2318-2323
126. Campbell, D.S. and Saville, B. (1967) Current principles and practices in elucidating structure in sulfur- vulcanized elastomers. Int. Rubber Conf., Brighton
127. Saville, B., and Watson, A.A. (1967) Structural Characterization of Sulfur-Vulcanized Rubber Networks. Rubber Chem. Technol. 40: 100-148
128. Campbell, D.S. (1969) Structural characterization of vulcanizates part X. Thiol-disulfide interchange for cleaving disulfide crosslinks in natural rubber vulcanizates. J. Appl. Polym. Sci. 13: 1201-1214
129. Boustani, A., Sahni, S., Gutowski, T., and Graves, S. (2010) Tire Remanufacturing and Energy Savings, MIT Sloan School of Management, MITEI-1-h-2010
130. https://www.annualreports.com/HostedData/AnnualReports/PDF/TSX_BOS_2019.pdf https://www.annualreports.com/HostedData/AnnualReports/PDF/TSX_BOS_2020.pdf
131. https://www.kaltire.com/en/retreading/

7

Stable Free Radical Assisted Shear Induced Devulcanization

Anu Mary Joseph[a], Benny George[b,*], K.N. Madhusoodanan[b] and Rosamma Alex[b]

[a]*Department of Chemistry, Newman College, Thodupuzha, Idukki, Kerala, India.*
[b]*Rubber Technology Division, Rubber Research Institute of India, Rubber Board, Kottayam, Kerala 686 009 India.*
**georgebenni@hotmail.com*

7.1 Introduction

The process of vulcanization crosslinks rubber chains to a three-dimensional network thereby transforming rubber from viscous to an elastic material. Vulcanized rubber is insoluble, infusible, and non-biodegradable under normal conditions facilitating its wide range of applications. On the contrary, the same crosslinks dispute with all environmentally benign recycling and disposal strategies of used rubber products. The major recycling strategies followed in developed countries are [1] [2] (i) energy recovery by incineration in cement kilns, urban heating and power plants and (ii) material recovery mostly by granulation to crumb rubber powder, the major share of which is used as infill in artificial football turf pitches and only a low percent of regenerated rubber is consumed by the rubber industry. It is also worth noting that many of the commercially practiced recycling processes are a compromise as most feasible practical solution to remove the menace of used tyres and are not environmentally the best solutions. In emerging economies the public concern on such matters are not as prominent as in developed countries so as to prompt the respective governments to formulate and implement appropriate legislations. This has resulted in lack of proper statistics of the fate of the end-of-life tyres and other used rubber products.

From a scientific and environmental perspective, the reuse of reclaimed rubber by the rubber industry itself is the best recycling approach. Devulcanization or decrosslinking, which involves selective scission of crosslinks to retrieve the polymer main chains with minimum scission and modification, is the appropriate strategy to achieve this goal. The key issue in reuse of the devulcanized rubber is the recovering of its original vulcanizate properties. Apart from the fact that the regenerated rubber is produced from scrap rubber products which have already undergone environmental ageing with consequent reduction in original properties, the current devulcanization technologies are not capable of completely removing the crosslinks and cannot fully avoid main chain scission. The net effect is that the regenerated rubber will have lower properties than the virgin compound, limiting its large-scale use in products like tyres. Incorporation of the devulcanized rubber compound, in most cases, negatively influences the compound viscosity, scorch and cure characteristics. As of today, no technology is available which can completely reverse the vulcanization process or de-crosslink a rubber product. Most of the work in the field is aimed to improve the efficiency of devulcanization by maximizing the scission of crosslinks with minimum main chain scission so that higher amount of regenerated rubber can be incorporated in to virgin rubber compounds, with minimal impact on mechanical properties [3]. This chapter describes a strategy to enhance the effectiveness of de-crosslinking that involves blocking the recombination of free radicals generated by the shear scission of cross links using a stable free radical [4]-[7].

7.2 Shear induced devulcanization

7.2.1 Mechanical and mechanochemical devulcanization

Shear mediated de-crosslinking or devulcanization of rubber vulcanizates at ambient or high temperatures envisages retrieval of polymer chain through selective crosslink scission with negligible main chain scission. The process is expected to generate better quality regenerated rubber compound that can be reused as a raw material for rubber industry. Devulcanization can be carried out by shearing crumb rubber powder in a two roll mill, at temperatures below 100^0C, or using specially designed extruders, at controlled temperatures, sometimes in the presence of a devulcanizing agent (mechanochemical devulcanization) or by shear force alone (mechanical devulcanization).

Mechano-chemical devulcanization of rubber vulcanizates is mostly carried out in a two-roll mill and is extensively practiced for its reliability. Here, devulcanizing chemical agents are likely to ensure (i) selective crosslink scission and (ii) stabilization of moieties produced *via* shear induced crosslink cleavage. The De Link process and the Lev Gum process, based on US Patent 5770632 A and US Patent 6387966 B1 respectively [8] [9] are the two most successfully commercialised devulcanization processes and both uses mechanical shear along with chemical modifiers to achieve devulcanization. These processes claim that the devulcanizing agents together with mechanical shearing selectively uncouple the sulphur crosslinks in a vulcanized network. Mechano-chemical shear-induced devulcanization and revulcanization of rubber vulcanizates are also described in the literature, with thiuram compounds (tetramethyl thiuram disulphide (TMTD) and tetra benzyl thiuram disulphide (TBzTD)) serving as both devulcanizing and curing agent [10]. Many similar processes making use of different devulcanization agents like disulphides, benzoyl peroxide (BPO) etc are widely reported [11] [12] [13] [14].

A careful review of reports on mechanochemical devulcanization processes carried out under ambient conditions suggests that dissimilar chemical modifiers follow similar mechanism to achieve crosslink scission and subsequent radical stabilization. A comparison of different mechanochemical devulcanization processes by Chapman et.al. employing different modifiers or chemicals showed comparable revulcanizate properties of devulcanized rubber irrespective of the devulcaniszation agents [15]. It is worth noting that, this study included both US Patent 5770632 A and US Patent 6387966 B1 along with mechanochemical devulcanisation using many novel combinations of chemicals, mainly various accelerator/activator combinations. Madhusoodhanan et.al. reported a simple mechanical devulcanisation process of ground tyre rubber (GTR) by shearing in a two-roll mill without using any devulcanising agent [16]. The revulcanizate properties obtained in this study were as good as the revulcanizate properties obtained in the comparative study by Chapman et al. These observations suggest that the devulcanising agents do not have an impact in shear induced crosslink scission in mechanochemical devulcanisation processes. This raises concerns regarding the function of the devulcanizing chemicals employed in mechanochemical devulcanization procedures. It might be assumed that, if dissimilar chemicals can bring about the same result through similar processes, it is the process (mechanical shearing) and not the chemical that bring about the result.

7.3 Stable free radical assisted shear induced devulcanization

In any shear induced devulcanization processes, it has to be assumed that some of the free radicals formed by the shear breaking of crosslinks might be stabilized by some species in the matrix and a significant number might recombine. The cumulative effect of these two opposing processes would result in a net lower percent of devulcanization. In this context, it has been hypothesised that the generation of free radical moieties via crosslink scission and the subsequent crosslink reformation via free radical recombination reactions are competitive. Therefore, any species that can arrest these free radical recombining reactions can significantly contribute towards the efficiency of the devulcanization process. Based on this hypothesis, a cyclic nitroxide stable free radical *viz.* 4-hydroxy 2, 2, 6, 6-tetramethylpiperidine 1-oxyl (4-hydroxy TEMPO abbreviated as 4HT) is introduced as a novel devulcanization aid for mechanochemical devulcanization [4,5]. 4HT is not a devulcanization agent but it assists devulcanization process by averting the recombination of shear-generated free radicals from crosslink scission thereby increasing the efficiency of devulcanization process.

Cyclic nitroxide free radicals belong to a set of free radicals that can remain stable at room temperature and at high temperatures. 2, 2, 6, 6-Tetramethyl piperidine-1-oxyl (I) is one such nitroxyl radical, abbreviated as TEMPO. In these nitroxides, the energy gain from oxygen –oxygen dimerization is lower than the loss of the resonance energy of the two aminoxyl groups, making O–O dimerization thermodynamically unfavourable [17]. 2, 2, 6, 6-tetramethylpiperidine 1-oxyl (TEMPO) and its derivatives (Figure 7.1) are widely used industrially as highly selective oxidation catalysts in pharmaceuticals, flavours and fragrances, agrochemicals, paints etc. [18]. In synthetic polymer chemistry, the capability of these stable free radicals to preferentially combine with carbon centred free radicals in presence of alkoxy radicals is utilised both to mediate living radical polymerisation of vinyl monomers [19] and to induce sulphur/accelerator like scorch control in peroxide vulcanization of rubbers [20].

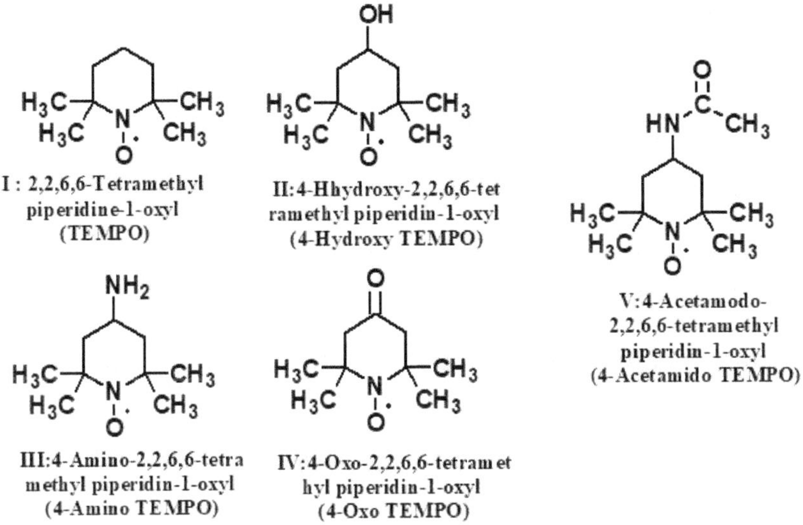

Figure 7.1 *Cyclic nitroxide stable free radicals.*

Mechanical devulcanization of carbon black filled natural rubber vulcanizates with thc assistance of 4HT was performed in a two-roll mixing mill of roller dimensions 15 cm x 30 cm and a friction ratio of 1:1.25 (12:15 rpm) at a nip gap of 0.05 mm (approx.). The thermal degradation of polymer due to rise in temperature during shearing was reduced by circulating water through the rolls. Carbon black filled natural rubber vulcanizates prepared in the laboratory (with known composition and vulcanizate properties) was used for devulcanization studies to make a reliable estimate of the efficiency of devulcanization. A comparative evaluation of the novel 4HT assisted devulcanization process for carbon black filled NR vulcanizates and contemporary mechanochemical devulcanization processes based on US Patent 5770632 A and US Patent 6387966 B1, designated as USP I and USP II respectively, was also carried out. Industrial feasibility of the novel method was then established via devulcanization of commercial waste rubber samples such as buffing dust and ground tyre rubber (GTR). A corresponding control experiment was also done in which devulcanization was carried out mechanically in the absence of the stable free radical. The sol-gel analysis of the devulcanized samples was used to reconfirm the percentage of the devulcanization results determined through crosslink density measurements. Horikx analysis was carried out to ascertain the extent of crosslink and main chain scission. Chemical probe analysis of the virgin and devulcanized samples was used to investigate the changes brought in by the devulcanization process in the crosslink distribution pattern. 10g of pre-masticated virgin natural rubber was blended with 100g of the devulcanized crumb for consistency and processing easiness before revulcanization. The mechanical properties of the revulcanized rubber were *determined as per the relevant ASTM standards.*

7.3.1 Stable free radical assisted shear induced devulcanization of carbon black filled rubber vulcanizates

Residual crosslink density (CLD) and percent devulcanization

The 4HT assisted devulcanization of carbon black filled NR vulcanizates, originally prepared using three different cure systems *viz.* conventional (CV), semi efficient (semi EV) and efficient (EV) are given in Figure 7.2 [5]. The positive influence of 4HT on shear-induced devulcanization is clear even at 1phr concentration. The influence of cure system of the original sample on the efficiency devulcanization is also evident from these observations. CV, semi EV and EV cured samples show unique responses towards mechanical devulcanization resulting in different percent devulcanization. Increasing concentrations of 4HT increased the percentage of devulcanization significantly. The changes observed in the residual crosslink density of devulcanized samples and the corresponding devulcanization percentage in response to the varying dosage of 4HT are summarized in Figure 7.2 and Table 7.1 respectively. It was observed that the optimum dosage of 4HT was 3phr for EV cured samples and 4-5phr for CV and semi-EV samples.

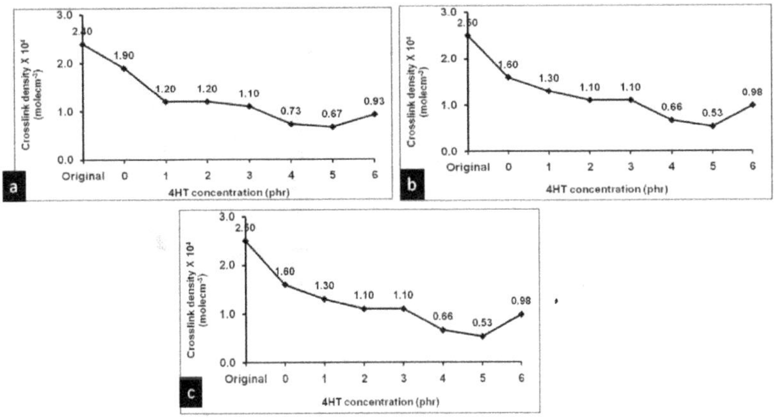

Figure 7.2 *Effect of 4-Hydroxy TEMPO on crosslink density of (a) conventionally (b) semi efficiently and (c) efficiently vulcanized NR samples [From reference 5 with permission].*

Table 7.1 *Effect of 4HT concentration on percentage of devulcanization [From reference [5] with permission]*

Sample	4HT, phr						
	0	1	2	3	4	5	6
CV	20.8	50	50	54.2	69.6	72.1	61.3
Semi EV	36	48	56	56	73.6	78.8	60.8
EV	40.5	77.1	79.4	80.6	76.5	73.5	-

The effect of particle size of carbon black used to reinforce the initial vulcanizates on the residual crosslink density of devulcanized samples is outlined in Figure 7.3. Devulcanization of carbon black filled NR vulcanizates prepared with 40phr N220 (Intermediate Super Abrasion Furnace black, ISAF), N330 (High Abrasion Furnace black, HAF), N550 (Fast Extrusion Furnace black, FEF) and N660 (General Purpose Furnace black, GPF) fillers indicated that:

(i) The residual CLD of 4HT assisted devulcanized samples are significantly lower than the corresponding values of mechanically devulcanized samples.

(ii) The type of carbon black has not significantly influenced the efficiency of devulcanization

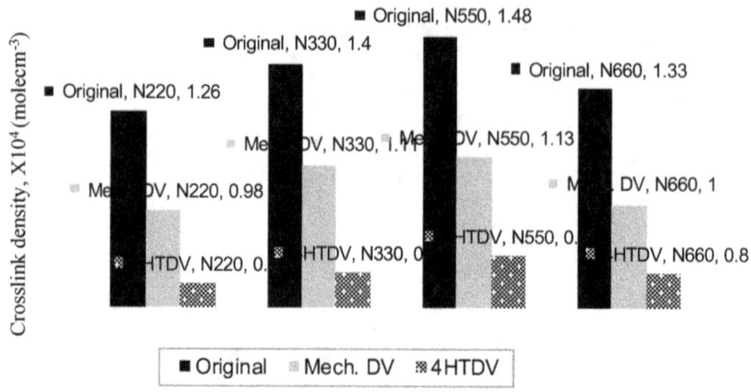

Figure 7.3 *Effect of 4HT on residual crosslink density of devulcanized NR vulcanizates reinforced with different carbon black fillers. [From reference 7 with permission].*

The trend observed in the residual CLD of NR vulcanizates containing different amounts of N330 (20-60phr) is given in Figure 7.4. As the CLD of the original vulcanizates increased with increase in filler content due to the strong interaction between rubber molecules and the surface of the carbon black particles, a corresponding increase in residual CLD of the 4HT assisted devulanized samples were also observed. Conversely, the percent devulcanization decreased as the CLD of the original sample increased with filler content. The percent devulcanization of samples containing 20, 30, 40, 50 and 60phr N330 was 56, 53, 53, 47 and 46 percent, respectively. This is in agreement with the earlier study that the residual crosslink density of the devulcanized sample prepared from vulcanizates of high crosslink densities will be high and vice versa [21].

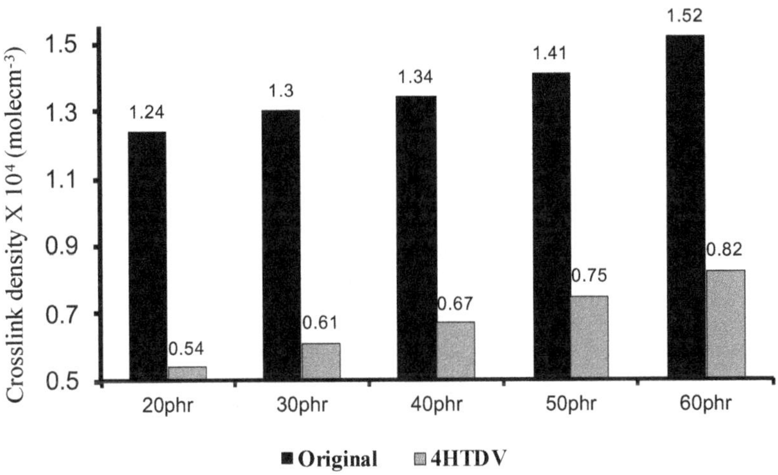

Figure 7.4 *Residual crosslink density 4HT assisted devulcanised NR vulcanizates reinforced with different amounts of N 330 carbon black filler. [From reference 7 with permission].*

In the case of NR/polybutadiene rubber (BR) blends, significantly lower residual crosslink densities and corresponding higher percent devulcanization were observed for 4HT-assisted devulcanization in comparison with mechanical devulcanization (Table 7.2). The samples showed a resistance towards mechanical devulcanization with increase in the BR content in the blend, making residual CLD determination practically impossible over 30 percent BR content. 4HT could resolve this issue to some extent which indeed suggests the capability of stable free radical as an efficient aid for shear induced devulcanization.

Comparing the results of 4HT-assisted mechanical devulcanization of carbon black-filled NR vulcanizates to those mechano-chemical devulcanization processes based on US Patents 5770632 A (USP I) and 6387966 B1 (USP II) allowed us to reconfirm our earlier observation regarding the effect of devulcanizing agents on the efficiency of devulcanization. The residual crosslink densities after devulcanization following mechanical, USP I protocol and USP II protocol, is given in Figure 7.5 [22]. The comparable residual cross link densities of mechanical, USP I and II samples substantiate our observation that these 'devulcanizing agents' have no role in crosslink scission, which is totally attributed to the shear applied by the two roll mill, and subsequent stabilisation of the free radicals generated. In contrast, samples that have undergone similar mechanical shearing in presence of the stable free radical has significantly lower residual crosslinks, which substantiate our hypothesis.

Table 7.2 *Crosslink densities of the original and devulcanised NR/BR blends with corresponding percent devulcanization*

NR/BR blend ratio	Crosslink density x 10^4, mole/cm^3			Percent devulcanization	
	Virgin	DV	4HT DV	DV	4HT DV
100/0	0.97	0.77	0.45	20.2	53.5
80/20	1.1	0.69	0.43	37.8	61.6
70/30	0.84	0.58	0.44	30.9	60
60/40	1.2	-	0.46	-	62
50/50	1.2	-	0.49	-	59.4

[From reference 6 with permission]

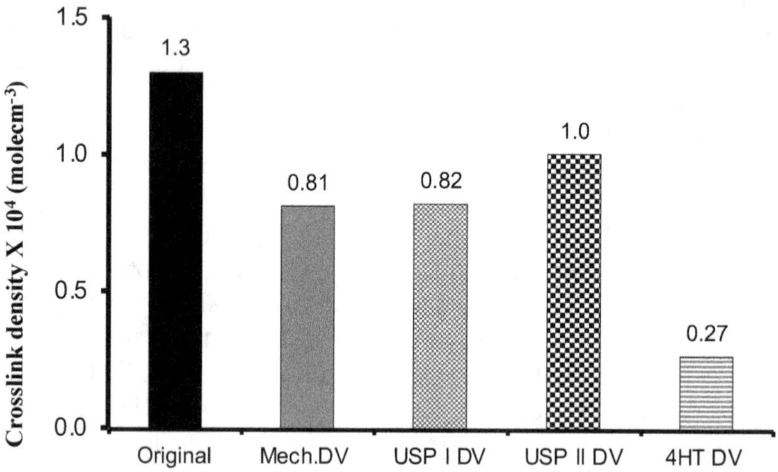

Figure 7.5 *Residual crosslink density of NR vulcanizates devulcanized by different methods. [From reference 22 with permission].*

Horikx analysis

Conceptual association between the sol fraction produced by polymer network degradation and the related decrease in crosslink density has been derived using Horikx equations [23]. Whether the soluble component is created by main-chain cleavage or crosslink cleavage, these equations hold true. In order to distinguish the contribution of crosslink and main chain scission towards the sol fraction of devulcanized rubber samples, Verbruggen et al. and Seghar et al. successfully compiled the theoretical Charlsbey model to validate the crosslink formation and main-chain scission happening during radiation vulcanization of rubber [24,25]. Using Horikx plots, the type of bond scission that the sample experienced throughout the devulcanization process is determined. Accordingly, the sample in question has undergone minimal main-chain scission if the experimental results associated with a certain crosslink density and accompanying sol fraction lie below the conventional main-chain scission curve. This observation in turn suggests that the sample has been devulcanized by a process involving selective breakage of cross-links. Experimental result either above or very close to the chain scission curve indicates deterioration of samples through significant main-chain scission induced by the devulcanization process.

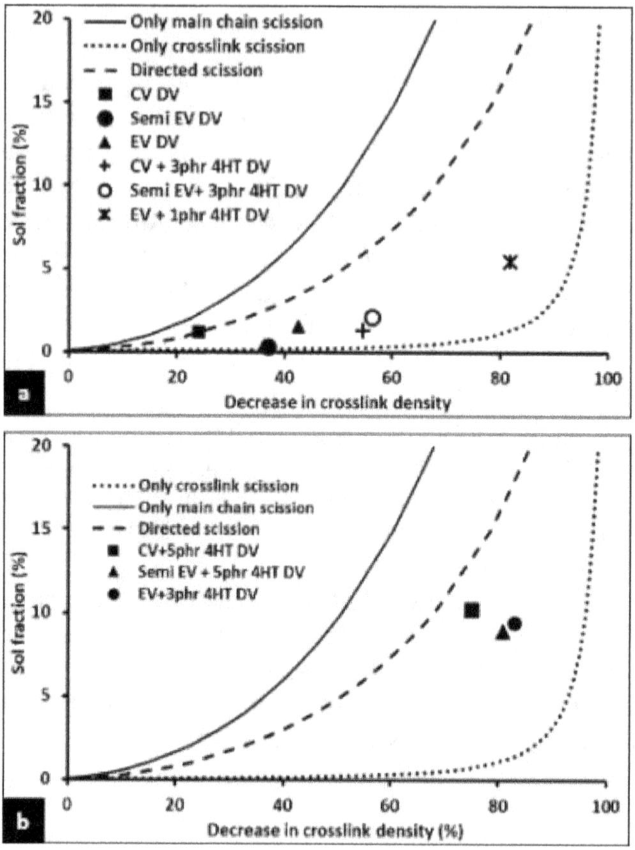

Figure 7.6 *Horikx plot of devulcanized NR vulcanizates (a) less than optimum amount of 4HT and (b) with optimum 4HT concentration. [From reference 5 with permission].*

Horikx curves suggested that stable free radical assisted shear induced devulcanization is highly likely to proceed via crosslink scission as the major bond scission route. Sol fraction of various devulcanized samples were plotted against corresponding percent devulcanization in a standard Horikx plot (Figure 7.6 a,b) The least percent devulcanization and higher sol fraction associated with mechanical devulcanization of CV cured samples suggests significant main chain scission (via directed scission). Horikx findings for mechanical devulcanization of semi-EV cured materials, on the other hand, indicate a predominance of cross-link scission. This observation is supported by higher percent devulcanization at lower sol fraction. It is interesting to note that mechanical devulcanization of EV-cured samples results in a drop in crosslink density and corresponding sol fraction values suggests a combination of cross-link and main chain scission. Recalling earlier publications that recommended establishing a balance between crosslink and main chain scission for the optimal devulcanization results, mechanical devulcanization of EV cured samples happens successfully to fulfil this need [26] [27].

Horikx analysis of devulcanized CV and semi-EV cured samples using lower than optimum amounts of 4HT shows lower residual crosslink density and low sol fraction than expected. These findings consistently point to a devulcanization process in which the main form of bond scission is crosslink scission. When 4HT is employed as a devulcanization aid at an optimum amount, the results suggest a balance between crosslink and main chain scission indicated by higher sol fraction at higher percent devulcanization. These findings suggest that 4HT increased devulcanization efficiency by striking a balance between main chain scission and crosslink scission.

Horikx study of the devulcanized NR vulcanizate with varying N330 black content (20-60phr) revealed that with increase in filler content in the sample:

(i) sol fraction of the devulcanized sample marginally increased
(ii) percent devulcanization decreased and
(iii) main chain scission (directed scission) increased.

The inverse relation between sol fraction and percent devulcanization of the samples suggests an increase in main-chain scission with increase in filler content. Increased amount of filler in the sample might be exerting higher stress to the sample during shearing leading to higher degree of main chain scission.

A comparative study of 4HT assisted shear-induced devulcanization and contemporary mechanochemical devulcanization processes resulted in a Horikx plot as given in Figure 7.7. Comparable sol fractions were found in materials devulcanized using mechanical, USP I, and USP II methods. The sample devulcanized using the USP II procedure had a substantially lower percentage devulcanization than the other two, though a larger sol percentage at a lower percent devulcanization is also accounted for by the data's location on the main chain scission curve. The results of samples devulcanized using mechanical (control) and USP I devulcanization techniques are shown alongside a directed scission curve, which suggests that main chain (directed scission) and crosslink scissions were involved in the devulcanization process. Devulcanization with 4HT again suggests equilibrium between main-chain and crosslink scission. It might be a worthy effort if some mathematical model could be developed, that could predict the optimum stress needed for the shear induced devulcanization to attain equilibrium between main-chain and cross-link scission. Anything higher than the optimum would be detrimental whereas lower stress is ineffective.

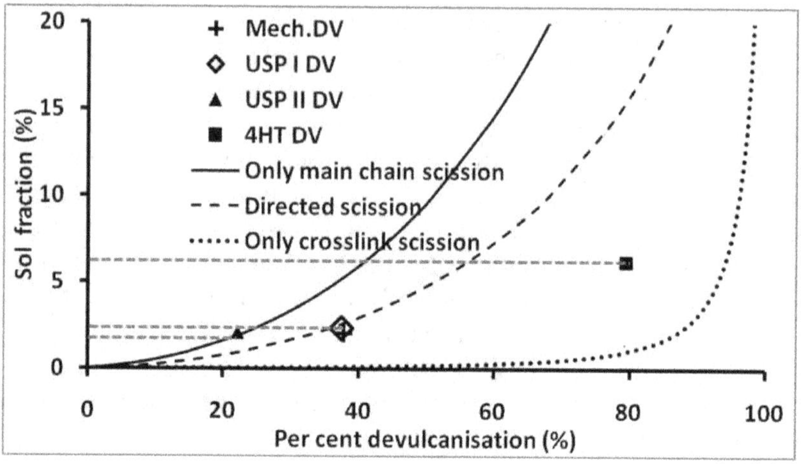

Figure 7.7 *Horikx plot of NR samples devulcanised by mechanical USP I, USP II and 4HT protocols. [From reference 22 with permission].*

Chemical probe analysis

Three forms of crosslinks-polysulphidic, disulphidic, and monosulphidic crosslinks-are created in sulphur-vulcanized rubber samples, and the concentration of each type is determined by the accelerator/sulfur ratio of the cure system used. The kind of bond scission involved in the devulcanization process can be determined by comparing the chemical probe analysis of a sample before and after devulcanization. The results of the chemical probe analysis of the CV, semi-EV and EV cured samples before and after devulcanization are graphically demonstrated in Figure 7.8 a,b and c, respectively. The percentage of devulcanization is most strongly influenced by the degree to which the predominant crosslink type (polysulphidic crosslinks in CV and semi-EV and disulphidic crosslinks in EV cured samples) has been reduced. This is demonstrated by the striking variation in residual crosslink densities between samples that were devulcanized without and with the use of 4-hydroxy TEMPO. The stable free radical assisted devulcanization process resulted in the scission of a significantly higher number of majority crosslink type as compared to the devulcanization process without 4HT.

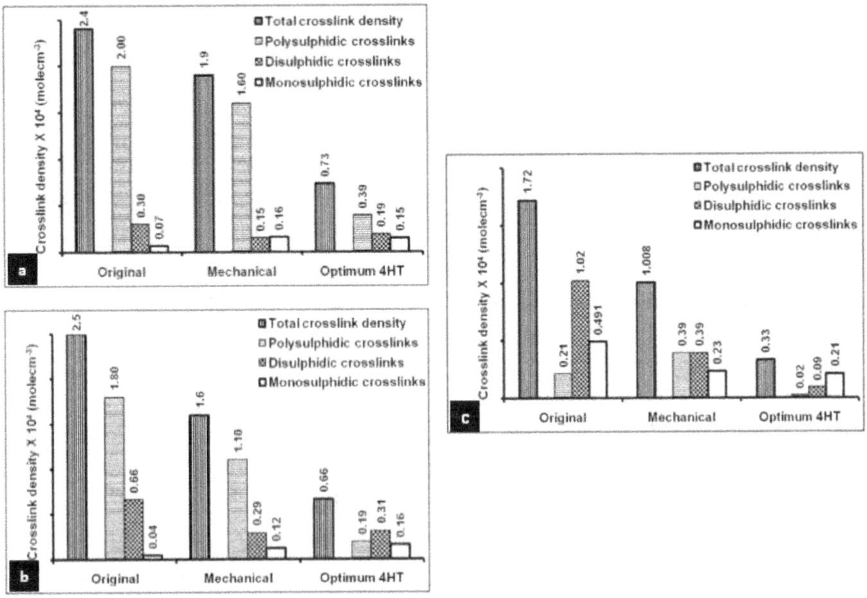

Figure 7.8 *Chemical probe analysis results of (a) CV, (b) semi-EV and (c) EV cured samples. [From reference 5 with permission].*

In case of samples cured using conventional and semi efficient cure systems, the number of polysulphidic and disulphidic crosslinks were reduced both by mechanical and 4HT mediated devulcanization but to different extent. Surprisingly, the absolute number of monosulphide crosslinks of the devulcanized samples was higher than that of the original vulcanizates in both cases. This might be explained by chain-shortening reactions that took place during shear-induced devulcanization at ambient temperatures, which are comparable to the crosslink desulphuration reactions that take place during post-crosslinking at vulcanization temperatures.

Post-devulvcanisation crosslink distribution pattern of EV-cured samples indicate significant increase of polysulphidic crosslinks after mechanical devulcanizantion, whereas only marginal increase in polysulphidic crosslinks was observed after 4HT-assisted devulcanization. This increase in polysulphidic crosslinks can only be accounted by the re–crosslinking of shear cleaved crosslinks during devulcanization. In 4HT-mediated devulcanization, it is highly likely that most of these free radicals generated by crosslink scission combine with the stable free radical and are stabilized right away during mechanical shearing. Because of this, the likelihood of their re-crosslinking is reduced, which is reflected in the negligible rise in residual polysulphidic crosslinks in these samples.

Another interesting aspect of post-devulvcanization crosslink distribution pattern of EV-cured samples is the absence of increase in the monosulphide CLD, as observed with CV and semi-EV cured samples. This might be as a result of the much smaller amount of polysulphidic crosslinks that undergo desulphuration reactions during shearing in EV samples compared to CV and semi-EV samples.
Apart from these observations, analysing the percentage of different bonds cleaved under mechanical and 4HT mediated devulcanization outlines a different picture as detailed in Table 7.3. In mechanically devulcanized CV and semi-EV samples, the percent reduction in number of polysulphidic crosslinks is comparatively less than that of the reduction in disulphidic crosslinks. 4HT assisted devulcanization reversed this trend in such a way that the percentage of cleaved polysulphidic bonds was well above that of shear cleaved disulphidic bonds. Owing to its lowest bond dissociation energy and abundance in the vulcanized network, polysulphidic crosslinks are expected to be the primary target towards shear induced bond cleavage. The recombination of crosslink fragments might be responsible for the effective low percent cleavage of polysulphidic crosslinks observed with mechanical devulcanization of CV and semi-EV cured samples.

Both in mechanical and 4HT mediated shear devulcanization of CV samples, the maximum shear stress will be concentrated on the most abundant polysulphidic crosslinks. These shear induced crosslink scissions will be simultaneously complemented by crosslink fragment stabilization by the stable free radical, restraining the possible re–crosslinking to a minimum. 4HT mediated devulcanized semi-EV samples also follow similar pattern of crosslink scission during devulcanization except for the higher numbers of disulphidic crosslinks cleaved initially along with polysulphidic crosslinks. This might be attributed to the higher number of disulphidic crosslinks present in the sample initially. The meagre number of monosulphide bonds in original semi-EV sample is consequently displayed as its significant increase after devulcanization owing to crosslink reformation during devulcanization process.

Table 7.3 *Comparison of percent change in crosslink density after devulcanization*

Crosslinks	Percent change in crosslink density after devulcanization					
	CV		**Semi – EV**		**EV**	
	Mechanical	Optimum 4HT	Mechanical	Optimum 4HT	Mechanical	Optimum 4HT
Total	-21	-70	-36	-74	-41.6	-80.8
Polysulphidic	-20	-80.5	-38.9	-89.4	+85.7	-90.1
Disulphidic	-50	-37	-56	-53	-61	-91
Monosulphidic	+128	+114	+200	+300	-53	-57

+ denotes an increment and – denotes a decrement

[From reference [5] with permission]

Since the number of polysulphidic cross-links in the original EV vulcanizate was low, the recombination of crosslink fragments during mechanical devulcanization surfaced as an increase in polysulphidic CLD. On the other hand, in the stable free radical assisted devulcanization of the EV sample, more than 90 percent reduction in the absolute number of polysulphidic crosslinks is observed. This again confirms that stable free radical assists shear induced devulcanization by blocking the possible re–crosslinking of bonds cleaved under shear.

The chemical probe analysis from comparative study of 4HT assisted devulcanization and contemporary mechanochemical devulcanization processes (as per USPI and USPII) of semi-EV cured NR samples is given in Figure 7.9. The samples devulcanized by process other than stable free radical assisted devulcanization is found to share same crosslink distribution pattern. As all other parameters, except the chemical aid, in these processes were the same, any difference might be attributed to the devulcanization aid used. The crosslink distribution pattern of the devulcanized samples points to a high level of residual polysulphidic crosslinks of about 1 x 10^{-4} mole/cm^3 in all samples, except for 4HT assisted devulcanized sample in which the value dropped to 0.2 x 10^{-4} mole/cm^3. Also, the contribution from disulphidic and monosulphidic crosslinks together is roughly the same in all the cases. These observations reconfirm that crosslink scission in shear induced devulcanization processes is an act of the mechanical shear force alone and the devulcanization agents employed in mechanochemical devulcanization processes, which are supposed to facilitate selective bond scission, have nothing to do with crosslink scission (except in influencing shear stress due to plasticising effect). The remarkably low crosslink densities attained by 4-hydroxy TEMPO assisted mechanical devulcanization can be very well correlated with the effective blocking of shear generated crosslink fragments by the stable free radical alone.

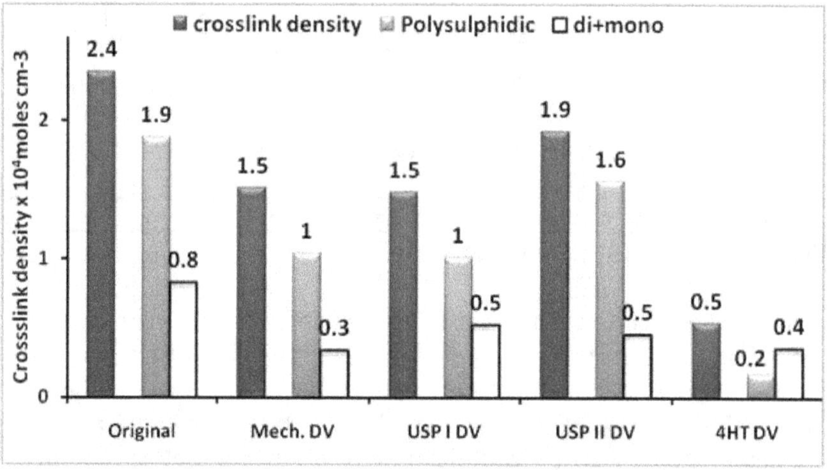

***Figure 7.9** Comparison of crosslink density and distribution pattern of the NR samples devulcanized by different methods. [From reference 22 with permission].*

Probable mechanism of stable free radical assisted mechanical devulcanization

The results of chemical probe analysis carried out on NR vulcanizates before and after devulcanization led to three inferences:

(i) Devulcanization efficiency of the process can be directly correlated with the net effective number of crosslinks of the majority kind.

(ii) Polysulphidic crosslinks present in the sample undergoes shear-induced desulphuration to monosulphidic crosslinks at ambient conditions resembling the post-crosslinking reactions during vulcanization.

(iii) Stable free radical significantly increased the efficiency of shear-induced devulcanization, most probably by blocking the re-crosslinking of shear broken crosslinks.

Figure 7.10 illustrates the probable mode of action of 4-hydroxy TEMPO in mechanical devulcanization of carbon black filled NR vulcanizates. Shear induced devulcanization of rubber vulcanizates generate crosslink fragments which are free radical species generated by crosslink scission. These free radicals may re-join to induce re-crosslinking, if not blocked by some species. 4-hydroxy TEMPO is a suitable candidate to stabilize these free radicals to prevent regeneration of crosslinks. This will in turn lead to lower residual crosslink density by 4-hydroxy TEMPO assisted mechanical devulcanization over mechanical devulcanization of the same sample.

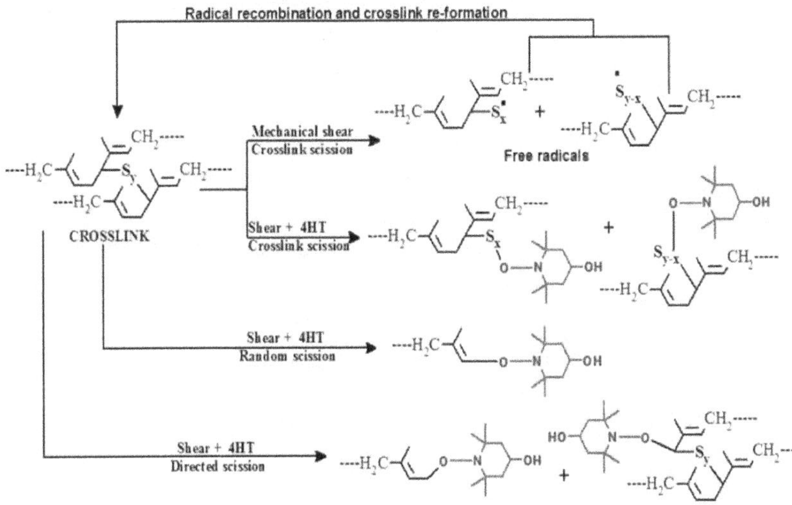

Figure 7.10 *Probable Mode of action of 4-Hydroxy TEMPO in mechanical devulcanization of carbon black filled NR vulcanizates [From reference 5 with permission].*

Two main observations support the proposed mechanism:
(a) Considerably low number of polysulphide crosslinks in conventionally and semi-efficiently vulcanized samples subjected to 4-Hydroxy TEMPO mediated devulcanization over analogous mechanical devulcanization,
(b) significantly higher percentage of polysulphidic cross links (more than 85%) in mechanically devulcanised EV sample than its original

sample whereas significantly lower percentage of (less than 10%) polysulphidic crosslinks in 4HT assisted devulcanised EV sample than present in the original sample.
Although the reaction of thiyl radical (RS·) and stable nitroxyl radical (>NO·) leading to the formation of an unstable adduct (>NO-SR) that can decompose into various products is reported [28], the fate of adduct formed by the 4HT and the thiyl radicals during devulcanization of NR vulcanizates is not clear.

Cure characteristics of devulcanized rubber

Revulcanization cure curves of the devulcanized NR vulcanizates and the cure curves of the corresponding virgin compounds (CV, semi-EV and EV systems) are given in Figure 7.11. The minimum torque of the 4HT assisted devulcanized samples was lower than that of the mechanically devulcanized samples, with consequent higher M_H-M_L value. Both these observations indirectly point to the higher devulcanization levels attained by the samples prepared by 4HT assisted mechanical devulcanization. The scorch and cure time of all devulcanized rubbers were comparable and were significantly lower than the corresponding virgin compounds. A recent study reported that low scorch times when revulcanizing devulcanized rubber or its blends with virgin compounds are not related to or caused by the devulcanization or revulcanization processes. Instead, they depend on the cure system used to vulcanize the original sample [29]. It was found that the low scorch safety could be traced back to the accelerated sulphur vulcanization chemistry involved in the preparation of original sample.

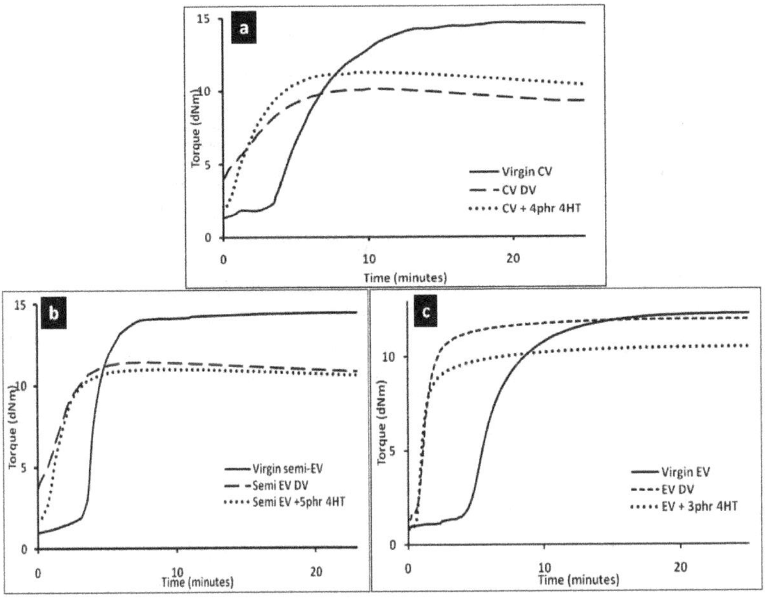

Figure 7.11 *Original and revulcanization cure curves of (a) conventional, (b) semi-Efficient and (c) efficient vulcanised samples. (Devulcanized rubber-100g, NR-10g, Antioxidant TDQ-1.5phr, Zno-5phr, Stearic acid-2phr, MBS-0.8phr, S-1.5phr, Cure temperature 150^{0}C) [From reference 5 with permission].*

Revulcanizate properties

The revulcanizate properties of carbon black-filled NR vulcanizates devulcanised using varying amounts of 4HT are given in Table 7.4. The mechanical properties of the revulcanized samples were significantly lower than those of the corresponding virgin samples. These properties got better as more 4-hydroxy TEMPO was added as a devulcanization aid, reaching their best level at the optimal 4HT dosage. The tear characteristics of the revulcanizates, as opposed to the tensile strength which levelled out at still low levels of 4HT, clearly demonstrate the systematic improvements in revulcanizate qualities with increasing amounts of 4HT up to its maximum.

The influence of the particle size of carbon black filler on devulcanisation efficiency and corresponding revulcanisate properties were studied using NR vulcanisates reinforced with N220 (ISAF), N330 (HAF), N550 (FEF) and N660 (GPF). In all the cases, the 4HT assisted devulcanized samples showed better revulcanizate properties when compared to the corresponding mehcanically devulcanized samples [**7**]. Irrespective of the particle size of the carbon black filler used for the reinforcement of the original samples, the revulcanizate properties of the 4HT DVR were comparable. Another interesting observation is that, irrespective of the type of filler, the stable free radical assisted devulcanization can provide revulcanizate properties almost comparable with that of the original N660 filled vulcanizate (Figure 7.12). This might be an indirect indication that, the undevulcanized parts present in the revulcanised sample might be acting as filler with a particle size comparable with the particle size of N660 black. Thus, the revulcanizate properties of the devulcanized samples could not rise above the vulcanizate properties attainable by N660 reinforced NR vulcanizates. In the case of mechanically devulcanized samples, these vulcanized moieties present in the devulcanized rubber might be still of higher size or in higher numbers so that these might not be properly dispersed in the matrix to exert any reinforcing effect. On the contrary, they might be acting as points of weakness and result in lower revulcanizate properties. The remarkably good revulcanizate properties obtained after 4HT aided devulcanization of N660 filled samples might be due to the uniformity in the size of fillers - both N660 and un-devulcanized moieties - present in the sample rendering effective reinforcement.

Table 7.4 *Revulcanizate properties of NR samples devulcanized with varying amounts of 4HT*

Vulcani-zate/ 4HT dosage		Tensile strength, MPa	Elongation at Break, %	M100, MPa	M200, MPa	M300, MPa	Tear strength (N/mm)
	Virgin	30 ± 0.88	671 ± 9.42	2.2	5.1	9.4	119 ± 3.63
	0phr	19.7 ± 0.94	533 ± 18.70	1.9	4.3	7.8	37 ± 3.33
	1phr	24.5 ± 0.37	617 ± 12.08	2	4.4	8	46 ± 2.02
	2phr	23.6 ± 0.7	613 ± 16.03	2	4.5	8	48 ± 2.63
CV	3phr	24.5 ± 0.97	616 ± 17.46	2.1	4.7	8.2	52 ± 1.74
	4phr	25.5 ± 0.8	668 ± 39.27	2.1	4.6	8.3	64 ± 3.00
	5phr	25 ± 0.70	683 ± 9.52	2	4.4	7.9	61 ± 2.40
	6phr	23 ± 1.50	618 ± 25.95	2	4.3	7.7	58 ± 1.16
	Virgin	31.7 ± 0.99	635 ± 2.75	2.58	6.4	11.52	110.5 ±1.5
	0phr	22.5 ± 0.91	579 ± 17.76	2	4.3	7.7	39 ± 3.69
	1phr	24.9 ± 0.54	627 ± 8.59	2	4.4	7.8	49 ± 3.41
	2phr	24.8 ± 0.48	637 ± 11.30	2	4.4	7.7	50 ± 2.75
Semi EV	3phr	24.9 ± 0.54	637 ± 6.05	2	4.4	7.8	55 ± 2.77
	4phr	26.6 ± 0.34	634 ± 8.16	2.1	1.5	8.2	60 ± 2.57
	5phr	27 ± 0.12	662 ± 9.43	2	4.4	7.9	65 ± 2.42
	6phr	22.4 ± 0.34	654 ± 18.16	1.7	3.8	6.7	54 ± 4.33
	Virgin	27 ± 0.65	699 ± 2.33	2	4.6	8	110 ± 5.5
	0phr	16 ± 0.95	455 ± 3.90	2.2	4.8	8.5	48 ± 2.97
	1phr	21 ± 1.30	574 ± 7.90	2	4.8	8	55 ± 2.54
EV	2phr	22.7 ± 1.42	584 ± 9.00	2	4.5	8	58 ± 2.46
	3phr	21.5 ± 0.81	581 ± 7.12	2	4.5	7.8	59 ± 3.18
	4phr	22 ± 0.71	593 ± 8.73	2	4.6	8	59 ± 3.18
	5phr	20 ± 1.69	587 ± 7.12	2	4.6	7.8	57 ± 2.90

[From reference No 5. With permission]

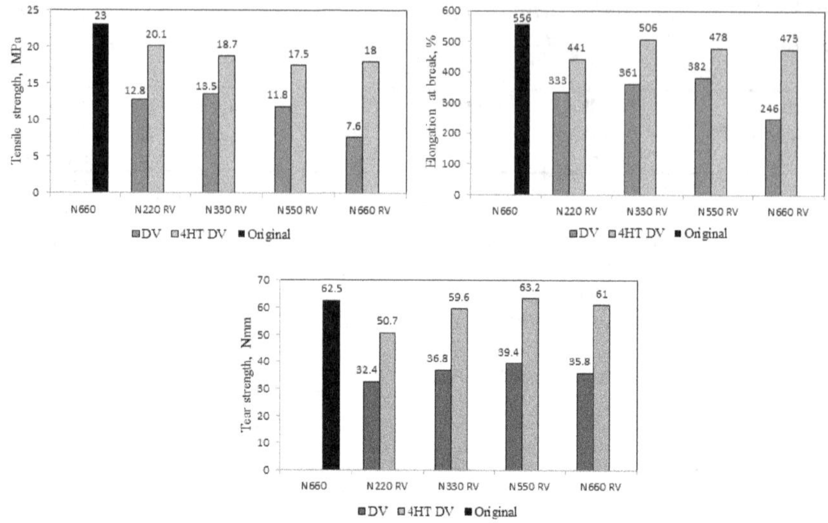

Figure 7.12 *Comparison of revulcanizate properties of the samples containing different carbon blacks with original properties of NR/N660 vulcanizate. [From reference 7 with permission].*

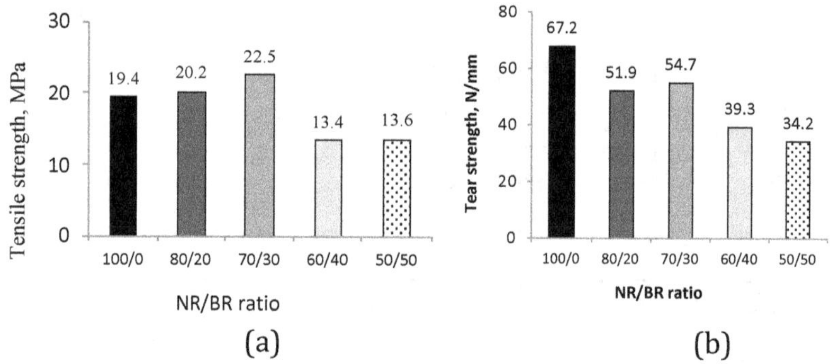

Figure 7.13 *Revulcanizate properties of NR/ BR blends [From reference [6] with permission].*

The revulcanizate properties of the devulcanised NR/BR blend samples throw light on the effect of BR content on the efficiency of shear-induced devulcanization. Revulcanizate properties of the mechanically devulcanized samples were significantly inferior to the corresponding virgin vulcanizates consistent with the higher residual crosslink density of these samples. The revulcanizate properties of the 4HT assisted devulcanised samples were significantly higher than the corresponding mechanically devulcanised samples. However, the positive influence of 4HT on revulcanizate properties significantly dropped when the BR content was 40 parts and above in the original sample (Figure 7.13).

In the case of devulcanization of NR material as per USP I and USP II protocols, the revulcanizate tensile strength and tear strength were comparable with those of mechanically devulcanized samples (Figure 7.14). The slight differences in the tensile and tear strengths between these revulcanizates might be attributed to the variations in the revulcanization formulations employed rather than the effect of the devulcanization process involved, which is well established by the residual crosslink density and crosslink distribution pattern of these devulcanized samples. The 4HT assisted devulcanized samples possessed a significantly higher tensile strength and tear strength after revulcanization further confirms effectiveness of stable free radical as a devulcanization aid.

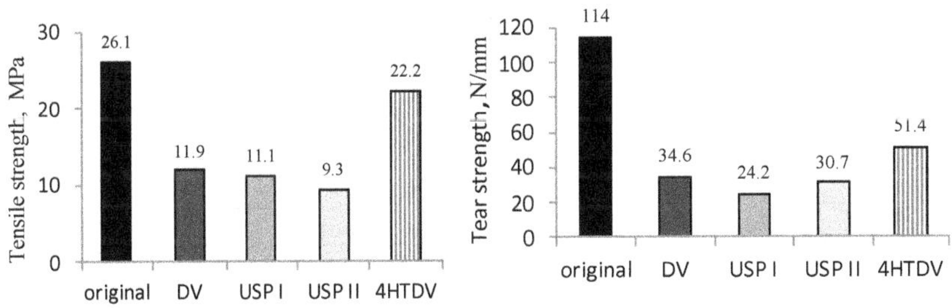

Figure 7.14 Comparison of revulcanizate properties of NR samples devulcanized by different methods. [From reference 22 with permission].

7.4 Applications of stable free radical assisted shear induced devulcanization

7.4.1 4HT assisted devulcanization of ground tyre rubber and tyre tread buffing

Ground tyre rubber (GTR) and tyre tread buffing (buffing dust) are important commercial sources of waste rubber. Stable free radical assisted shear induced devulcanization of these commercial samples substantiated its potential for commercial practice. The positive impact of 4HT on shear-induced devulcanization of GTR and buffing dust is evident from the reduction in the required number of passes for effective devulcanization. The required number of passes through the tight nip reduced from 30 to 20 and 50 to 42 in the case of buffing dust and GTR respectively when 1phr 4HT was used as a devulcanising aid. Lesser number of passes indicates reduction in time and power requirement, which directly add on to the commercial and practical feasibility of the method. Residual crosslink density of mechanically and 4HT assisted devulcanized buffing dust were 0.54 x 10^{-4} mole/cm^3 and 0.25 x 10^{-4} mole/cm^3 respectively, suggesting an additional 53.7 percent reduction in residual crosslink density for 4HT mediated devulcanization. The revulcanizate properties of 4HT assisted devulcanized samples were invariably better than the corresponding mechanically devulcanized samples (Figure 7.15). Many moulded general rubber goods, like mats, can be directly prepared using devulcanized buffing dust or GTR.

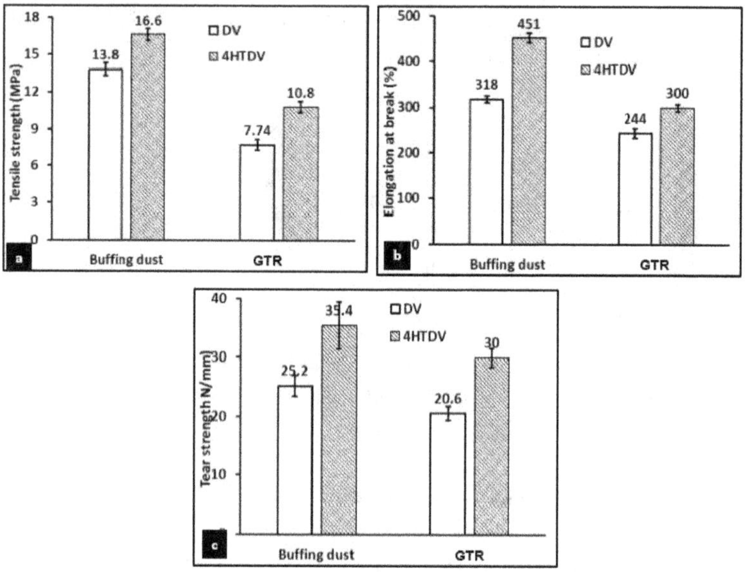

Figure 7.15 *Revulcanizate properties of devulcanized buffing dust and GTR. [From reference 22 with permission].*

7.4.2 Blends of virgin rubber compound (NR/BR 70/30) and devulcanized buffing dust

Virgin rubber (NR/BR 70/30) and devulcanized buffing dust were blended in the ratios 95/5, 90/10, 85/15, 80/20 and 75/25 and compounded following the formulation used for preparing the pre-cured tread from which the buffings are sourced. The cure characteristics of the control and the blends were comparable except for the scorch and cure times. The scorch time of the blends decreased as the devulcanized rubber content in the blend increased. The mechanical properties of the blends showed notable difference with the control mostly with respect to tear strength and abrasion (Table 7.5). 4HT assisted devulcanization could resist the substantial deterioration of tear strength of the blends when devulcanized rubber is incorporated as clearly inferred from Table 7.5.

Table 7.5 *Mechanical properties of blends of virgin rubber (NR/BR 70/30) and devulcanized buffing dust*

Vulcanizate	Hardness (Shore A)	Tensile strength (MPa)	EB (%)	M300 (MPa)	Tear strength (N/mm)	Abrasion loss (mm³)	Heat build-up (°C)
Control, 100/0	63	20.6	512	10.7	92.7	57	17
NR/DV (95/5)	63	21.3	580	9.9	87.8	54	18
NR/4HTDV (95/5)	63	21.8	582	9.7	90	54	19
NR/DV (90/10)	62	22.2	594	9.3	68	72	18
NR/4HTDV (90/10)	63	21.4	580	9.3	82.3	72	19
NR/DV (85/15)	62	22	581	9.4	63.8	72	18
NR/4HTDV (85/15)	62	21.3	522	9.2	79.8	72	19
NR/DV (80/20)	61	20.5	557	9.2	60	72	19
NR/4HTDV (80/20)	62	19.3	527	9.3	73.5	72	19
NR/DV (75/25)	61	20.1	563	8.9	49.6	90	17
NR/4HTDV (75/25)	61	19.9	545	9.2	54	81	18

[From reference No 22 with permission]

The original sample and the blends with devulcanized rubber were subjected to ageing at 70°C for 7 days and the properties were compared with the corresponding un-aged samples. The original blend could retain 100 per cent tensile strength, an 18 percent increase in M300 and 92 per cent retention in elongation at break but could only retain 64percent of tear strength after ageing. NR/4HTDV blends retained mechanical properties to a greater extent after ageing in comparison with corresponding NR/DV blends. It could retain above 80 percent of tensile strength, above 70 percent of elongation at break and tear strength. It may be noted that the absolute value of tear strength after ageing was comparable to that of the control sample (Figure 7.16).

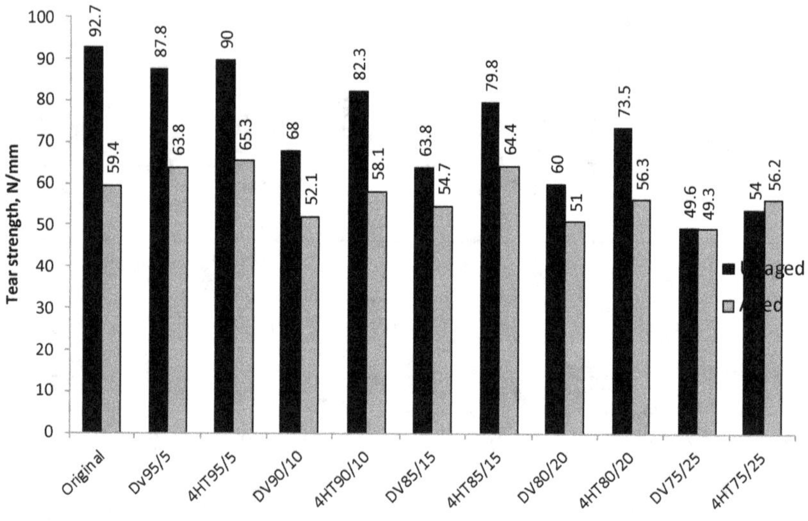

Figure 7.16 *Effect of ageing (70°C/ 7 days) on blends of virgin rubber and devulcanized buffing dust [From reference 22 with permission].*

7.4.3 Second round devulcanization of commercial devulcanized rubber

A commercial devulcanized rubber (CDR) sample with an approximate residual crosslink density of 2 x10^{-4} mole/cm^3 was subjected to second round devulcanization. CDR was further devulcanized at the tight nip of a two-roll mill without and with the assistance of 2phr 4HT. The residual crosslink density of CDR was reduced to 1.26 x 10^{-4} mole/cm^3 (37% reduction) after second round mechanical devulcanization whereas it further lowered to 0.85 x 10^{-4}

mole/cm^3 after 4HT assisted devulcanization. The revulcanizate tensile strengths and tear strengths of CDR, mechanically devulcanized CDR and 4HT assisted devulcanized CDR are given in Figure 7.17. Over 40 percent increase in tensile strength and over 100 percent increase in tear strength in stable free radical assisted devulcanized sample are noteworthy.

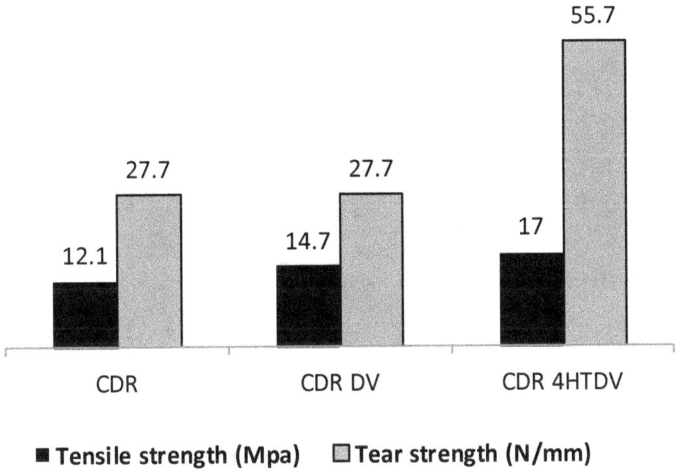

Figure 7.17 *Revulcanizate properties of commercial devulcanised rubber (CDR) after second round devulcanization [From reference 22 with permission].*

7.4.4 Factory scale devulcanization of tyre tread buffing

The positive results of 4HT-assisted devulcanization from several lab scale trials are further validated through factory scale trial (lager volumes) using tyre tread buffings. Lab scale devulcanization of the same sample was carried out as the control experiment. The factory scale trials could reproduce the results obtained in the lab-scale trials forecasting the 4HT assisted mechanical devulcanization process as a promising method for maximum property retention through efficient devulcanization (Figure 7.18).

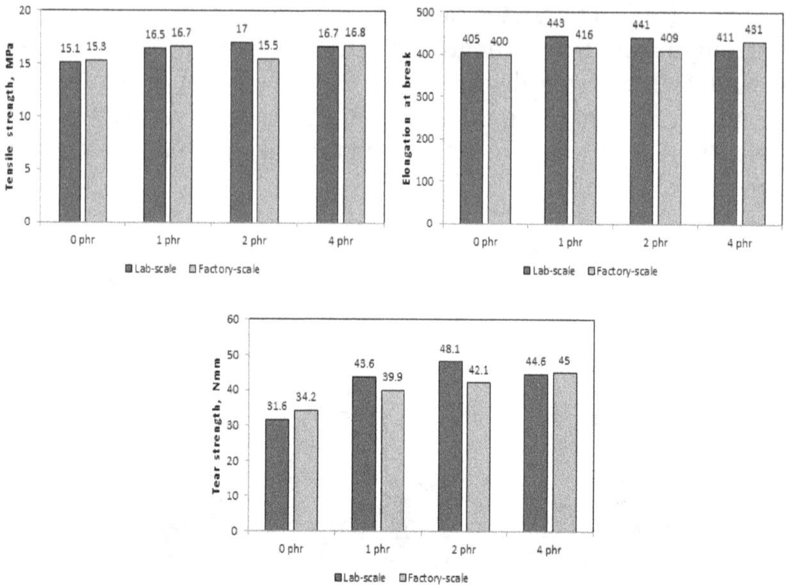

Figure 7.18 *Revulcanizate properties of factory scale devulcanised buffing dust in comparison with laboratory scale devulcanised sample. [From reference 22 with permission].*

7.5 Concluding remarks

Devulcanization of crumb rubber powder sourced from scrap rubber products, especially tires, by a simple shear process, like in a two-roll mill at ambient temperature and pressure, using devulcanizing agents is well known in rubber industry. Comparison of two patented and industrially practiced mechanochemical devulcanization process with simple shear devulcanization without the devulcanising agents demonstrated that these additives did not have an impact to increase the efficiency of shear-induced devulcanization. These results led us to hypothesise that crosslink breaking and reformation are competitive in shear-induced devulcanization and any chemical that can arrest the re-combination of the free radical species formed by crosslink scission can significantly improve the efficiency of devulcanization. Accordingly, cyclic nitroxide stable free radical viz. 4-hydroxy TEMPO was used as a "devulcanization aid" for shear induced devulcanization at ambient conditions.

The efficacy of stable free radical as an efficient devulcanizing aid was demonstrated by 4HT assisted shear induced devulcanization of

NR vulcanizates, with known formulation and properties, at ambient temperature in comparison to shearing in a two roll-mill without 4HT. Similarly, NR vulcanizates reinforced with four different carbon blacks with varying particle sizes, NR/BR blend vulcanizates with varying BR content were also devulcanized. Further, two industrially practiced mechanochemical devulcanization protocols were compared with the shear devulcanization with and without the presence of the stable free radical. The devulcanized rubbers were characterised in terms of residual crosslink density and estimation of percent devulcanization, Horikx analysis and crosslink distribution by chemical probe analysis. Based on these studies, a mechanism which involves reaction of sulphur free radicals formed by crosslink scission with stable free radical is recommended. Low residual crosslink density in general and low residual polysulphidic crosslink density in particular observed with 4-hydroxy TEMPO assisted devulcanization and higher revulcanizate properties as a result strongly suggest that the proposed reaction is possible.

The percent devulcanization was greatly affected by the cure system (CV, semi-EV, and EV) and therefore the crosslink distribution of the original sample. When compared to mechanical devulcanization, the stable free radical allows for a high percentage of devulcanization, and at the optimum concentration, the percentage of devulcanization in both conventional and semi-EV samples is the same. The EVcured samples with the smallest amount of polysulphidic crosslinks achieved the highest percentage of devulcanization. The addition of 4-hydroxy TEMPO was observed to aid devulcanization of conventionally cured NR samples (with the highest proportion of polysulphidic crosslinks) the most in relation to increase in percent devulcanization. All of these results imply that the percent devulcanization depends equally on the bond dissociation energy and the rate of bond reformation via crosslink fragment recombination.

Among the revulcanizate properties of devulcanised rubber, tear strength was significantly sensitive to percent devulcanisastion. Mechanical devulcanization with 20-40 percent devulcanization could retain 60-70 percent of the original tensile strength. However, at the same level of devulcanization, the revulcanizate tear strength is only 30-43 percent of the original value. Stable free radical assisted devulcanization increased the percent devulcanization to 70-80 percent with consequent increase in percent retention of tensile strength to 85 percent and tear strength to 53-59 percent in comparison with the original sample.

Devulcanization of NR vulcanizates reinforced with four different carbon black fillers namely N220, N330, N550 and N660 of different particle sizes showed that the type of filler present in the original sample did not have a decisive role on the revulcanizate properties when stable free radical was used as devulcanizing aid. It is worth noting that the revulcanizate properties of all the carbon black filled NR vulcanizates devulcanized with the assistance of 4HT were comparable with the original vulcanizate properties of N660 reinforced NR disregarding the type of black filler. At fixed filler concentration, high polybutadiene content reduced the devulcanization efficiency of stable free radical. The percent retention of revulcanizate properties considerably dropped when the polybutadiene content was 40 parts or above in NR/BR blends.

Comparison of residual crosslink density, cross link distribution and revulcanizate properties of two industrially practiced mechanochemical devulcanization protocols with simple mechanical devulcanization further proved that the "devulcanization agents" used in these processes did not play any role in increasing the efficiency of devulcanization. The 4HT assisted devulcanization of commercially available crumb rubber powder and tire buffing dust and its blends of the respective devulcanised samples with virgin rubber compounds indicated the feasibility of using stable free radical as an effective devulcanization aid with commercial potential.

References

1. Tire Recycling Riding on (2019), *Global Recycling,* **5** (2), 4-8. https://www.eu-recycling.com/flips/gr22019/gr22019.html#4 Accessed on 22/12/2020
2. Recycled rubber infill material has a role to play in a Circular Economy (2020), *ETRMA updates, 2* September; https://www.etrma.org/library/recycled-rubber-infill-material-has-a-role-to-play-in-a-circular-economy/ Accessed on 22/12/2020
3. Fazli A, Rodrigue D (2020) Recycling waste tires into ground tire rubber (GTR)/Rubber Compounds: A Review. J Comp Sci 4:1-43. htpps://doi:10.3390/jcs4030103

4. George B, Joseph AM, Madhusoodanan KN, Alex R, Jacob J. (2017) Stable free radical assisted devulcanization of rubber. Indian Patent 370393, 19 May 2017.
5. Joseph AM, Madhusoodanan KN, Alex R, George B (2018) Stable free radical assisted mechanical devulcanization of carbon black filled natural rubber vulcanizates. *Rub Chem Technol* 91: 469. htpps://doi:10.5254/rct.18.81633
6. Joseph AM, Madhusoodanan KN, Alex R, George B (2019) Stable free radical assisted mechanical devulcanization: Devulcaanisation of NR/BR blends. Rub Sci 32: 68-82
7. Joseph AM, Madhusoodanan KN, Alex R, George B (2020) Stable free radical assisted mechanical devulcanization of carbon black filled natural rubber vulcanizates: Role of carbon black filler. Rub Sci 33: 59-73.
8. Sekhar BC, Kormer VA, Sotnikova EN, Mironyuk VP, Turnova LN, Nikitina NA (1998) Reclaiming Elastomeric Materials, US patent 5770632 A.
9. Goldstein V, Kopylov M (2002) Method and Composition for Devulcanization of Waste Rubber, US patent 6387966 B1.
10. Mandal SK, Alam N, Debnath SC (2012) Reclaiming of ground tyre rubber by safe multifunctional rubber additives: I. Tetra benzyl thiuram disulphide. *Rub Chem Technol* 67:629-644. htpps://doi:10.5254/rct.12.88949
11. Jana GK, Das CK (2005) Devulcanization of automobile scrap tyres by a mechanochemical process. *Prog in Rub Plast and Polym Recycl Tech* 21: 319-331. https://doi.org/10.1177/147776060502100405
12. Rooj S, Basak GC, Maji PK, Bhowmick AK (2011) New route for devulcanization of natural rubber and the properties of devulcanized rubber. *J Polym Envir* 19: 382-390. htpps://doi:10.1007/s10924-011-0293-5
13. De D, Maiti S, Adhikari B (2000) Reclaiming of rubber by a renewable resource material (RRM) III. Evaluation of properties of NR reclaim. *J Appl Polym Sci* 75: 1493-1502. https://doi.org/10.1002/(SICI)1097-4628(20000321)75:12<1493::AID-APP8>3.0.CO;2-U
14. Premachandra JK, Edirisinghe DG, De Silva MIA (2011) A novel reclaiming agent for ground rubber tyre (GRT) Part I: Property evaluation of virgin natural rubber (NR)/ novel reclaimed GRT blend compounds. *Prog in Rub Plast and Polym Recycl Tech* 27: 31-47. http://dl.lib.mrt.ac.lk/handle/123/12471

15. Chapman A (2007) Recycling of the tyre rubber into new rubber products through efficient devulcanization, Report of Waste and Resources Action Programme (WRAP), UK
16. Madhusoodanan KN, Mohandas KS, Kuriakose B, Thomas KT, Mathew NM (1998) *An efficient mechanical method for reclamation of vulcanized scrap rubber.* International Rubber Conference (IRC), Chennai, India.
17. Bohle DS (2010) The Nitrogen Oxides: Persistent Radicals and van der Waals Complex Dimers. In: Stable Radicals: Fundamentals and Applied Aspects of Odd-Electron Compounds, Hicks, R.G. (ed.), 1st edition, Wiley, United Kingdom, pp.199.
18. Ciriminna R, Pagliaro M (2010) Industrial oxidations with organocatalyst TEMPO and its derivatives. *Org Proc Res Develop* 14: 245-251. https://doi.org/10.1021/op900059x
19. Cunningham MF (2008) Controlled/living radical polymerization in aqueous dispersed systems. *Prog in Polym Sci* 33: 365-398. https://doi:10.1016/j.progpolymsci.2007.11.002
20. George B, Alex R. (2014) Stable free radical assisted scorch control in peroxide vulcanization of EPDM. Rub Sci 27: 135-145.
21. Joseph AM, Prasannan N, Madhusoodanan KN, Alex R, George B (2017) Mechanical devulcanization of carbon black filled natural rubber vulcanizates : Effect of crosslink density. *Rubber* Science 30: 181-192.
22. Joseph AM, Stable free radical assisted mechanical devulcanization of carbon black filled rubber vulcanizates, Ph.D. dissertation, Cochin University of Science and Technology, India, 2016
23. Horikx M M (1956) Chain scissions in a polymer network. *J Polym Sci* 19(93): 445-454. https://doi.org/10.1002/pol.1956.120199305
24. Verbruggen M A L, van der Does L, Dierkes W K, Noordermeer J W M (2016) Experimental validation of the Charlesby and Horikx models applied to devulcanization of sulfur and peroxide vulcanizates of NR and EPDM. *Rub Chem Technol* 89: 671-688. https://doi.org/10.5254/rct.16.83776
25. Seghar S, Asaro L, Aït Hocine N (2019) Experimental Validation of the Horikx Theory to be Used in the Rubber Devulcanization Analysis. *J Polym Envir* 27: 2318–2323. ttps://doi.org/10.1007/s10924-019-01513-z

26. Saiwari S, Dierkes WK, Noordermeer JWM (2013) Devulcanization of whole passenger Car tire material. *Kaut Gummi Kunst* 66: 20-25.
27. Levin VY, Kim SH, Isayev AI (1997) Vulcanization of ultrasonically devulcanized SBR elastomer. *Rub Chem Technol* 70: 120-128. https://doi:10.5254/1.3538412
28. Busfield KW, Aldabbagh F, Jenkeins DI (2011) The reactivity of sulfur-centred radicals towards TMIO (1, 1, 3, 3-Tetramethyl-2,3-dihydroisoindol-2-yloxyl). A new type of radical fragmentation reaction. Aus. J. Chem., 54: 313-318. https://doi.org/10.1071/CH01100
29. Joseph AM, George B, Madhusoodanan KN, Alex R (2017) Cure characteristics of devulcanised rubber: The issue of low scorch. *Rub Chem Technol* 90: 536-549. https://doi.org/10.5254/rct.17.83737

8

Rubber Recycling by Microwave Devulcanization: Effect of Ionic Liquids and Properties of Recycled Composites

Nourredine Aït Hocine[a,*], Lucia Asaro[b], Karima Aoudia[c], Said Seghar[d] and **Saïd Azem[c]**

[a]*INSA CVL, Univ. Tours, Univ. Orléans, LaMé, 3 rue de la Chocolaterie, CS 23410, 41034 Blois Cedex, France.*
[b]*Institute of Materials Science and Technology (INTEMA), University of Mar del Plata and National Research Council (CONICET), Av. Colón 10850, 7600 Mar del Plata, Argentina.*
[c]*LEC2M, Université Mouloud Mammeri de Tizi-Ouzou B.P 17, Hasnaoua 15000 Tizi-Ouzou, Algérie.*
[d]*RubberGreen Industrie, Route de Bavay 44, Frameries 7080, Belgium.*
**nourredine.aithocine@insa-cvl.fr*

8.1 Introduction

Storage and elimination of used tires and rubber wastes have become an urgent environmental concern. In fact, each year, hundreds of used tires and industrial rubber wastes are accumulated all over the world.

Rubber recycling is a challenge because the material is vulcanized. Vulcanization is a known thermo-chemical process which consists of incorporating sulfur into a mixture of rubber to generate a 3D cross-linked network, providing the desired properties [1]. During this treatment, Sulfur atoms are chemically linked to the rubber molecules to form Carbon-Sulfur-Carbon (C-S-C) and Carbon-Sulfur-Sulfur-...-Carbon (C-Sx-C) bonds. The vulcanization process is irreversible at standard atmospheric conditions of temperature and pressure. Therefore, the recycling of this kind of material is not easy.

One of the proposed methods to recycle rubbers is the devulcanization. This treatment consists of the breaking of C-S and/or S-S bonds without affecting the main chain Carbon-Carbon (C-C) of the polymer, i.e. without degradation of the material. Several devulcanization methods have been proposed during the last decades, including: thermomechanical [2-5] mechanochemical [6-8], high power ultrasound [9], biological [10-12], and microwave [13-15].

The microwave devulcanization process was developed by Goodyear in 1978 and successfully implemented by Novotny et al. [16]. It consists of exposing the waste rubber, for a short time, to controlled microwave irradiation. Actually, microwave heating is based on the interaction between an oscillating electrical field of microwaves and the molecular dipoles and/or charged ions existing in the material. It is generated by three main mechanisms: dipolar polarization, ionic conduction and interfacial polarization. Dipoles are heated by the dipolar polarization (dielectric heating), whilst samples with free ions or ionic species are heated by the ionic conduction mechanism. Among several techniques, microwave devulcanization presents some advantages that makes it one of the most promising methods for rubber recycling [13]. These advantages include the possibility of continuous processing, the easy adjustment of the process parameters, and the ability to treat large amounts of material. As the microwave radiation can penetrate the materials, heat can be generated throughout the volume of the samples and it is possible to achieve a rapid and uniform heating of thick materials [17]. This devulcanization method enables to apply large amounts of energy rapidly and uniformly to the rubber, using a little quantity of chemical agents or without using agents during the process [18], which makes it eco-friendly [19]. However, several authors have found that when the process parameters are not optimized, for example when using long microwave exposure periods, many structural changes and breaking of the main polymer chains take place leading to a partially degraded material [20-23]. This phenomenon can make it difficult to create new crosslinks in devulcanized ground tire rubber (DGTR) during its revulcanization. That is why the microwave process has to be deeply studied and analyzed.

Numerous studies regarding microwave devulcanization are recently presented in the literature. An investigation conducted by Goodyear society on the devulcanization of ethylene propylene diene rubber (EPDM) by microwaves showed that properties of the treated material were not considerably altered after treatment [14,16]. It was pointed out that although the EPDM is a non-polar rubber, the microwave energy was transferred in the whole volume of the material through the carbon black that was initially introduced as a reinforcing filler. Bani et *al.* [24] also performed similar studies in waste rubber and observed that the microwave devulcanization process was effective only beyond the heating temperature of 300 °C. Landini et *al.* [25] performed microwave devulcanization in bromobutyl rubber (BIIR). The power and time of treatment were identified as the most important factors to reach high devulcanization degrees. Scagliusi et *al.* [26] devulcanized chloroprene rubber by microwave irradiation. After the treatment, the obtained material was revulcanized. Authors observed that the microwave irradiation time affects the properties of the recycled material, hardness and fracture resistance were observed to decrease, whereas rupture strain increased. Vega et *al.* [27] studied microwave devulcanization combined with the effect of a devulcanization reagent, diphenyl disulfide (DPDS). The mentioned chemical agent was supposed to improve and facilitate the devulcanization process. Authors highlighted the beneficial effects of DPDS on the efficiency of microwave devulcanization. In another study, Hong et *al.* [19] compared the performance of microwave and ultrasonic devulcanization procedures, and concluded that the microwave technique was more effective than ultrasonic method.

Although the process has favorable results, one of the strategies developed to improve the microwave devulcanization efficiency and also to reduce the energy consumption consists of the impregnation of the waste rubber with a solvent before treatment. In fact, the impregnation of the waste rubber could promote heating when exposed to microwave radiation. Ionic liquids (ILs) are considered as an alternative to conventional organic solvents. They were already used to improve the dispersion of nanosized zinc oxide (ZnO) and silica in nitrile butadiene rubber (NBR) [28]. Actually, ILs are molten salts composed of a large variety of organic cations and organic/inorganic anions [29]. These liquids present several interesting properties, as: excellent solvent for several organic/inorganic materials, negligible vapor pressure, high thermal and chemical stability, non-flammability, high ionic conductivity, low viscosity and large electrochemical capacitance [29,30]. The ionic conduction mechanism is the most important effect in the heating of ILs by microwave; it increases their dielectric loss factor. For the aforementioned, ILs are potential solvents to be used in microwave devulcanization.

Once rubber is recycled, it has to be used in new materials or formulations. The addition of recycled rubber in polymer matrices and composite materials is a great option to consume a non-negligible amount of waste rubber. Epoxy resins are well known to have good properties (physicochemical, mechanical and thermal) [31]. To improve these properties, it is common to fill the matrices with particles including silica, rubber [32], minerals, carbon or glass fibers [33], and others [34-38]. He et al. [35] studied the influence of the addition of rubber (acrylic) particles on epoxy fracture behavior. Fracture results showed that toughness increased to a given value and then decreased as the concentration of rubber particles was increased, suggesting an optimum concentration for toughening. These results were supported by micrograph observations. Yagneswaran et al. [37] investigated the interaction of mechanochemical devulcanized GTR powder with an epoxy matrix. Structural and thermal analysis were performed by several techniques (SEM-EDX, FTIR, DSC and TGA). Results mainly showed an increase in thermal stability of the epoxy with increasing GTR content. Good dispersion and interaction of GTR in the epoxy matrix were found. A dominant catalyzing effect of GTR on the curing reaction was highlighted since it was observed that the presence of GTR promotes the curing of epoxy resin. In the work reported by Piszczyk et al. [38], flexible polyurethane foams were modified with two kinds of GTR, untreated and thermomechanically reclaimed GTR. The impact of the rubber treatment on the mechanical and thermal properties of the resulting polyurethane composites was analyzed. It was found that the incorporation of rubber particles into polyurethane foam caused significant changes in the structure of the matrix. It particularly enhanced its cross-link density and had positive influence on its thermal stability and its compressive strength. However, to obtain materials with improved properties, it seems useful to perform devulcanization treatment before adding the rubber to the epoxy matrix [39-41].

The aim of this chapter is to describe and analyze the effect of microwave devulcanization process on the effectiveness of waste rubber recycling. Thus, Styrene butadiene rubber (SBR) and waste ground tire rubber (GTR) were devulcanized using microwave irradiation and the performance of the devulcanization treatment were analyzed in terms of crosslink scission. The effect of ionic liquid ([Pyrr][HSO4]) as a medium for microwave devulcanization is studied. Finally, the incorporation of the devulcanized rubber in an epoxy matrix is investigated and the changes in mechanical properties of the obtained material are discussed.

8.2 Materials

Two types of rubber were investigated in this study, one is a styrene butadiene rubber "SBR" which is an industrial rubber, and another is ground tire rubber "GTR" coming from waste tires. For circumstantial facts, each of the two rubbers is studied in a given laboratory, at a certain period, that is why the achieved characterizations are not the same for both rubbers.

The SBR was a mixture of two grades of SBR: 50 phr of SBR1500 with a styrene content of 23.5% and 50 phr of SBR1900 with a styrene content of 75%. The rubber was vulcanized at 170 °C for 7 min, under compression in a hydraulic molding press (Agila, Kortrijk-Bissegem, Belgium), and then ground to particles with size ranged from 1 to 4 mm. The glass transition temperature, T_g, was -48 °C for the raw SBR and at -43 °C for the vulcanized SBR. The increase of T_g from -48 °C to -43 °C suggested the successful vulcanization of the studied material. An ionic liquid (IL) "pyrrolidinium hydrogen sulfate [Pyrr][HSO_4]", which is a green solvent, was used as devulcanization medium during the SBR treatment. The C_p of [Pyrr][HSO_4] was constant up to 120 °C and it was proved that was three times smaller than that of distilled water, which suggests that it presents high ability to convert microwave energy in heat. The onset degradation temperature of the IL measured by thermogravimetry was ~289 °C, proving that this IL has a competitive thermal stability. More details about the rubber and IL characterization are presented in Seghar et al. [42].

The GTR was kindly provided by Phenix Technologies Company (Sancheville 28800, France). It is originated from tire sidewalls and the material particle size is ranged from 250 to 500 μm.

Composites were prepared by adding 10 wt.% of GTR and DGTR particles to a mixture of a Bisphenol-A-(Epichlorhydrine) epoxy resin (molecular weight max 700) and bisphenol-F epoxy resin (molecular weight ≤ 700/C13/C15 Alkyl glycidyl ether), provided by SOLOPLAST (Germany). The hardener used has the following composition: 3 aminomethyl-3,5,5-trimethyl-cycloexyl amine, Bisphenol A-epoxy resin (molecular weight < 700) and alcohol benzyl. Unfilled material was also prepared by mixing the epoxy resin with 45 wt.% of hardener, during 15 minutes, in an internal propeller mixer.

8.3 Experimental methods

8.3.1 Devulcanization process

Domestic commercial microwave was used to perform rubber devulcanization. For that, rubber (SBR or GTR) was placed in a 250 ml beaker exposed to microwave irradiations during different times t. In the case of SBR the microwave power, P, was ranged from 350 to 900 Watts. In the case of GTR the magnetron power was 656 Watts. The corresponding microwave specific energy E (J/kg) was evaluated from Equation (1):

$$E = \frac{P.t}{M} \tag{1}$$

In order to improve the efficiency of the rubber microwave devulcanization, the SBR powder was mixed with 10 wt.% of IL [Pyrr][HSO_4]. The mixture was left to stand for 1 hour, prior to microwave exposure. The SBR soaked in [Pyrr][HSO4] is denoted SBR-P in this chapter. SBR and GTR microwave devulcanization conditions are summarized in Table 8.1.

Table 8.1 *Material coding and devulcanization conditions*

Material	Time of treatment (s)	Microwave power (W)	Energy (Wh/kg)
SBR	0	0	0
SBR-110	30	656	110
SBR-220	60	656	220
SBR-330	90	656	330
SBR-440	120	656	440
SBR-P	0	0	0
SBR-P-110	30	656	110
SBR-P-220	60	656	220
SBR-P-330	90	656	330
SBR-P-440	120	656	440
GTR1	15	350	0.15
GTR2	15	500	0.21
GTR3	15	650	0.28
GTR4	15	750	0.32
GTR5	15	900	0.38
GTR6	60	350	0.60
GTR7	60	500	0.85
GTR8	60	650	1.11
GTR9	60	750	1.28
GTR10	60	900	1.54

8.3.2 Evaluation of rubber devulcanization degree

The soluble fractions of the virgin and devulcanized rubbers were determined by extraction in a Soxhlet extractor for 24 h in acetone to remove low-molecular weight polar substances like remains of accelerators and curatives. This operation was immediately followed by an extraction for 72 h in toluene to remove the non-polar components such as oil and non-crosslinked polymer residues or soluble polymer chains released from the network by the devulcanization process. The gel fraction was weighed. Then the samples were dried in a vacuum oven at 50 °C for 24 h and finally weighed.

The soluble fraction S of the material was evaluated using Equation (2):

$$S\ (\%) = \frac{M_i - M_f}{M_i} \times 100 \qquad (2)$$

where M_i and M_f are the SBR rubber weight before and after extraction operation respectively.

The apparent crosslink density v_c was determined according to Equation (3) [43]:

$$v_c = \frac{v_r + \chi\, v_r^2 + \ln(1 - v_r)}{v_s\left(0.5\, v_r - v_r^{\frac{1}{3}}\right)} \qquad (3)$$

with:

$$v_r = \frac{m_r/\rho_r}{\frac{m_r}{\rho_r} + \frac{m_s}{\rho_s}} \qquad (4)$$

In Equations (3) and (4), v_r is the volume fraction of the polymer in the swollen specimen, v_s is the molar volume of the solvent, χ is the Flory-Huggins polymer solvent interaction parameter, m_r is the mass of the dry rubber, m_s is the mass of the solvent absorbed by the rubber, ρ_r and ρ_s are the densities of the dry rubber and solvent respectively.

The devulcanization degree was calculated according to the standard ASTM D6814-02, using Equation (5):

$$devulcanization\ (\%) = \left(1 - \frac{v_f}{v_i}\right) \times 100 \qquad (5)$$

where ν_i and ν_f are the crosslink densities of the samples before and after devulcanization, respectively.

It is known that a successful devulcanization process results in crosslink scission, however, actual processes also cause breaking of the main chain of the polymer. The devulcanization quality was evaluated using Horikx's theory [44]. This theory correlates the rubber soluble fraction and the relative decrease in crosslink density. When only main-chain scission takes place, the relative decrease in the apparent crosslink density is given by Equation (6):

$$1-\frac{v_f}{v_i}=1-\left[\frac{(1-\sqrt{S_f})^2}{(1-\sqrt{S_i})^2}\right] \tag{6}$$

where S_i and S_f are the soluble fraction of the untreated vulcanizate and reclaimed vulcanizate respectively, v_i is the crosslink density of the untreated vulcanizate and v_f is the crosslink density of the reclaimed vulcanizate. When pure crosslink scission occurs, the soluble fraction is related to the relative decrease in crosslink density by Equation (7):

$$1-\frac{v_f}{v_i}=1-\left[\frac{\gamma_f(1-\sqrt{S_f})^2}{\gamma_i(1-\sqrt{S_i})^2}\right] \tag{7}$$

where γ_i and γ_f are the average numbers of crosslinked units per chain before and after treatment respectively. These parameters were determined from the relation between soluble fraction S and the number of crosslinked units per chain γ, given by Charlesby [45, 46] under the following equation (Equation 8):

$$S=\frac{1}{(1+\gamma-\gamma s)^2} \tag{8}$$

The Horikx diagram allows to carry out an accurate analysis of the performed devulcanization. In fact, in this chart, the devulcanization degree and the soluble fraction of the rubber are related, so that both the degree and quality of devulcanization, in terms of crosslink against main chain scission, can be qualitatively estimated. Theoretical Horikx curves are presented in Figure 8.1. The analysis consists of considering that the experimental points positioned near the lower curve (dash line) correspond to a more selective sulfur scission, and those located near the upper curve (solid line) correspond to main chain scission and polymer degradation. The points between the two curves undergo reclaiming, a combination between devulcanization (sulfur bond scission) and degradation (main chain scission).

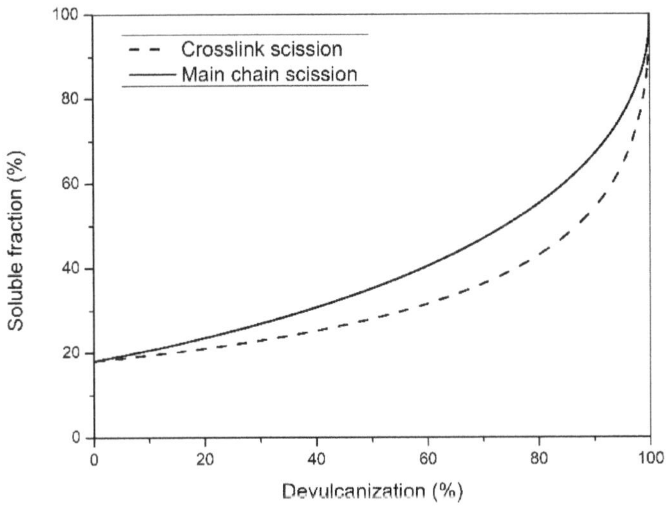

Figure 8.1 *Theoretical Horikx curve for the studied SBR rubber.*

8.3.3 Mooney viscosity

The viscosity of the devulcanized SBR rubber was measured with a Mooney Viscometer from Gibitre Instruments (Bergamo, Italy), according to the ASTM D1646 Standard. The test consists of the slow and continuous rotation of a disk in a heated chamber filled with rubber, for a given time. The resistance to this rotation induced by the rubber is measured, in arbitrary torque units, as the Mooney viscosity of the tested material. In this study, samples were preheated during 1 min, and the viscosity was measured after 4 min of testing, at 100 °C. The results are presented as ML (1 + 4) 100 °C. At least, 3 measurements were carried out for each material.

8.3.4 Mechanical characterization of epoxy-based composites

Mechanical properties of epoxy/GTR and epoxy/DGTR composites were determined by means of three point bending tests, using a Zwick Z20 testing machine. Parallelipipedic specimens of 80 mm x

10 mm x 4 mm were cut from the plates of material using a diamond blade. Distance between sills was fixed at 64 mm. Tests were carried out at ambient temperature and at a machine crosshead speed of 2 mm min^{-1}. Stress-strain curves were recorded up to total breaking of the specimens. Young modulus, strength at break and strain at break of the neat matrix and composites were deduced from these stress-strain curves.

8.3.5. Scanning electron microscopy

Morphological analysis of GTR was achieved by Scanning Electron Microscopy performed in a SEM, Philips ESEM XL30, equipped with a tungsten filament and an EDS-X (Energy Dispersive Spectrometry) system. The working accelerating voltage was 20 KV. GTR powder was also analyzed with EDX technique in order to identify its components and evaluate their proportion. Also, post-mortem fracture surfaces of bending samples were observed to aid in the understanding of mechanical results.

8.4 Results and discussion

8.4.1 SBR

Figure 8.2 presents the crosslink densities (Equation (3)) and the soluble fraction (Equation (2)) of SBR and SBR-P samples, as a function of the microwave energy. The crosslink density decreases slightly and slowly with the microwave energy up to 220 Wh/kg, critical value beyond which the crosslink density decreases significantly for the SBR-P rubber. The soluble fraction increases with the microwave energy up to 440 Wh/kg in both cases, with more significant increase for the SBR-P rubber beyond the critical value of 220 Wh/kg. These results confirmed that the liquid ionic [Pyrr][HSO_4] favored the microwave devulcanization process.

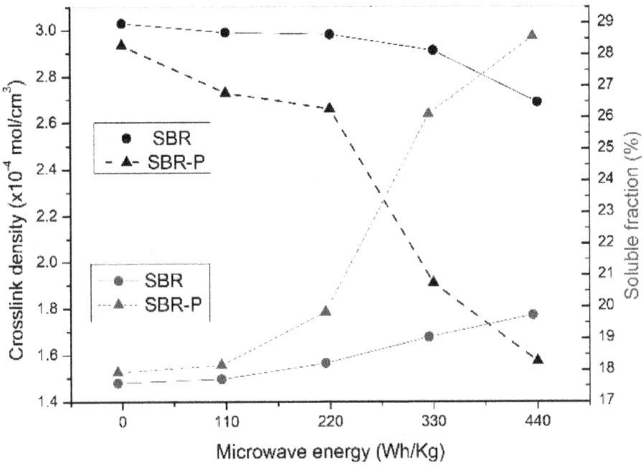

Figure 8.2 *Crosslink density and soluble fraction of SBR ground rubber with and without ionic liquid as a function of the microwave energy.*

Devulcanization fractions are represented in Figure 8.3 as a function of the average temperature generated in the rubber powder. The temperature was measured with an infrared sensor during microwave irradiation. The rubber powder was stirred during the devulcanization treatment. The experimental data of both SBR and SBR-P are fitted by power laws. It is apparent that for a given microwave energy value, average temperature reached by SBR-P was much higher than the one reached by dry SBR.

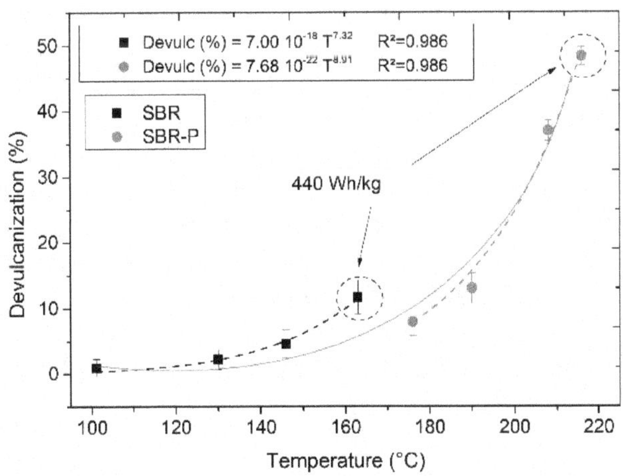

Figure 8.3 *Devulcanization of SBR ground rubber with and without ionic liquid as a function of the temperature reached during microwave treatment.*

To further analyze the rubber devulcanization by microwave process, the distribution of the treatment temperature was recorded with an infrared camera FLIR JADE-MWIR, in both SBR and SBR-P, right after microwave treatment. The obtained cartographies are shown in Figure 8.4, as an example for a treatment energy of 40 Wh/kg. The temperature distribution is more homogeneous in the SBR-P powder than in the dry SBR powder. This result proved that microwave irradiation produced efficient internal heating in IL medium, resulting in higher and homogeneous average temperature in the treated material.

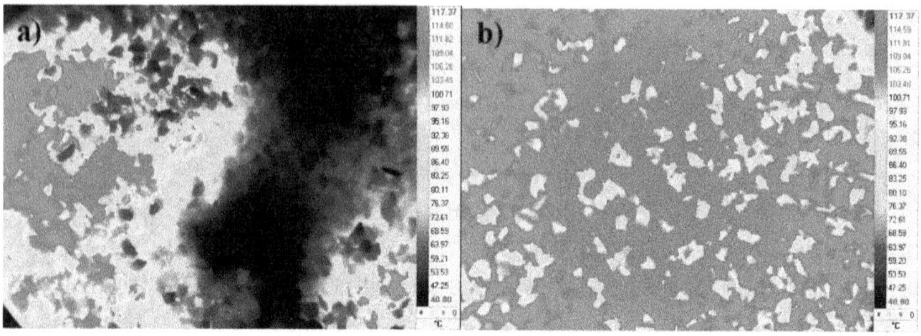

Figure 8.4 *Temperature of the SBR ground rubber (a) without ionic liquid and (b) with ionic liquid, measured with a thermal camera.*

Figure 8.5 (a) and (b) show theoretical curves of Horikx given by Equations (6) and (7), and the experimental points corresponding to the soluble fractions of SBR and SBR-P, as a function of devulcanization degree evaluated using Equation (5). The experimental data fits well with the Horikx's curve corresponding to the crosslink scission (Equation (7)), suggesting that the devulcanization process was successful i.e. that is occurred by rather selective breakage of crosslinks. However, percentage of this devulcanization was higher as the microwave energy was increased, for SBR and SBR-P but more significantly for SBR-P. For instance, for the same microwave energy of 440 Wh/kg, apparent devulcanization of SBR-P reached ~50%, which is four times greater than the value obtained for SBR without IL.

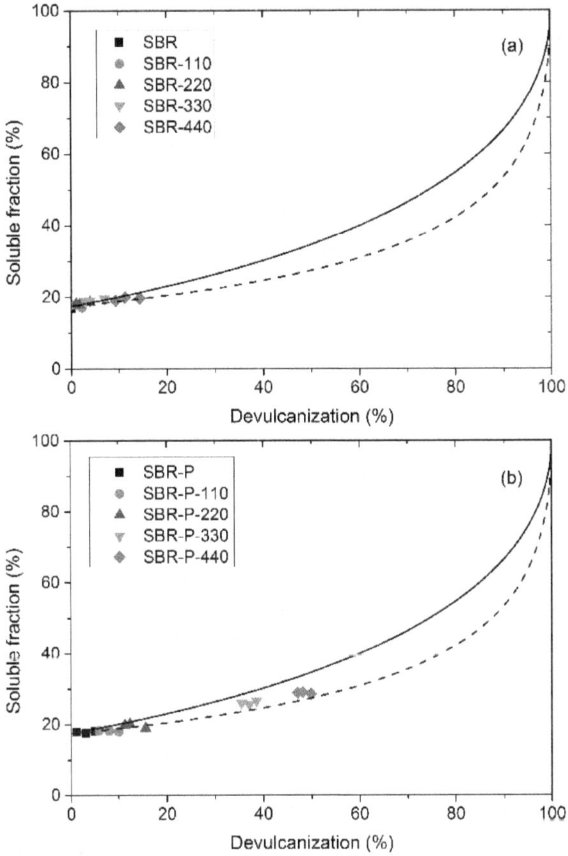

Figure 8.5 *Horikx diagram for SBR ground rubber (a) without ionic liquid and (b) with ionic liquid.*

Gums were processed with the devulcanized SBR and SBR-P materials, treated with microwave energy of 440 Wh/kg. A laboratory two-roll mill was used with the friction ratio of 1.3, at temperature of 50 °C and mixing time of 10 min for each blend. The soluble fractions and the devulcanization proportions of the obtained materials are reported in the Horikx diagram in Figure 8.6. Data of both microwave treated SBR powders are also plotted in this figure. It can be observed that experimental points of gums shifted from 13% to 38% and from 50% to 65% of devulcanization for SBR and SBR-P respectively, while remaining well fitted by the main-chain scission theoretical curve of Horikx. This result indicated that the mechanical shearing of rubber favored the devulcanization process, suggesting that combining mechanical loading with microwave energy and IL could be a most profitable procedure allowing an optimal devulcanization of rubbers.

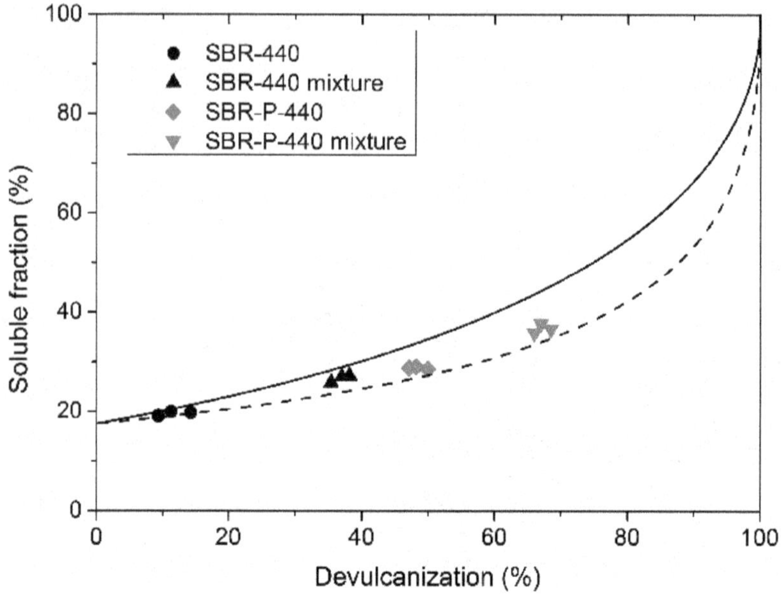

Figure 8.6 *Horikx diagram for SBR powder and SBR mixture.*

Figure 8.7 shows the Mooney viscosity and Mooney relaxation of virgin and reclaimed rubber obtained after ground rubber mixing, confirming the earlier results. In fact, it is well known that more is the degree of devulcanization, lower is the Mooney viscosity [47]. The specimen generated from the ground rubber previously treated by microwave in the presence of IL exhibited Mooney viscosity more close to that of the virgin rubber, proving high devulcanization of this specimen.

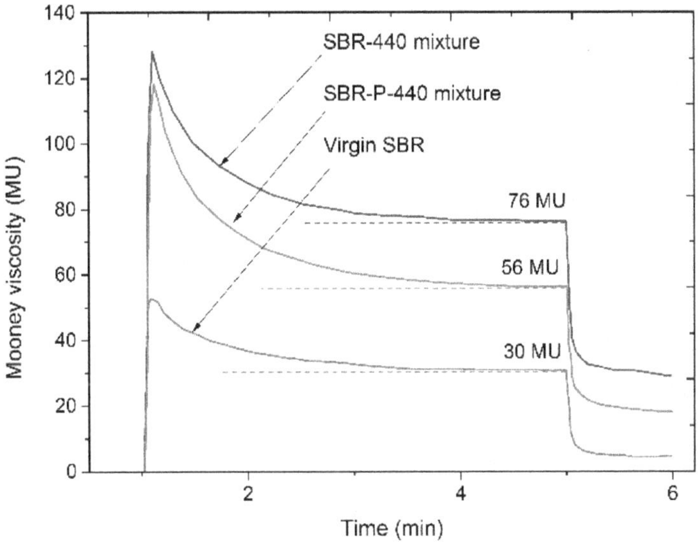

Figure 8.7. *Mooney viscosity ML (1+4) 100°C for unvulcanized raw rubber and different recycled rubbers (MU is Mooney unit of viscosity: 1 MU = 0.083 N·m).*

8.4.2 GTR

Characterization of devulcanized GTR

Figure 8.8 shows FTIR spectra of DGTR exposed to microwaves at different power values for 15 seconds along with the untreated GTR spectrum. Main transmission peaks of GTR spectrum are listed in Table 8.2, with their attributions, in accordance with the literature [48, 49]. Figure 8.8 shows that the material undergoes many structural changes during the devulcanization treatment. Indeed, the occurrence of new chemical links and the disappearance of some peaks can be observed. To better analyze these chemical links and functions, zooms in specific zones of Figure 8.8 were performed and depicted in Figure 8.9.

Table 8.2 *Main FTIR characteristic peaks of the GTR*

Wavenumber (cm^{-1})	Attribution
3250	O-H
3150-2850	C-H
1870-1650 (1750)	C=O
1680-1600 (1610)	C=C
1340	C-H deformation
1300-1000 (1270)	C-O
1300-1000 (1180)	S-O-C
1080	C-O-C
1280-1150 (1045)	S=O
980	C-C /C-O
880	C-O
820	C-H deformation
775-570 (730, 540, 530)	C-S
485	S-S

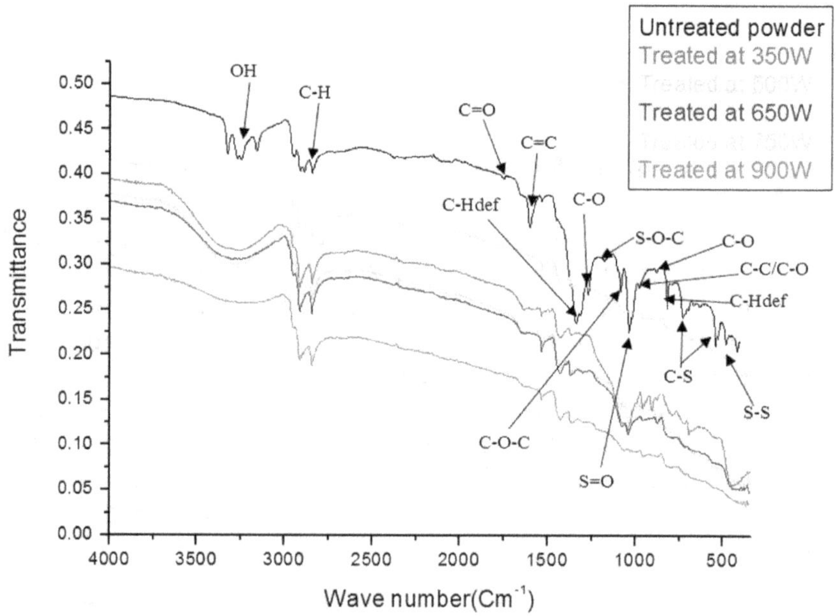

Figure 8.8 *Superposition of the FTIR spectra of GTR and DGTR treated at different powers for 15 seconds.*

Figure 8.9a shows a zoom in the range from 3750 to 3000 cm^{-1} representing the hydroxyl functionality. Three small initial peaks can be observed in the untreated rubber around 3250 cm^{-1}, and a wide band for the DGTR treated at 350, 500 and 650 W, corresponding to the formation of hydroperoxides (–OOH). This large band blunts for DGTR treated with a power of 750 and 900 W, confirming the well-known thermal instability of such hydroperoxides. Peak of C-H link located in the range from 3000 to 2800 cm^{-1} (Figure 8.9b) does not show differences between GTR and DGTR, suggesting that the hydrocarbon backbone of GTR was not degraded by microwave irradiation, even at high power levels. The region of the FTIR spectrum corresponding to the carbonyl, between 1800 and 1500 cm^{-1}, shows that important modifications occurred during the devulcanization treatment, as shown in Figure 8.9c. Indeed, peaks corresponding to C=C and C=O links vanish in DGTR spectra, whatever the treatment microwave power. These results agree with those reported in literature that conclude long microwave exposure periods result in many structural changes and in breaking of the main polymer chain [20] [21] [22]. The rupture of C=C and C=O bonds would lead respectively to lower macromolecular mass fractions and to the formation of, among others, sulfur dioxide (SO_2).

In summary, results exhibited in Figures 8.9b and 8.9c suggest that the principal macromolecular backbone of the rubber was only partially degraded by the microwave devulcanization process: C-H bonds remained intact but C=C links were partially broken. Figure 8.9d shows important modifications of the FTIR spectra in the range from 1500 to 750 cm^{-1}. The bands corresponding to C-O, S-O-C, C-O-C, S=O, C-C bonds undergo a strong decrease for the treatment power of 350, 500 and 650 W and form a single band around 1050 cm^{-1} assigned to sulfonyl groups such as sulfone -R-SO_2-R- and sulfoxide R-S=O [48]. These bands disappear for the treatment power of 750 and 900 W, suggesting that the elements containing these bonds reacted with oxygen and hydrogen to form gas. These results confirm that an oxidative degradation of principal chain occurred simultaneously with the rubber devulcanization during the microwave treatment. This oxidative phenomenon seems to be more significant at high microwave powers.

FTIR spectra zoomed in the range from 750 to 400 cm^{-1} revealed a strong decrease of peak heights of C-S and S-S bonds, for the treatment power of 350 W (Figure 8.9e). These bands disappeared in the DGTR treated with the higher microwave powers, confirming the breaking of a large proportion of the vulcanization links i.e. the success of the material devulcanization.

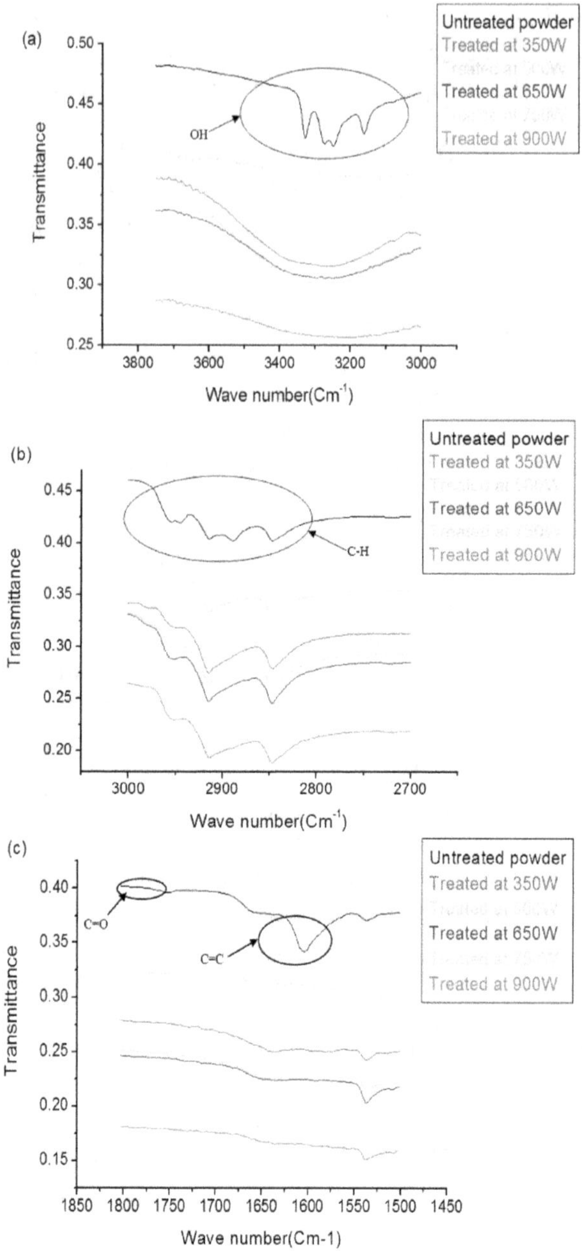
(a)
Untreated powder
Treated at 350W
Treated at 650W
Treated at 900W
OH
Transmittance
0.50
0.45
0.40
0.35
0.30
0.25
3800
3600
3400
3200
3000
Wave number(Cm-1)
(b)
Untreated powder
Treated at 350W
Treated at 650W
Treated at 900W
C-H
Transmittance
0.45
0.40
0.35
0.30
0.25
0.20
3000
2900
2800
2700
Wave number(Cm-1)
(c)
Untreated powder
Treated at 350W
Treated at 650W
Treated at 900W
C=O
C=C
Transmittance
0.40
0.35
0.30
0.25
0.20
0.15
1850 1800 1750 1700 1650 1600 1550 1500 1450
Wave number(Cm-1)

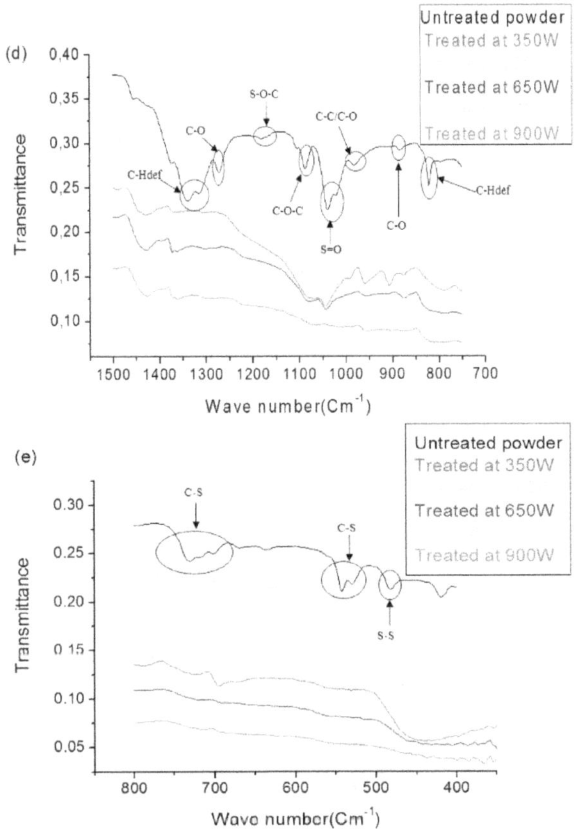

Figure 8.9 *Zooms of FTIR spectra of Figure 8.3 in the range: (a) from 3750 to 3000 cm^{-1}; (b) from 3000 to 2700 cm^{-1}; (c) from 1800 to 1500 cm^{-1}; (d) from 1500 to 750 cm^{-1}; (e) from 800 to 350 cm^{-1}.*

FTIR spectra of DGTR exposed to microwaves for 60 s, at different power values along with GTR spectrum, are shown in Figure 8.10.

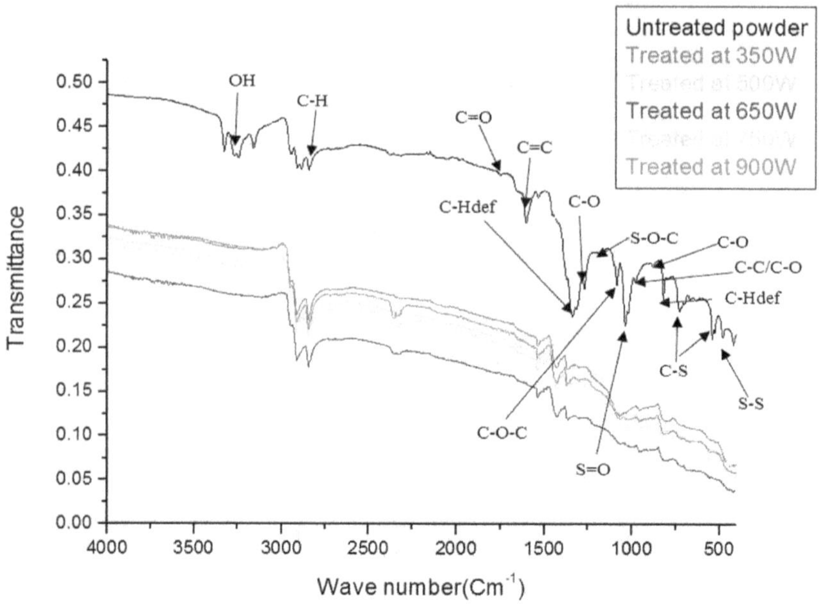

***Figure 8.10** Superposition of the FTIR spectra of GTR and DGTR treated at different powers for 60 seconds.*

Figure 8.11 shows enlargements and superposition of various zones for better analysis and understanding of physical-chemical occurred during the devulcanization treatment. Results exhibited similar trends to those corresponding to treatments of 15 seconds. Indeed, Figure 8.11a shows the disappearance of the hydroxyl group for all the microwave power values applied in the devulcanization treatment. The peak of C-H stretching vibration does not show differences between the GTR and DGTR for all microwave power values of treatment, as depicted in Figure 8.11b, meaning that the hydrocarbon backbone is not degraded, whatever the microwave power level applied. The disappearance of the bands assigned to the C=C and C=O vibrations is observed in the region of the carbonyl located between 1800 and 1500 cm^{-1} (Figure 8.11c), whatever the microwave power value of the treatment, which argues in favor of the rubber degradation. Figure 8.11d shows significant modification of the FTIR spectra in the range from 1500 to 750 cm^{-1}, mainly characterizing the C-O, S-O-C, C-C and S=O bonds. It is easily seen that these bands disappear or, at least, undergo a strong decrease. This indicates that simultaneously with the rubber devulcanization, occur two phenomena during the treatment: degradation which is revealed by the breaking of C-C bonds and deoxidation by eliminating oxygen. Figure 8.11e shows the disappearance of FTIR C-S and S-S bonds in the range from 750 to 400 cm^{-1}, for all the applied microwave powers, suggesting the success of the devulcanization treatment. In conclusion, FTIR analysis shows that microwave irradiations of GTR induced a double effect: devulcanization and partial degradation of the ground rubber. This phenomenon is commonly called regeneration.

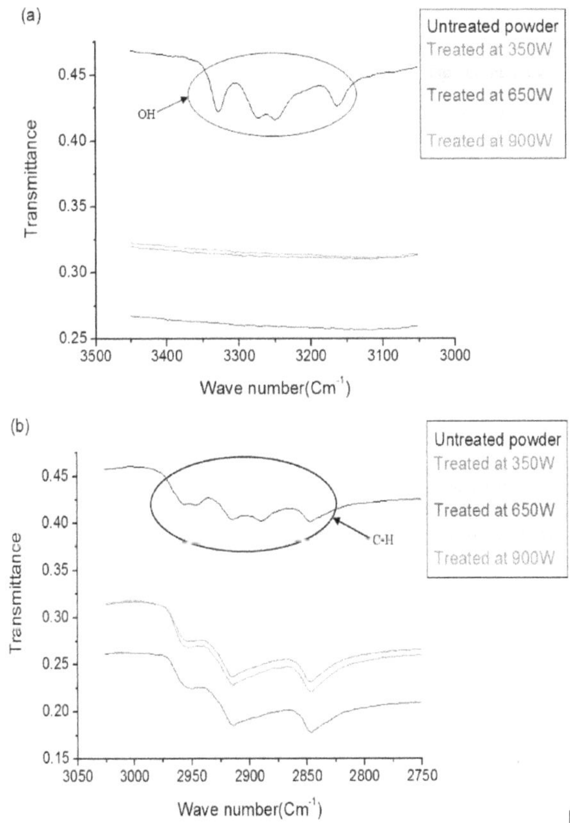

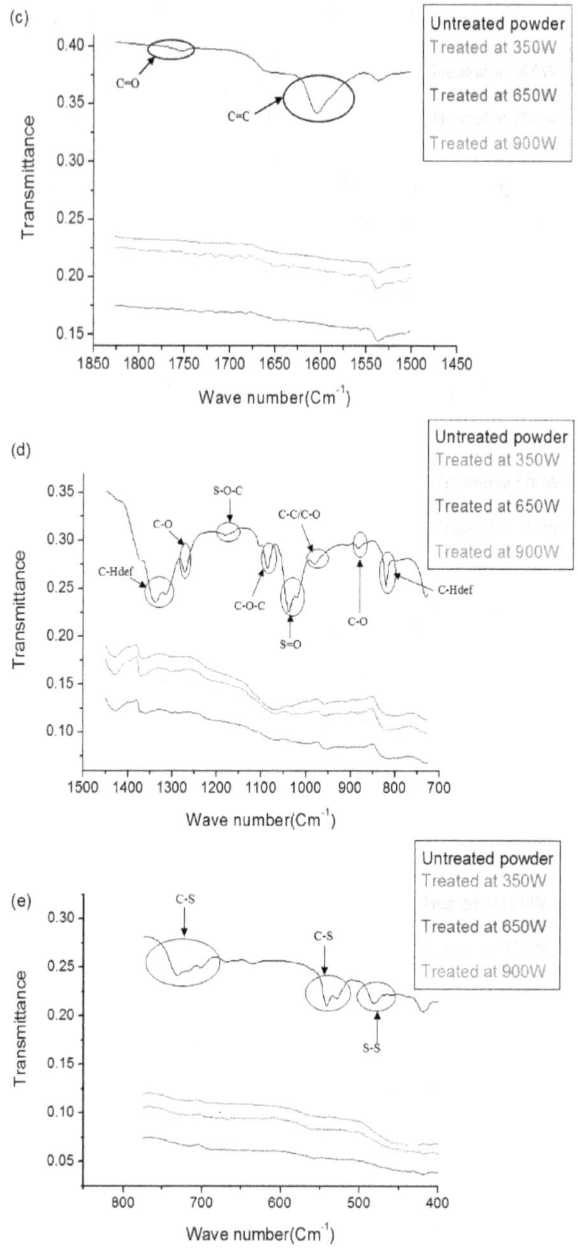

***Figure 8.11** Zooms of FTIR spectra of Figure 8.5 in the ranges: (a) from 3450 to 3050 cm^{-1}, (b) from 3025 to 2750 cm^{-1}, (c) from 1825 to 1500 cm^{-1}, (d) from 1450 to 750 cm^{-1}, (e) from 800 to 400 cm^{-1}.*

Swelling values obtained from the swelling test, are shown in Figure 8.12 as a function of microwave energy. It can be clearly seen that the swelling degree sharply increases with increasing the treatment specific energy, reaching 63.22 % for 1.389 kJ kg^{-1}. Beyond that critical energy level, it tends to stabilize i.e. specific energy should be highly increased to obtain a significant gain in the swelling value. Devulcanization degree is also reported in Figure 8.12 as a function of microwave energy. Devulcanization degree exhibits the same trend as swelling degree, confirming existence of a relationship between the two quantities. In fact, it is found that devulcanization degree increases with energy employed in microwave treatment up to 1.389 kJ kg^{-1}. Beyond this threshold value, only a little improvement in devulcanization degree is observed, since it approaches asymptotically to 95 %. These results suggest that a weak energy is enough to accomplish the devulcanization or regeneration process. Moreover, the high measured values of devulcanization support FTIR outcomes, i.e. a decrease of rubber crosslink density or a strong regeneration of DGTR.

Based on these observations, DGTR particles treated with energy levels smaller than 2.16 kJ kg^{-1}, i.e. treated during 15 s, where chosen to be used as fillers in the epoxy matrix.

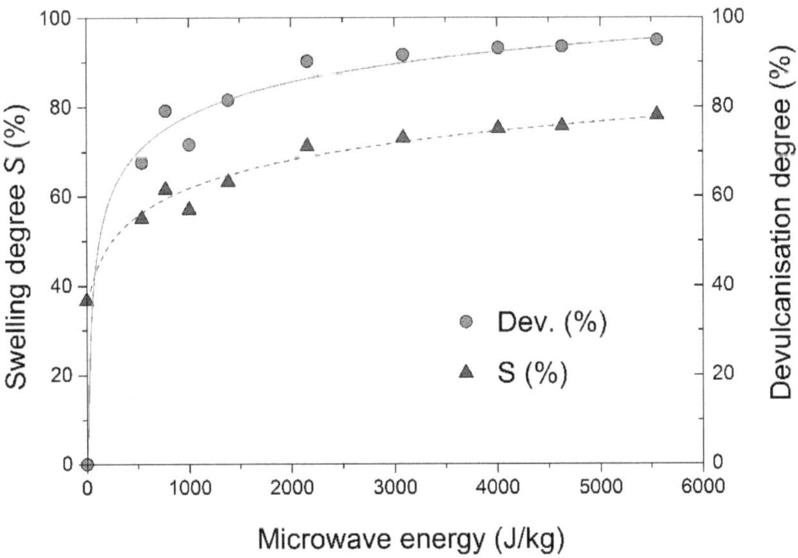

Figure 8.12 *Swelling degree and devulcanization degree as functions of treatment energy.*

Properties of GTR and DGTR epoxy composites

Stress at break, strain at break and Young's modulus of epoxy obtained from three-point bending tests are respectively 21.35 MPa, 0.044 and 723.73 MPa. Composites' mechanical properties also obtained from these tests are reported in Figure 8.13 as a function of the treatment microwave energy. It appears that incorporation of GTR or DGTR in the epoxy matrix resulted in a decrease of its mechanical properties. Nevertheless, despite experimental discrepancies, particularly visible in the stress at break values (Figure 8.13), results show that microwave treatment of GTR induced a relative improvement in the flexural stress and strain at break of the composites (Figure 8.13). This suggests that DGTR particles' surfaces are more chemically reactive and have better adhesion with epoxy matrix than GTR particles. This interaction seems to be higher for DGTR powder treated with 1157 J kg^{-1} i.e. at 750 Watts during 15 s, which leads to an increase of ~20 % in flexural stress and strain at break of the epoxy/DGTR composite, compared to the epoxy/GTR composite. However, such increase is less significant for composite filled with DGTR particles treated with 1389 J kg^{-1} i.e. at 900 Watts during 15 seconds, which could be related to a partial degradation of the DGTR treated with high microwave energy value.

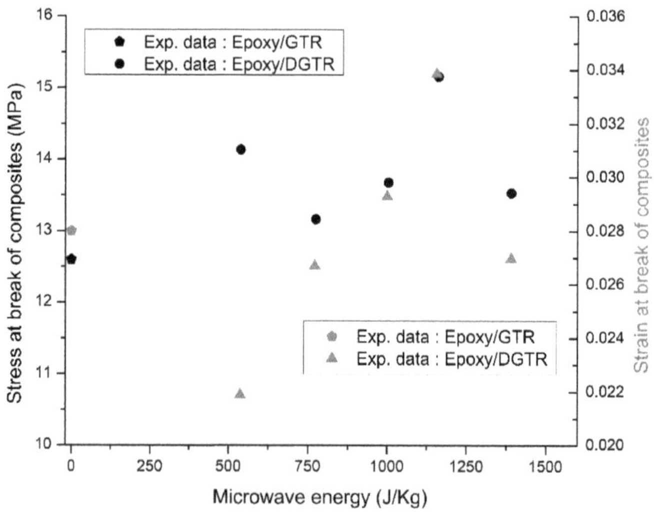

Figure 8.13 *Stress at break and strain at break of composites as a function of treatment energy.*

Generally, when a rubber modifier is added to a thermoset resin, its toughness increase is accompanied by a significant decrease in Young's modulus due to the less rigidity of elastomers. In our case, a significant decrease in composites Young's modulus is observed with the increase of the treatment microwave energy, which may be explained by the rubber rigidity diminution with increasing devulcanization degree. One can expect that this affects the composites Young's modulus more significantly as the forces at the epoxy/DGTR interface are stronger. It is also interesting to note that the decrease of the composites Young's modulus follows a linear regression as a function of the treatment microwave energy (Figure 8.14).

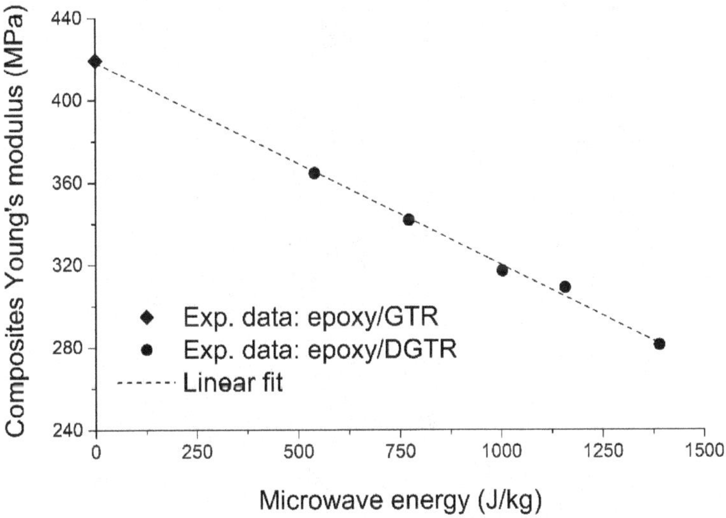

Figure 8.14 *Composites Young's modulus as a function of treatment energy.*

SEM observations of fracture surfaces were performed to analyze the effect of the microwave treatment on composites' microstructure, particularly on the epoxy/rubber interface. Micrographs are shown in Figure 8.15. A noticeable gap is highlighted in Figure 8.15a between the untreated GTR and the matrix, suggesting a weak adhesion of the GTR to the epoxy matrix. This led to an easy pull of the GTR from the matrix during flexural tests, in accordance with composites mechanical properties discussed above. Failure surfaces of composites with DGTR (Figures 8.15b–f) show more continuous interfaces than that obtained with GTR. This effect seems to be more marked as microwave energy increases, which is probably due to good physicochemical interactions generated at the interfaces by functional groups created on DGTR surface during devulcanization treatment. Such interactions result in an increase of ultimate stress and strain of epoxy/DGTR composites, with increasing microwave energy up to 1157 J kg^{-1}. Beyond this limit, mechanical properties decrease despite the apparent better interface between epoxy and DGTR (Figure 8.15f). This decrease is probably related to the pronounced degradation of DGTR which prevails over the devulcanization effect.

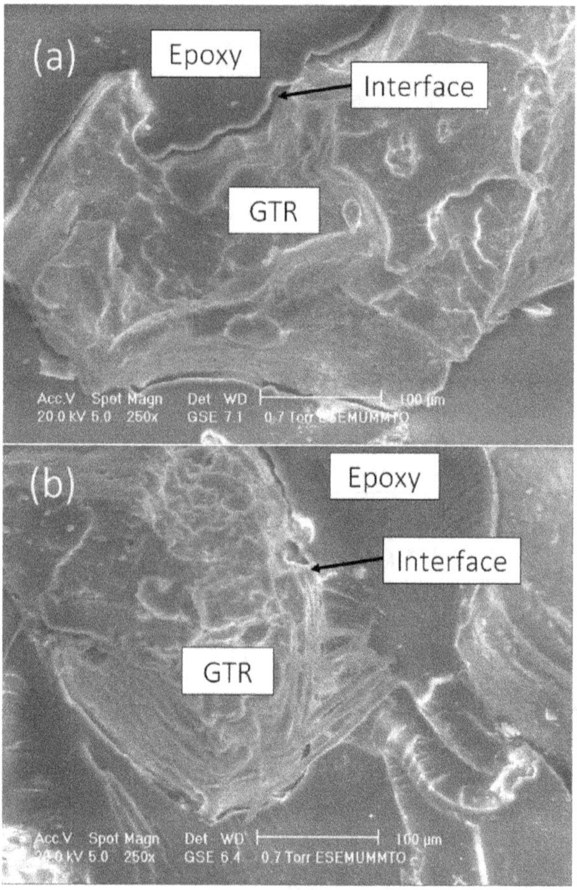

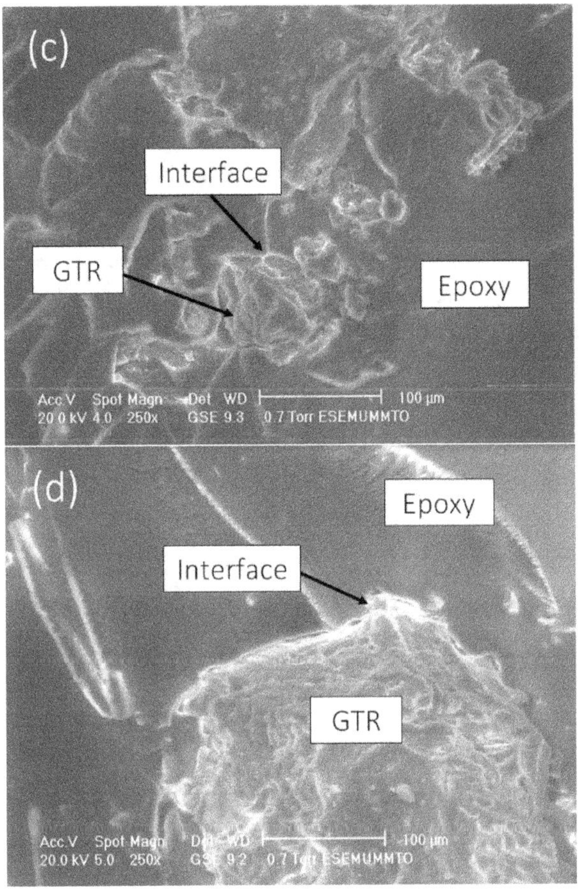
(c)
Interface
GTR
Epoxy
Acc.V Spot Magn Det WD 100 µm
20.0 kV 4.0 250x GSE 9.3 0.7 Torr ESEMUMMTO
(d)
Epoxy
Interface
GTR
Acc.V Spot Magn Det WD 100 µm
20.0 kV 5.0 250x GSE 9.2 0.7 Torr ESEMUMMTO

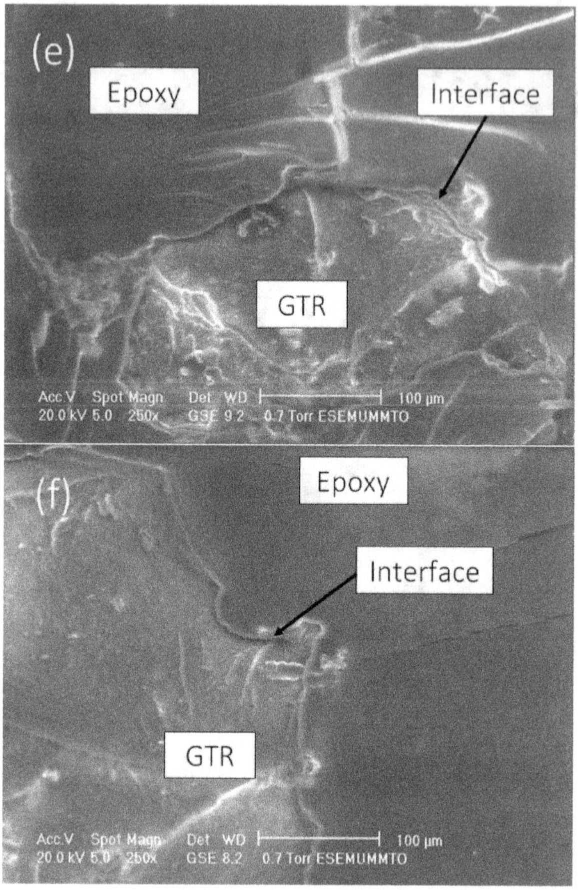

Figure 8.15 *SEM micrographs of failure surfaces of epoxy composites filled with: (a) GTR; (b) DGTR treated with 540 J kg^{-1}, (c) DGTR treated with 772 J kg^{-1}, (d) DGTR treated with 1003 J kg^{-1}, (e) DGTR treated with 1157 J kg^{-1}, (f) DGTR treated with 1389 J kg^{-1}.*

8.5 Conclusion

Devulcanization of styrene butadiene rubber (SBR) and Ground Tyre Rubber (GTR) by microwave irradiations has been studied. The effect of IL (pyrrolidinium hydrogen sulfate [Pyrr][HSO_4]) on this devulcanization was particularly analyzed. High ability of IL to convert microwave energy in heat has been proved. The devulcanization evaluated from equation (5) was higher as the microwave energy was increased and when the SBR was impregnated in IL, confirming again the positive role of IL in the devulcanization treatment. Analysis of the temperature distribution through the ground rubber proved that microwave irradiation produced efficient internal heating when microwave energy was coupled with the IL. The experimental data fitted well with the Horikx's curve corresponding to the crosslink scission, suggesting that devulcanization process occurred in all microwave treated materials by rather selective breakage of crosslinks. The magnitude of this devulcanization was higher as the microwave energy was increased and when the rubber was imbibed in IL.

When the GTR is concerned, FTIR analysis combined with swelling measurements of the virgin and devulcanized powders were performed to check the effectiveness of the devulcanization treatment. Results showed that microwave irradiations induced a so-called regeneration phenomenon i.e. the combined devulcanization and degradation of the rubber material. A dependence between devulcanization degree and microwave energy was observed, concluding that the maximum energy to be used should be 1.389 kJ kg^{-1}, since higher energies did not produce major changes in the degree of devulcanization. Composites were obtained by mixing GTR or DGTR particles in an epoxy matrix. Epoxy composites filled with DGTR had better mechanical properties than those filled with GTR, at the same loading. Microscopic study of failure surfaces showed that DGTR/epoxy adhesion seemed to be better than that obtained with untreated GTR, confirming the successful microwave treatment in recycling of waste rubbers.

As a main conclusion, the use of IL exhibits strong potential towards successful and efficient devulcanization of rubbers and reduce the energy required for devulcanization treatment. Moreover, optimization of the devulcanization parameters, particularly identification of an adequate energy value, will lead to the successful devulcanization i.e. low material degradation.

References

1. Coran AY (1995) Vulcanization: Conventional and Dynamic. Rubber Chem. Technol. 68:351–375
2. Kumar CR, Fuhrmann I, Karger-Kocsis J (2002) LDPE-based thermoplastic elastomers containing ground tire rubber with and without dynamic curing. Polym Degrad Stab 76:137–144.

https://doi.org/10.1016/S0141-3910(02)00007-1

3. Zhang SL, Zhang ZX, Pal K, et al (2010) Prediction of mechanical properties of waste polypropylene/waste ground rubber tire powder blends using artificial neural networks. Mater Des 31:3624–3629. https://doi.org/10.1016/j.matdes.2010.02.039
4. Zhou X, Wang F, Yuan X, et al (2015) Usage of slurry oil for the preparation of crumb-rubber-modified asphalt emulsions. Constr Build Mater 76:279–285. https://doi.org/10.1016/j.conbuildmat.2014.11.064
5. Sutanto P, Picchioni F, Janssen LPBM (2006) Modelling a continuous devulcanization in an extruder. Chem Eng Sci 61:7077–7086. https://doi.org/10.1016/j.ces.2006.07.012
6. Yagneswaran S, Tomar N, Smith DW (2013) Non-isothermal curing kinetics of epoxy/mechanochemical devulcanized ground rubber tire (GRT) composites. Polym Bull 70:1337–1351. https://doi.org/10.1007/s00289-013-0926-z
7. Sutanto P, Laksmana FL, Picchioni F, Janssen LPBM (2006) Modeling on the kinetics of an EPDM devulcanization in an internal batch mixer using an amine as the devulcanizing agent. Chem Eng Sci 61:6442–6453. https://doi.org/10.1016/j.ces.2006.05.024
8. Dubkov KA, Semikolenov S V, Ivanov DP, et al (2012) Reclamation of waste tyre rubber with nitrous oxide. Polym Degrad Stab 97:1123–1130. https://doi.org/10.1016/j.polymdegradstab.2012.04.006
9. Li J, Liang M, Guo S, Lin Y (2004) Studies on chain scission and extension of polyamide 6 melt in the presence of ultrasonic irradiation. Polym Degrad Stab 86:323–329. https://doi.org/10.1016/j.polymdegradstab.2004.04.021
10. Li Y, Zhao S, Wang Y (2012) Microbial Desulfurization of Ground Tire Rubber by Sphingomonas sp.: A Novel Technology for Crumb Rubber Composites. J Polym Environ 20:372–380. https://doi.org/10.1007/s10924-011-0386-1
11. Bredberg K, Andersson BE, Landfors E, Holst O (2002) Microbial detoxification of waste rubber material by wood-rotting fungi. Bioresour Technol 83:221–224
12. Yao C, Zhao S, Wang Y, et al (2013) Microbial desulfurization of waste latex rubber with Alicyclobacillus sp. Polym Degrad Stab 98:1724–1730. https://doi.org/10.1016/j.polymdegradstab.2013.06.002

13. Adhikari B, De D, Maiti S (2000) Reclamation and recycling of waste rubber. Prog Polym Sci 25:909–948. https://doi.org/10.1016/S0079-6700(00)00020-4
14. Fix SR (1980) Microwave devulcanization of rubber. Elastomerics 112:38–40
15. Simon DÁ, Pirityi DZ, Bárány T (2020) Devulcanization of ground tire rubber: microwave and thermomechanical approaches. Sci Rep 10:1–13. https://doi.org/10.1038/s41598-020-73543-w
16. Novotny DS, Richard L M, FranK C M, David N T (1978) Microwave Devulcanization of Rubber
17. Thostenson ET, Chou T-W (1999) Microwave processing: fundamentals and applications. Compos Part A Appl Sci Manuf 30:1055–1071
18. Isayev AI (2013) Recycling of Rubbers, Fourth Edi. Elsevier Inc.
19. Yeo Joo Hong, Keuk Min Jeong, Prosenjit Saha, Jaechan Suh JKK (2015) Processing and Characterization of Microwave and Ultrasonically Treated Waste-EPDM/LDPE Polymer Composites. https://doi.org/DOI 10.1002/pen.23916
20. Garcia PS, de Sousa FDB, de Lima JA, et al (2015) Devulcanization of ground tire rubber: Physical and chemical changes after different microwave exposure times. Express Polym Lett 9:1015–1026. https://doi.org/10.3144/expresspolymlett.2015.91
21. Karger-Kocsis J, Mészáros L, Bárány T (2013) Ground tyre rubber (GTR) in thermoplastics, thermosets, and rubbers. J Mater Sci 48:1–38. https://doi.org/10.1007/s10853-012-6564-2
22. Scuracchio CH, Waki DA, Da Silva MLCP (2007) Thermal analysis of ground tire rubber devulcanized by microwaves. J Therm Anal Calorim 87:893–897. https://doi.org/10.1007/s10973-005-7419-8
23. Isayev AI, Sujan B (2006) Nonisothermal vulcanization of devulcanized GRT with reversion type behavior. J Elastomers Plast 38:291–318. https://doi.org/10.1177/0095244306067424
24. Bani A, Polacco G, Gallone G (2011) Microwave-induced devulcanization for poly (ethylene--propylene--diene) recycling. J Appl Polym Sci 120:2904–2911
25. Landini L, de Araújo SG, Lugão AB, et al (2007) Preliminary

analysis to BIIR recovery using the microwave process. Eur Polym J 43:2725–2731. https://doi.org/10.1016/j.eurpolymj.2007.03.017

26. Scagliusi SR, Araújo SG, Landini L, Lugão AB (2009) Study of Properties of Chloroprene Rubber Devulcanizate by Radiation in Microwave. In: International Nuclear Atlantic Conference--INAC Rio de Janeiro, RJ, Brazil
27. B. Vega, L. Montero, S. Lincoln, N. Agullo′ SB (2008) Control of Vulcanizing/Devulcanizing Behaviorof Diphenyl Disulfide with Microwaves as theHeating Source. J Appl Polym Sci 108:1969–1975
28. Maciejewska M (2014) Effect of ionic liquids on the dispersion of zinc oxide and silica nanoparticles, vulcanisation behaviour and properties of NBR composites. Express Polym Lett 8:932–940. https://doi.org/10.3144/expresspolymlett.2014.94
29. Armand M, Endres F, MacFarlane DR, et al (2009) Ionic-liquid materials for the electrochemical challenges of the future. Nat Mater 8:621–629
30. Silvester DS, Rogers EI, Compton RG, et al (2008) Technical Aspects. Electrodepos. from Ion. Liq. 287–351
31. Chozhan CK, Alagar M, Sharmila RJ, Gnanasundaram P (2007) Thermo mechanical behaviour of unsaturated polyester toughened epoxy-clay hybrid nanocomposites. J Polym Res 14:319–328. https://doi.org/10.1007/s10965-007-9114-x
32. Jia QM, Zheng M, Xu CZ, Chen HX (2006) The mechanical properties and tribological behavior of epoxy resin composites modified by different shape nanofillers. Polym Adv Technol 17:168–173. https://doi.org/10.1002/pat.710
33. Wingard CD (2000) Characterization of prepreg and cured epoxy/fiberglass composite material for use in advanced composite piping systems. Thermochim Acta 357–358:293–301. https://doi.org/10.1016/S0040-6031(00)00400-7
34. Chekanov Y, Arrington D, Brust G, Pojman JA (1997) Frontal curing of epoxy resins: Comparison of mechanical and thermal properties to batch-cured materials. J Appl Polym Sci 66:1209–1216. https://doi.org/10.1002/(sici)1097-4628(19971107)66:6<1209::aid-app20>3.3.co;2-7
35. He J, Raghavan D, Hoffman D, Hunston D (1999) The influence of elastomer concentration on toughness in dispersions containing preformed acrylic elastomeric particles in an epoxy matrix. Polymer (Guildf) 40:1923–1933.

https://doi.org/10.1016/S0032-3861(98)00436-4

36. Harsch M, Karger-Kocsis J, Holst M (2007) Influence of fillers and additives on the cure kinetics of an epoxy/anhydride resin. Eur Polym J 43:1168–1178. https://doi.org/10.1016/j.eurpolymj.2007.01.025
37. Yagneswaran S, Tomar N, Smith DW (2013) Non-isothermal curing kinetics of epoxy/mechanochemical devulcanized ground rubber tire (GRT) composites. Polym Bull 70:1337–1351. https://doi.org/10.1007/s00289-013-0926-z
38. Piszczyk Ł, Hejna A, Formela K, et al (2015) Effect of ground tire rubber on structural, mechanical and thermal properties of flexible polyurethane foams. Iran Polym J (English Ed 24:75–84. https://doi.org/10.1007/s13726-014-0301-4
39. Oliphant K, Baker WE (1993) The use of cryogenically ground rubber tires as a filler in polyolefin blends. Polym Eng Sci 33:166–174. https://doi.org/10.1002/pen.760330307
40. Naskar AK, De SK, Bhowmick AK (2002) Thermoplastic elastomeric composition based on maleic anhydride-grafted ground rubber tire. J Appl Polym Sci 84:370–378. https://doi.org/10.1002/app.10348
41. Naskar AK, Bhowmick AK, De SK (2001) Thermoplastic elastomeric composition based on ground rubber tire. Polym Eng Sci 41:1087–1098. https://doi.org/10.1002/pen.10809
42. Seghar S, Aït Hocine N, Mittal V, et al (2015) Devulcanization of styrene butadiene rubber by microwave energy: Effect of the presence of ionic liquid. Express Polym Lett 9:1076–1086. https://doi.org/10.3144/expresspolymlett.2015.97
43. Flory PJ (1941) Molecular size distribution in three dimensional polymers. III. Tetrafunctional branching units. J Am Chem Soc 63:3096–3100
44. Horikx M (1956) Chain scissions in a polymer network. J Polym Sci Part B Polym Phys 19:445–454. https://doi.org/https://doi.org/10.1002/pol.1956.120199305
45. Charlesby A (1953) Solubility and molecular size distribution of crosslinked polystyrene. J Polym Sci 11:513–520. https://doi.org/https://doi.org/10.1002/pol.1953.120110601
46. Charlesby A (1954) Gel formation and molecular weight distribution in long-chain polymers. Proc R Soc London Ser A Math Phys Sci 222:542–557

47. Sutanto P (2006) Development of a Continuous Process for EPDM Devulcanization in an Extruder
48. Lambert, J.B., Shurvell, H.F., Cooks RG (1987) Introduction to Organic Spectroscopy. Macmillan Publ New York
49. Pavia, D.L., Lampman, G.M., Kriz G. (1979) Introduction to Spectroscopy: A Guide for Students of Organic Chemistry. WB Saunders Co, Philadelphia

9

On the Use of Ground Tire Rubber as Thermoplastic Polymer Modifiers

Caren Rosales and **Valeria Pettarin***

Institute of Materials Science and Technology (INTEMA), University of Mar del Plata-CONICET, Av. Colon 10850, 7600 Mar del Plata, Argentina
*pettarin@fi.mdp.edu.ar

9.1 Introduction

The huge quantity of waste rubber is mainly derived from the continue development of the automotive market. In fact, approximately 80 % of waste rubber comes from end of life tires [1]. Land filling was the first way of dispose waste tires. However, it causes serious environmental troubles. The impermeability and hollowed shape of tires allow them to serve as water holders, providing sites for mosquito larva to grow, these being vectors of deadly diseases. Furthermore, discarded tires could produce fires that are difficult to extinguish [2]. The global concern about the continue accumulation of discarded tires and the developing ecological awareness demand new ways and regulations to solve this problem. The developed regulations promote the research of the scientific community to explore alternative solutions to reuse this waste material [1, 3].

However, the reuse or recycling of tires is not a simple task. Tires are a complex and sophisticated assemblage of components: high mechanical standard requirements like flexibility, forces and adhesion make the design of this engineering structure to be constructed from several materials. For example, automobile tires are made of more than ten elements, and tires used for trucks have near twenty different elements. Even tough, vulcanized rubbers constitute the major relative component of tires [2]. Vulcanization is an irreversible process wherein rubber crosslinking happens through the formation of sulfur bonds [4]. As a result, a cross-linked structure is obtained due to transverse bonds that connect different rubber chains. This vulcanized structure makes rubber an elastic thermoset material that cannot be reprocessed by melting as thermoplastic polymers due to its insoluble and infusible characteristics [2].

It is therefore of crucial relevance to find methods for tires recycling that combine both low cost and technological feasibility. One proposed way is to devulcanize the rubber and reuse it. Although such processes work out well, they are expensive and consequently they are not frequently applied in industrial scale [5]. Another industrial recycling method of tires is their down-sizing or down-cycling. It consists on disassembling tires and grinding rubber into particles achieving a powder know as ground tire rubber (GTR) [1, 3]. Even though this is a technologically complex process that involves specific equipment [2], it is already well established in industry. GTR is actually used in surfaces of athletic tracks or games, or in combination with other materials such as asphalts [6–8], concrete [9–11] and mortars [12].

On the other side, a commonly used practice in industry to improve mechanical behavior and toughness of a polymer is to incorporate rubber particles into it. Then, an attractive option seems to be to add GTR to a polymeric material, so competitive products can be shaped. The polymer matrix could be thermoplastic, thermoset or even rubber[5]. Compound of waste tire rubber with polymers allows to lower the cost of the final products while contributing to the world's 3R (reduce, reuse and recycle) concept: reducing the volume of used virgin polymer, reusing end-of-life rubbers, and recycling the waste rubber [2]. The incorporation of GTR to thermoplastic polymers is especially attractive due to different reasons. The incorporation of only 10% of GTR in thermoplastics, would mean a large use of recycled GTR. In addition, the thermoplastic material can act as a matrix for dispersed GTR, or as a binder in a blend where GTR is the main component [5].

It is well known that the mechanical performance of composite materials depends on several factors such as quantity and behavior of each component, size, shape, geometrical arrangement and interfacial adhesion of present phases, as well as on the way they are obtained and prepared [13, 14]. In this sense, one of the main problems when compounding GTR and polymers is their incompatibility, and then, it is necessary to promote interfacial adhesion between GTR and polymer matrix. Moreover, GTR obtained by standard industrial grinding process presents high particle size (between 400 and 600 μm approximately), being this a drawback since it is well known that the addition of such large particles in a thermoplastic matrix results in composite materials with brittle behavior [3]. Different strategies to compatibilize GTR with polymers and to diminish GTR's particle size have been reported [3]. Sometimes, the search of compatibility involves an increase in costs that damages the initial premise of obtaining a material at an acceptable cost.

Through this chapter, results obtained from the incorporation of GTR in thermoplastic polymers are reviewed. Strategies applied to improve performance of the obtained composites are also explained, and their results summarized.

9.2 Unmodified GTR

As previously mentioned, GTR particles can be dispersed in a thermoplastic matrix, as toughening agent. In that sense, incorporation of GTR into a virgin polymer should result in an upcycling process. However , when GTR is simply added to a polymer matrix, the blend's mechanical properties are generally poor due to the lack of interfacial interactions between thermoplastic matrix and cross-linked structure of GTR particles [15, 16]. To improve GTR/polymer compatibility it is necessary to modify the GTR powder. An initial approach to enhance mechanical performance of these blends is to control the size and morphology of GTR particles.

Regarding mean particle size, Mujal et al. [17] found in PP/GTR composites that composites with particle diameter lower than 200 μm showed improved mechanical properties when compared with composites with particles in the range between 200 and 500 μm, while the worst mechanical properties were observed for composites with particle diameter larger than 500 μm. This is easy to explain taking into account that the smaller the particle size, the larger the specific surface area of each single particle. Therefore, contact surface between GTR powder and matrix is enlarged for low diameter particles, increasing the interaction between components [18].

The grinding process controls relevant factors such as particle shape and the specific surface area of the powder. The effect of these factors was reported by some researchers. Goncharuk et al. [19] found that composites with particles obtained by roller grinding exhibited better properties that those with particles obtained by cryogenic grinding and elastic-strain grinding. They ascribed this result to a higher specific surface area generated in these grinding processes. Liu et al. [20] found that the ductility of composites based on PP was higher for particles grounded at room temperature than for composites made with a cryogenic ground powder. They explained this finding by the less developed particles surface in the latter case.

Another important feature is the quantity of GTR to be incorporated to the thermoplastic matrix. Colom et al. [21] carried out a study in which different amounts of GTR (size 400-600 microns) were incorporated into an HDPE matrix. The compound with 20% untreated GTR was found to be the best compromise between mechanical performance and rubber content, while samples richer in GTR showed a decrease in tensile properties. Marín-Genescà et al. [22] analyzed the influence of GTR on several thermoplastic polymers – PVC, HDPE, PP, EVA, ABS, PA and PS, varying the content of GTR (200 μm diameter) between 0 and 70%. They found that the composites with high number of particles, between 40 and 70%, exhibited deteriorated mechanical properties. This was attributed to the formation of cracks and subsequent matrix failure, because of poor interfacial adhesion between GTR and matrixes that induces detachment and voids that act as stress concentrators. On the other side, when the GTR concentration was low (between 5% and 20%), the interfacial cohesion remained acceptable. Within these filling values, they confirmed that HDPE, PS, ABS and PVC based composites behaved in a similar manner, while composites based on EVA and PP exhibit a more flexible performance than the above mentioned composites, evidenced by larger elongation at break [22].

GTR can be also added to blends to obtain ternary composites. One option of ternary composites is those made of two thermoplastics and GTR. Following this idea, we incorporated GTR to a LLDPE/PP blend, with the aim of obtaining a completely recycled composite. GTR particles in different size ranges – lower than 300 μm, 300-500 μm and 500-800 μm – and in various weight percentages – 5, 10 and 15% w/w– were incorporated to a 75% w/w LLDPE / 25% w/w PP blend. We found that GTR particles did not affect the microstructure and the crystallinity of the thermoplastic matrix. The tensile behavior of the ternary materials was slightly affected by the composition of the mixture: composites became less rigid (Figure 9.1a) and more ductile (Figure 9.1b), this being directly related to the elastic nature of the rubber particles. A clear relationship between the tensile parameters and the sizes and percentages studied was not evidenced.

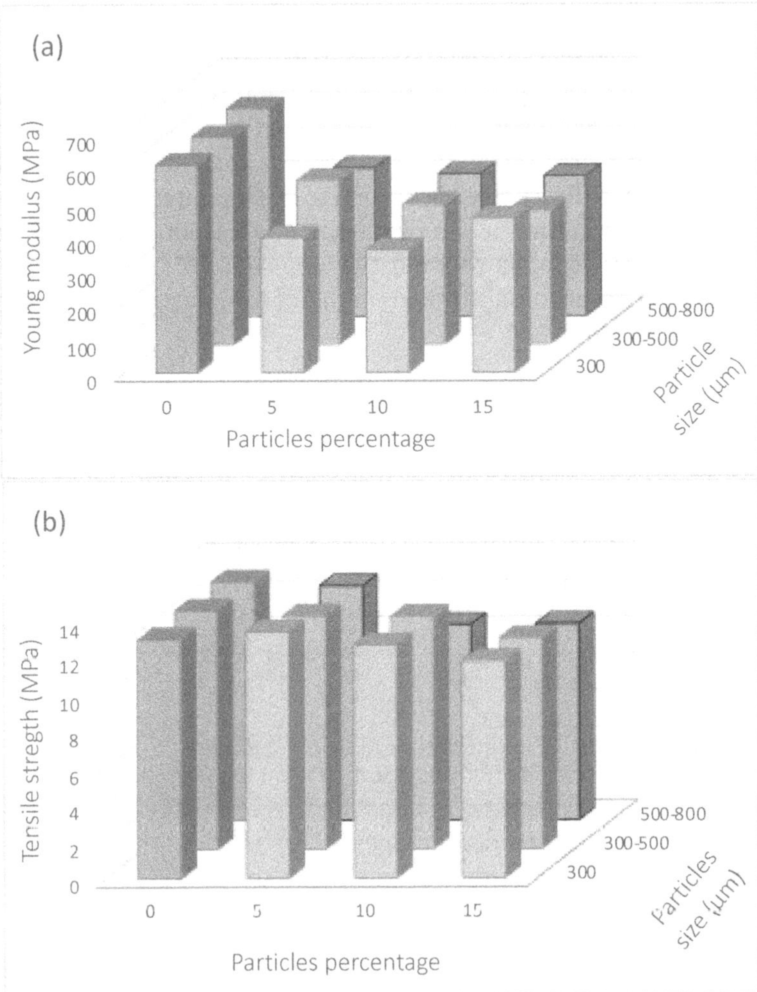

***Figure 9.1** a) Young modulus and b) tensile strength for ternary composites LLDPE/PP/GTR.*

However, large differences were found in the mechanical performance at fracture for the ternary mixtures, in comparison with the matrix blend (Figure 9.2 and Table 9.1). Fracture performance was evaluated by tensile tests of double-edge notched samples. Composites exhibited a completely ductile behavior, with stable crack propagation, with a smooth load decrease. As an example, curves corresponding to composites with 10% GTR is shown in Figure 9.2, along with macroscopic photos of damage in which ductile behavior is confirmed. Curves were analyzed by means of the EWF theory. According to it, the total work required to fracture a notched specimen can be separated in two parts that are associated with two different zones [23]. The energy associated with the first part – the essential work of fracture (**W_e**) – is linked to a process zone in which the crack is running and is directly proportional to the ligament section *lt*. The energy involved in the second part – the non-essential work of fracture (**W_p**) – corresponds to a plastic zone surrounding the process zone, that is proportional to the volume of the plastically deformed region. Therefore:

$$W_f = W_e + W_p = w_e . lt + \beta w_p . l^2 t \qquad (1)$$

where **w_e** is the specific essential work of fracture (per unit ligament area), **w_p** is the specific non-essential work of fracture (per unit volume), *l* is the ligament length, *t* is the specimen thickness and β is the plastic zone shape factor. To obtain the specific work of fracture, w_f, eq (1) is divided by the ligament area *lt*.

$$w_f = {}^{W_f}/_{lt} = w_e + \beta w_p \qquad (2)$$

GTR greatly influenced both energies involved in crack propagation (as evidenced in the propagation parameter **βw_f** in Table 9.1). In addition, different mechanical behaviors were observed that varied according to the relative percentage of the particles and their size. In Figure 9.3 it is observed that even though the LLDPE/PP blend exhibits a ductile fracture with stable crack propagation, the incorporation of GTR particles induced an even more ductile behavior, as evidenced by the highly fibrillated and plastically deformed fracture surface. Cavitation and plastic deformation of the matrix material are the main sources of energy dissipation during failure of ternary systems, regardless of the size and percentage of particles incorporated. Cavities are generated by the interfacial separation of GTR particles from the matrix and are elongated by the application of the tensile load. It is therefore evident that the sole inclusion of GTR particles, with no additional treatment or surface modification, can induce fracture mechanisms effective enough to improve both fracture toughness and energy involved in fracture propagation.

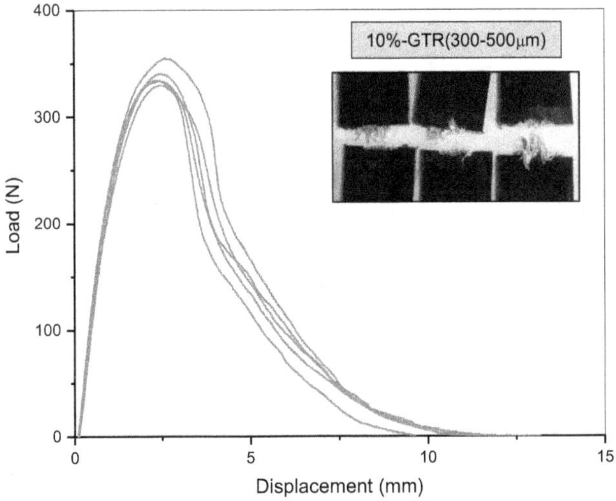

Figure 9.2 *Load-displacement curves and macroscopic damage for LLDPE/PP/10% GTR (300-500 microns).*

Table 9.1 *Essential work of fracture parameters for matrix and ternary composites*

Material	Particles percentage (%)	w_e (KJ/m²)	βw_f (MJ/m³)
7525		14.1 ± 1.6	0.6 ± 0.1
300	5	14.9 ± 5.1	1.8 ± 0.1
	10	26.0 ± 4.2	2.0 ± 0.2
	15	3.2 ± 3.0	3.3 ± 0.3
300-500	5	9.6 ± 6.1	3.2 ± 0.7
	10	16.4 ± 5.3	2.3 ± 0.4
	15	29.5 ± 4.0	2.8 ± 0.4

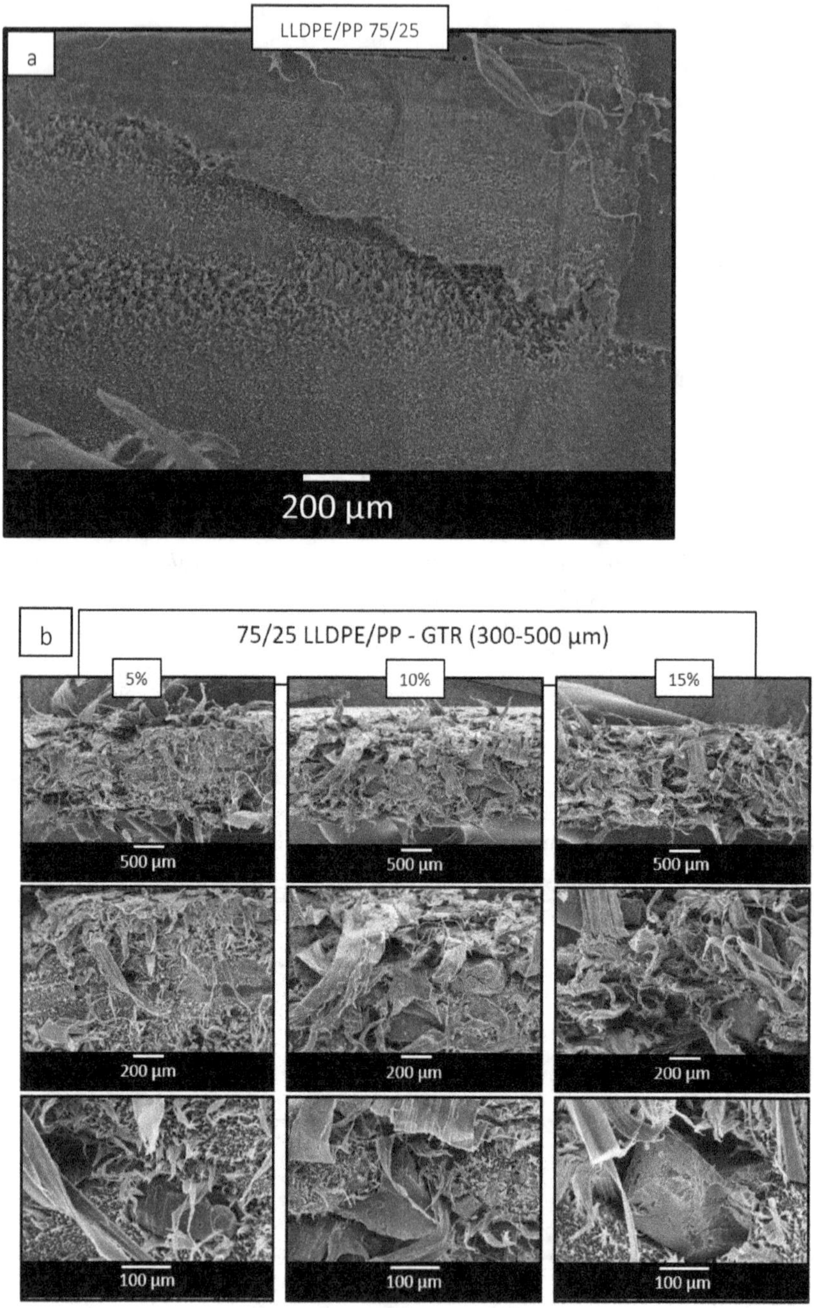

Figure 9.3 *SEM micrographs of fracture surfaces.*

Ternary composites can be also made of one thermoplastic and two rubbers. In this sense, García et al. [24] analyzed the rheological behavior of ternary mixtures based on PP for injection molding purposes. Specifically, he studied the feasibility of replacing rubbers like EPDM or EPR with recycled GTR. They found that the use of high-flow PP is a convenient strategy to counteract the GTR-induced increase in viscosity. Even the presence of a third rubber component such as EPR further reduces the viscosity of the mixture. He concluded that injection molding is a feasible strategy for large-scale recycling of this type of material.

Formela et al. [25] provided a deeper understanding about the mechanism of GTR interaction with other components in LDPE based ternary composites. They added two commercial non-polar elastomer to GTR. The addition of these compounds generated a surrounding layer of elastomer phase that encapsulated GTR. This behavior provides better thermal and mechanical properties by means of improved interfacial interaction between components. In accordance with other research, the authors found the best mechanical properties for a GTR content of 25% w/w.

Another approach is to use GTR particles as the majority component and add a thermoplastic polymer to behave as a binding agent. This is an interesting approach since it implies that simple processing techniques of thermoplastics, such as extrusion or injection molding, can be used to obtain "rubbery" pieces. In this sense, Basso et al. [1] studied GTR/PP composites with different GTR particle sizes – coarse (200 - 800 μm), intermediate (<425 μm), and fine powder (<74 μm) – with varying composition. The study demonstrated the possibility of use injection molding up to GTR content of 80% w/w. Following this approach, it is possible to reduce the use of virgin polymer and improve the content of recycled material. They also noticed an uneven dispersion of the powder in the injected parts, through the detection of sulfur. This resulted in distribution of surface finishes (larger roughness in the presence of a larger amount of GTR) and of mechanical properties (for example, lower hardness for larger GTR content).

Based on the previous findings, there is a general agree among researchers regarding the need to carry out chemical modifications on GTR to achieve better interaction or adhesion with polymer matrix in order to improve the mechanical performance of the final composites. The first approach is to modify the surface of GTR particles. Details on the GTR surface modification techniques and their results are discussed in the following section.

9.3 Surface activated GTR

Based on what was described in previous sections, it is evident how necessary it is to solve the problem of insufficient interfacial bonding and adhesion between GTR and polymer matrices, in order to achieve good mechanical resistance and long term stability [2]. Even though GTR is a vulcanized rubber, it presents 90% of the double bonds of the unvulcanized rubber [26], which provide places for chemical reactions such as grafting and/or cross- linking that would improve GTR/thermoplastic matrix compatibility. Moreover, additional active sites are formed on GTR surfaces during the grinding processes, that would be used to further improve GTR/thermoplastic compatibility [27]. Then, GTR surface modifications have been proposed as a possible solution. They produce interaction sites between phases present in composites, avoiding GTR particle agglomeration during processing [28], leading to better compatibility and smooth interfacial stress transfer, finally giving place to high mechanical properties. These compatibilization approaches are generally classified as physical methods and chemical methods. Among chemical methods, there are both reactive and non-reactive methods [29]. Some reported results of these methods are depicted in what follows.

Physical compatibilization is a method that aims to improve roughness and surface activity of GTR in order to create a thick interphase and to improve wetting by thermoplastic matrix. This is achieved by applying external forces in the form of mechanical or thermo-mechanical shearing [30], microwave [31], ozone [28], γ radiation [32], ultrasonic waves [33], ultraviolet (UV) light [34], plasma [35], and corona discharge [36] treatments. For example, Zhang et al. [30] reported mechanical milling as a cheap method to recycle waste tires. The pan-mill method involves high strong shear forces giving place to a pulverization like process of waste tires in which scission of the cross-linked GTR structure occurred, leading to better surface activation and dispersion of particles. They found that the number of groups with oxygen on the GTR particle surface increased with the number of milling cycles due to the reaction between free radicals generated during milling and oxygen from the atmosphere. They also stated that a high content of groups containing oxygen, i.e.

high number of polar groups, implies better interfacial interactions between GTR and matrix [37].

In chemical methods, reactive compatibilization involves the addition of a reactive molecule during mixing with the aim of produce a chemically linked interphase, and non-reactive compatibilization involves the addition of molecules such as copolymers (block or graft) and/or different nanoparticles (NP) such as clay, among others, to improve the blend compatibility [37–39].

Among reactive compatibilization, some methods aim to oxidize the GTR surface. Chemical modification, radiation induced modification, graft polymerization, and gas modification are some od these methods. Different chemicals, such as hydrogen peroxide, potassium permanganate, sulphuric acid or nitric acid can be used to oxidize GTR particles. During oxidation, polar groups are formed on GTR surface. These groups not only enhance the interaction with polar polymers but also they can interact with reactive compatibilizers such as maleic anhydride grafted polymer to improve the interaction between GTR particles and thermoplastic polymer matrix [2]. Another promising approach is to chemically modified GTR surface by grafting unsaturated monomers and oligomers, which in turn may both crosslink with virgin rubber or entangle with macromolecules of the thermoplastic matrix. Several researchers have followed this approach making use of various grafting procedures with styrene, GMA, and MAA. As an example, Naskar et al. [41] used a process based on a free radical process performed in an internal mixer at 160°C to graft MA on GTR surface (GTR-g-MA). They found better properties for the TPE/GTR-g-MA blend than for the TPE/GTR reference sample. Coiai et al. [42] showed that on the GTR surface there are reactive double bonds which are available sites for grafting. They made use of a free radical polymerization to graft styrene onto the GTR surface. The percentages of grafted styrene over the total weight of polymer formed reached different levels depending on the initiator used (dibenzoyl-peroxide or azobisisobutyronitrile). Authors attributed this fact to the differences in the H-abstraction capabilities of radicals formed from the different initiators. Zhang et al. [43] followed a similar approach by grafting styrene onto GTR surface. Tolstov et al. [44] made use of free radical and c-irradiation reactions to graft MA and acrylamide onto GTR, in order to add grafted GTRs to TPE, exploiting MA groups reactions with amine ($-NH_2$) functional groups. It was observed that the tensile properties of TPE based composites with reactively compatibilized GTR were superior as compared to those with non-compatibilized GTR. The best improvement in mechanical performance due to enhanced compatibility was observed for TPE that contain HDPE and GTR functionalized via grafting of reactive monomers. Shanmugharaj et al. modified GTR surface by allylamine UV-induced photografting [46, 47], and obtained better mechanical properties for PP/PP-g-MA/GTR composites than for non-functionalized GTR composites. They explained their findings by the chemical reaction between $-NH_2$ and anhydride groups of the graft and PP-g-MA compatibilizer, respectively [31, 46]. Another example is the one reported by Du et al. [47] who used thermally and UV-induced polymerization to graft bismaleimide onto GTR, aiming to produce new elastomers from NR/GTR blend with high content of GTR. They found that the best performance was observed for GTR/NR/bismaleimide (85/15/5 part) system when the sample was treated for 10 min at T = 110°C. Abdel-Bary et al. [48] reported the use of c-irradiation to graft GTR with several vinyl monomers such as acrylic acid, acrylamide and crylo- nitrile, attaining different graft yield depending on the monomer type. Yamazaki et al. [49] applied a heat treatment at 300°C in air and N_2 to polyisoprene, and vulcanized natural rubber with and without carbon black, and obtained lower molecular weight compounds. Afterward they used free radical initiation to copolymerize these rubbers with styrene. Fan and Lu [28] reported a two-step compatibilization of GTR: in a first stage they used ozonization to add "immobile" reactive hydroperoxide groups on the GTR surface, and then they graft methyl methacrylate. Obtained grafting degree depended on differences in both the polymerization temperature and the ozone treatment. Shahidi et al. [51, 52] produced a semi-interpenetrating network (semi-IPN) in GTR particles, by an "impregnation" process of polymerization of a reaction mixture composed of toluene (swelling agent), acrylic acid monomer, and an oil-soluble radical initiator. They used the obtained final product in wastewater purification. Sonnier et al. [16] applied a different modification treatment to oxidize GTR surface and improve interfacial compatibility. They used potassium permanganate and γ-irradiation to promote interfacial adhesion between GTR and PE matrix. However, there was no significant improvement in mechanical properties when comparing blends with and without this oxidation treatment of GTR particles. Egodage et al. [27] investigated the effect of using maleimide like curing system, in presence of a organic peroxide activator, in a PP/GTR blend with the aim of increase the curing of the rubber to obtain good aging characteristics. Physical properties of the rubber cured with a maleimide curing system resulted independent of the temperature of curing, and the capability of rubbers to crystallize when they were stretched was not markedly affect. They found that maleimide caused an increase in tensile and impact properties. However, above of 3 parts per hundred of maleimide caused a viscosity increased that difficult the processing, without further improvement in properties. The incorporation of peroxide activator induced an increase in interfacial cross-link formation, resulting in a homogeneous blend with further improved mechanical properties. Looking for simplifying surface modification process, through making it faster and cheaper, and more industrial viable, Kocevski et al. [6] tried an approach in which not initiator was added for reaction. Surface treatment of GTR powder was carried out by bulk polymerization of acrylic acid. They studied the effect of acid monomer concentration, reaction time and temperature. They obtained an enhancement in viscosity values and rheological properties, which was assigned to the improvement in interaction between components, attributed to the larger surface area of GTR particles and the anhydride formation on this surface.

Among researchers who applied non-reactive methods, Colom [21] studied the influence of modify GTR surfaces with three different acids (H_2SO_4, HNO_3 and $HClO_4$), when particles were added to a HDPE matrix. From the analysis of mechanical properties, he concluded that compatibility improvement was achieved when H_2SO_4 or HNO_3 was applied, producing a rougher surface with improved mechanical adhesion. The addition of above 10 % GTR modified with $HClO_4$ decreased Young's modulus and elongation at break in comparison with pure HDPE, which was attributed to weak interfacial adhesion. H_2SO_4 or HNO_3 treated GTR improved stiffness of HDPE due to the rigidification of the rubber after the extraction of additives produced by the treatment. Sulfuric acid was the most effective treatment on GTR particles' surfaces, reaching better mechanical properties because of more intense and bigger porosity at GTR surface that favors interfacial adhesion between components. This porosity is present as holes, providing a morphology in which the HDPE is capable to enter, then contributing to a extensive interlock with treated GRT. Following this favorable results, Hernandez et al [52] incorporated sulfur modified GTR to a PP matrix, in different concentrations until 50 wt% of GTR. Sulfuric acid drastically increased the surface roughness in 600% in comparison with untreated GTR. Consequently, a significant improvement on mechanical properties (especially Young modulus) was observed, even though crystallinity percentage was reduced.

From all the above results, it is clear that surface modification of GTR can insure acceptable interfacial interaction between GTR and a thermoplastic matrix. However, described techniques are still in a laboratory level and to the knowledge of the authors, they are no available commercially.

9.4 Devulcanized and reclaimed GTR

Another approach to improve the compatibility between GTR and thermoplastic matrixes is to diminish the degree of cross-links in GTR by applying methods of devulcanization or reclaiming [46, 54, 55]. These methods aim to compound a rubber that can be revulcanized, alike to a pristine rubber. It should be highlighted that they are related methods but rather distinct processes, because they have different targets of the chemical attack. On one side, devulcanization attacks the sulfuric crosslinks of vulcanized rubber, and it selectively cleaves S–S and C–S bonds, requiring high energy to split them (–C-S-C– 285 kJ/mol, –C–S–S–C– 268 kJ/mol and –C–S_x–C– 251 kJ/mol) [55]. On the other side, in reclaiming not only sulfuric crosslinks are broken but also considerable scission of C-C bonds along the polymeric chains usually occurs, giving place to fractions of low molecular weight. Despite of this clear difference between reclaiming and devulcanization, both procedures – in which GTR bonds cleavage occur – can be classified as thermochemical, thermomechanical, microwave, ultrasonic, and other processes [5].

Before discussing about the incorporation of devulcanized or reclaimed rubber in thermoplastic, it is important to describe the importance of thermoplastic vulcanizated (TPV) materials. TPVs are a type of rubber/thermoplastic blend in which the selective cross-linking of the dispersed elastomeric phase is produced during the mixing with the molten thermoplastic. Through this dynamic devulcanization process, a material in which cross-linked rubber particles are dispersed in a thermoplastic matrix is obtained. This final TPV blend presents the processability of the thermoplastic matrix along with the elasticity at room temperature given by the cross-linked elastomer particles. The most common TPVs are those based on EPDM and HDPE or PP. Their common applications are in the automotive industry, in pieces such as hoses or bumpers, but they are also used in various applications such as pipes, covering wires, handle tools, boots, etc. [56]. The rheological and physical properties of these materials are due to the above-described morphology, beyond being governed by the characteristics and processing conditions of the constituting materials. During processing, the difference between coalescence and break-up of the particles of the elastomeric phase is determinant of the TPV final morphology. There are some studies about size reduction, compatibility, mechanical performance and phase structure of thermoplastic vulcanizates (TPVs) based on polyolefins and GTR [57]. In this particular case in which the dispersed rubbery phase of TPV is a devulcanized elastomer, both devulcanization and revulcanization contribute to the refinement of the morphology, devulcanization by increasing the break-up ability of rubber droplets and revulcanization by decreasing the coalescence of those particles. Another positive effect of adding devulcanized GTR is that high amounts of recycled rubber can be incorporated in the TVP blend without compromise its properties, since the devulcanization process favors the interaction, i.e. the adhesion, between rubber particles and the thermoplastic matrix [58]. Among reported attempts to use GTR in TPV's, are the works of Kumar et al. [59] who studied the viability of producing TPE with pristine rubber (SBR, NR or EPDM), mechanical degraded GTR and LDPE; Punnarak [60] who presented the evaluation of TPV composites based on reclaimed tire rubber (RTR) – dynamically vulcanized with sulphur, MA and DCP – and HDPE; and Naskar [61] who reported a TPE in which the elastomeric phase was composed of EPDM, cryogenic GTR and ethylene-co-acrylic acid, vulcanized with DCP. From these works, it is established that GTR particles are encapsulated by the fresh EPDM – that present similar surface energy to the matrix, and then good compatibility with it – therefore giving place to a co-continuous phase and improving the adhesion between rubbery and thermoplastic phases. Nevertheless, during last year the price of EPDM has rising thus leading to its substitution by ethylene-octene and ethylene-butene elastomers based on metallocenes. Then, the effort to incorporate as much GTR as possible instead of EPDM in TPV formulations seems to be a good alternative for cost reduction of the final materials and as an upcycling recovery of GTR [3]. It is then evident that devulcanization and/or reclaiming are essential if this replacement is intended.

9.4.1 Themomechanical processes

In thermomechanical processes, rubber particles are treated in equipment such as mills, twin-screw extruders, etc. to subject them to elongational and/or shear stresses, at room or high temperature. In such processes, even at room temperature, an important decrease in the molecular weight of rubber may result due to the self-heating generated. Final rubber properties are highly dependent of the processing equipment, its characteristics, and processing parameters such as residence time, temperature and shear speed. At first, discontinuously operating open mills were used [55], but nowadays continuous operations using extruders are preferred [62]. However, the design and fabrication of devulcanization extruders is an developing research [63].

9.4.2 Themochemical processes

A thermochemical process is a thermomechanical process in which a reclaiming agent is present. Reclaiming compounds can be both inorganic and organic, and there is a wide range of them [64] that work exclusively at elevated temperatures. The easiest rubbers to devulcanize are those that were vulcanized by conventional curing systems, being conventional curing systems those with high sulfur/accelerator ratio giving place mainly to longer flexible polysulfidic crosslinks. Also "softeners" (such as asphalts [66, 67], bitumens [68, 69] and more recently oils from renewable resources [64]) are frequently used in GTR reclaiming to reduce the mechanical energy required for mixing by stretching the bonds between rubber chains. Both batch and continuous processes are used to produce the partial devulcanization of GTR. Batch processes result in powder GTR that can be added to thermoplastics and rubbers [70, 71]. On the other hand, continuous processes result in a plastic mass that can be either added to pristine curable rubbers or revulcanized [72–74].

Cavalieri et al. [74] performed devulcanization in solid state by using high energy ball milling (HEBM). They observed that HEBM produced the preferential scission of sulfur crosslinks and obtained good results when they grafted natural rubber onto the surface of GTR by applying this mechano-chemical milling.

Colom et al. [21] designed and performed four different experimental sets to accurately study TPV mechanical properties with GTR adding. They studied the effect of temperature on curing, which is related to peroxides activity. They obtained a very promising value of elongation at break (297%, keeping in mind that for a thermoplastic elastomer the elongation at break is always larger than 100%) with an appreciable quantity of 30% GTR particles. Further increase of the GTR particles amount to 40% leads to drastic drop of the elongation at break, down to 121%. It seems that the optimum amount of GTR particles in the composite is 30% if the objective is a balance between maximum quantity of GTR particles and good mechanical performance.

Fukumori et al. [75] obtained a thermoplastic elastomer (TPE) by the combination of devulcanization with dynamic vulcanization on line continuously. They applied a process consisting on successive steps of pulverization, then devulcanization, followed by blending with a thermoplastic polymer (such as PP), and finally dynamic vulcanization. Dynamic curing refers to the selective curing of the rubber in the molten thermoplastic under exhaustive shearing, i.e. within a thermoplastic matrix phase and dispersed submicron-sized crosslinked rubber particles. This so obtained thermoplastic dynamic vulcanizate (TDV) is another element of the TPE group.

One of the processes that are industrially available for the reclaiming and modification of GTR is to apply high pressure and temperature to produce the reactive sintering of GTR. It can be performed either without the incorporation of additives or with the incorporation of adhesive binders or curing systems. However, it is still limited to the production of simple shapes products with low-quality requirements, and therefore low-cost [5]. The modification or functionalization of GTR seems to be an interesting and promising approach to enhance the reactive sintering efficiency and therefore to increase possible applications of GTR. In GTR reclaiming it is crucial to find an equilibrium between selective rupture of cross-linking bonds and the undesired scission of main polymeric chains, which degrade the final performance of the obtained reclaimed rubber [77, 78]. This could be attained by controlling the process parameters. Moreover, there are alternative procedures to enhance the reclaiming efficiency, such as the adding of reclaiming agents that facilitate the cross-linking scission process. Some of the common additives used in polymer industry can be used as curing additives. Peroxides are therefore possible to be used as a free radicals source, which would enable the generation of new chemical bonds between polymer chains, at the same time that is degrading the main polymeric chains [78].

9.4.3 Ultrasonic, microwave

The energy of S–S bond is on the same level that the one of ultrasonic weaves [79]. Then, use ultrasound to break the crosslinked bonds of rubber seems to be a valid alternative (ultrasound low frequency range between 20 kHz and 500 MHz). In fact, ultrasonic devulcanization has been successfully used as a recycling strategy for rubber. For example, Isayev and coll. [80] investigated the recycling of various rubbers by continuous devulcanized them by using an extruder connected to an ultrasonic generator. Even though the crucial processing parameters are the ultrasound amplitude and the residence time, the temperature and the pressure also affect the rate of the devulcanization. It was also shown that after revulcanization, mechanical performance of these rubbers are near to the one of precedent rubber [81, 82]. Moreover, Isayev and co-workers successfully used ultrasonic treatment to obtain reclaimed silicon rubber, and incorporated it in both pristine and ultrasonically decomposed SBR with the aim to modify its surface and bulk properties [82].

Nowadays, the most promising technique of rubber devulcanization is the one that use microwaves due to the good properties observed in the devulcanized material and the high productivity potential. The technique is based on the volumetric heating of the rubber that is produced by microwaves which promotes a more uniform heating than the one reached by other heating methods based on convection and/or conduction [83]. The energy required for GTR devulcanization (around 180 Wh/kg) is easily obtained and controlled using microwaves [84]. One of the requirements to generate the necessary heat for GTR devulcanization by microwave technique is that the waste rubber must be polar. If CB is present, it assures the collection of internal energy and its even distribution in GTR, due to its high thermal conductivity and heat capacity [84]. It has been reported that microwave devulcanized GTR presents properties rather different of those of the original vulcanized rubber [85]. Among these different properties, the most outstanding is the ability to flow, i.e. to be remolded. It is then evident, that this ability to be remolded along with the possibility of its revulcanization, makes microwave devulcanization a technique with wide applicability. Furthermore, it is been demonstrated the ease of microwaves application and its many advantages, among which it is the high heating rate without any need of additional mechanical or chemical treatments [86]. Usually, microwave technique is coupled with extrusion and mixing equipment, in a way such as to produce a refined, devulcanized material rapidly and in a cost-efficient manner. Moreover, the obtained devulcanized material may be used as pristine material for the same product that has been produced from [87].

There are several works that reported the use of microwaves to devulcanize GTR. For example, Pistor et al. [84] applied microwave devulcanization to EPDM compounds in the presence of an apolar paraffinic oil. They arrived to the conclusion that if paraffinic oil is extracted from the rubber and compounds are subjected to microwaves for short time periods, devulcanization is best controlled. Zanchet et al. [88] added to a sulfur-curable SBR, up to 80 phr of SBR waste powder devulcanized by microwaves, and investigated its effect on the mechanical thermooxidative and photooxidative properties of the resulting composites. They observed that after 3 min of microwave treatment in a modified domestic-type oven with 900 W magnetron capacity, the sol content of the SBR waste powder increased from 7 to 67 %. De Sousa [58] used microwaves devulcanization of GTR followed by dynamically revulcanization on GTR/HDPE (80-20) blends, with constant microwaves power, at various exposure times. Poor mechanical properties were observed, which decrease with the increase in exposures times to microwaves. Additionally, blends with GTR untreated and vulcanized with additives showed better mechanical properties in comparison to that without additives.

9.4.4 Others

There are several other techniques to devulcanize and reclaim GTR. As an example, the work of Song et al. [57] could be cited. They studied the addition of a compatibilizier (HDPE-g-MA) to GTR devulcanized by light pyrolysis process, in which a decrosslinking mechanism occurred. The authors investigated different degrees of degradation on LLDPE based composite due to the increase in extrusion temperature. As the melting temperature increase, a more uniform dispersion was observed by SEM. Interfacial interaction was enhancement by degradation growth. The addition of a compatibilizer allowed a smaller and homogeneous dispersion. Elongation at break and tensile strength enhanced significantly because of better compatibility.

Another example is the work of Zedler and Łukasz [78], who studied a mechano-chemical process to modify GTR with the presence of bitumen and two different peroxides. Processing sintering efficiency was improved by the incorporation of bitumen as plasticizer, due to the decrease in consumed energy by the process. The added amount of peroxides have a significant influence of curing qualities, at the same time that significantly improved the processability of GTR, being this last one of the critical characteristics for GTR recycling. They found that if lower forces were applied on the motor due to the presence of bitumen, the amount of energy put into the process decreased, resulting in a noticeable diminish of the processing cost. They also found that curing characteristics and physico-mechanical properties were highly dependent on the type and amount of added peroxides; obtaining a product with satisfactory properties. They concluded that the peroxides structure is crucial in the creation of crosslink bonds, and this affect the physico-mechanical properties of final material.

9.4.5 Conclusions

Through this chapter, results obtained from the incorporation of GTR in thermoplastic polymers were reviewed along with the strategies applied to improve performance of the obtained composites. Even though some good results were obtained with the sole incorporation of unmodified GTR to thermoplastic matrices, it is clear that to achieve better overall performance it is necessary to promote better interfacial interaction and/or adhesion between GTR and polymer matrix. The efficiency of the compatibilization method is reflected on the composites' mechanical properties. Different compatibilization methods were depicted, like GTR surface activation, devulcanization and dynamic revulcanization. With this last technique, materials such as thermoplastic elastomers can be obtained, with promising mechanical properties and the obvious advantage of upcycling GTR. However, all compatibilization methods imply higher production costs, and go against the first objective of compound waste tire rubber with polymers, i.e. to lower the cost of final materials while contributing to the world's 3R (reduce, reuse and recycle) concept. It is then anticipated that in the near future, scientific and industrial communities will make efforts to develop cost effective compounds based on thermoplastics and GTR, diminishing compatibilization costs, and/or developing low-cost green methods.

References

[1] A. Basso, Y. Zhang, L. Linnemann, and H. Nørgaard Hansen, "Study of the distribution of rubber particles in ground tire rubber/polypropylene blends," *Mater. Today Proc.*, vol. 34, pp. 311-316, 2021, doi: 10.1016/j.matpr.2020.05.362.

[2] S. Ramarad, M. Khalid, C. T. Ratnam, A. L. Chuah, and W. Rashmi, "Waste tire rubber in polymer blends: A review on the evolution, properties and future," *Prog. Mater. Sci.*, vol. 72, pp. 100–140, 2015, doi: 10.1016/j.pmatsci.2015.02.004.

[3] X. Colom and X. Cañavate, *New Routes to Recycle Scrap Tyres.* 2011.

[4] V. V. Rajan, W. K. Dierkes, R. Joseph, and J. W. M. Noordermeer, "Science and technology of rubber reclamation with special attention to NR-based waste latex products," *Prog. Polym. Sci.*, vol. 31, no. 9, pp. 811–834, 2006, doi: 10.1016/j.progpolymsci.2006.08.003.

[5] J. Karger-Kocsis, L. Mészáros, and T. Bárány, *Ground tyre rubber (GTR) in thermoplastics, thermosets, and rubbers*, vol. 48, no. 1. 2013.

[6] S. Kocevski, S. Yagneswaran, F. Xiao, V. S. Punith, D. Smith, and S. Amirkhanian, "Surface modified ground rubber tire by grafting acrylic acid for paving applications," *Constr. Build. Mater.*, vol. 34, pp. 83–90, 2012, doi: 10.1016/j.conbuildmat.2012.02.040.

[7] Y. V. Vazquez and S. E. Barbosa, "Process Window for Direct Recycling of Acrylonitrile-Butadiene-Styrene and High-Impact Polystyrene from Electrical and Electronic Equipment Waste," *Waste Manag.*, vol. 59, pp. 403–408, 2017, doi: 10.1016/j.wasman.2016.10.021.

[8] X. Zhou, F. Wang, X. Yuan, M. Kuang, Y. Song, and C. Li, "Usage of slurry oil for the preparation of crumb-rubber-modified asphalt emulsions," *Constr. Build. Mater.*, vol. 76, pp. 279–285, 2015, doi: 10.1016/j.conbuildmat.2014.11.064.

[9] K. Afshinnia and A. Poursaee, "The influence of waste crumb rubber in reducing the alkali-silica reaction in mortar bars," *J. Build. Eng.*, vol. 4, pp. 231–236, 2015, doi: 10.1016/j.jobe.2015.10.002.

[10] B. S. Thomas, R. C. Gupta, and V. John Panicker, "Experimental and modelling studies on high strength concrete containing waste tire rubber," *Sustain. Cities Soc.*, vol. 19, pp. 68–73, 2015,

doi: 10.1016/j.scs.2015.07.013.

[11] B. S. Thomas, R. C. Gupta, and V. J. Panicker, "Recycling of waste tire rubber as aggregate in concrete: Durability-related performance," *J. Clean. Prod.*, vol. 112, pp. 504–513, 2016, doi: 10.1016/j.jclepro.2015.08.046.

[12] A. Popovici *et al.*, "Modern mortars with electronic waste scraps (glass and plastic)," *Mater. Plast.*, vol. 52, no. 4, pp. 588–592, 2015.

[13] J. Móczó and B. Pukánszky, "Polymer micro and nanocomposites: Structure, interactions, properties," *J. Ind. Eng. Chem.*, vol. 14, no. 5, pp. 535–563, 2008, doi: 10.1016/j.jiec.2008.06.011.

[14] S. Y. Fu, X. Feng, B. Lauke, and Y. W. Mai, "Effects of particle size, particle/matrix interface adhesion and particle loading on mechanical properties of particulate-polymer composites," *Compos. Part B Eng.*, vol. 39, no. 6, pp. 933–961, 2008, doi: 10.1016/j.compositesb.2008.01.002.

[15] S. P. Khatiwada *et al.*, "Morphology and Mechanical Properties of Star Block Copolymer Modified Epoxy Resin Blends," *Mater. Today Proc.*, vol. 4, no. 4, pp. 5734–5742, 2017, doi: 10.1016/j.matpr.2017.06.038.

[16] R. Sonnier, E. Leroy, L. Clerc, A. Bergeret, and J. M. Lopez-Cuesta, "Polyethylene/ground tyre rubber blends: Influence of particle morphology and oxidation on mechanical properties," *Polym. Test.*, vol. 26, no. 2, pp. 274–281, 2007, doi: 10.1016/j.polymertesting.2006.10.011.

[17] R. Mujal-Rosas, J. Orrit-Prat, X. Ramis-Juan, M. Marin-Genesca, and A. Rahhali, "Study on dielectric, mechanical and thermal properties of polypropylene (PP) composites with ground tyre rubber (GTR)," *Polym. Polym. Compos.*, vol. 20, no. 9, pp. 797–808, 2012, doi: 10.1177/096739111202000905.

[18] X. Lu, W. Wang, and L. Yu, "Waste ground rubber tire powder/thermoplastic vulcanizate blends: Preparation, characterization, and compatibility," *J. Appl. Polym. Sci.*, vol. 131, no. 3, pp. 1–9, 2014, doi: 10.1002/app.39868.

[19] G. P. Goncharuk, M. I. Knunyants, A. N. Kryuchkov, and E. S. Obolonkova, "Effect of the specific area and the shape of rubber crumb on the mechanical properties of rubber-filled plastics," *Polym. Sci.*, vol. 40, no. 5–6, 1998.

[20] H. S. Liu, C. P. Richard, J. L. Mead, and R. G. Stacer, "Development of Novel Applications for Using Recycled Rubber in

Thermoplastics, Chelsea Center for Recycling and Economic Development." p. 18, 2000.

[21] X. Colom, F. Carrillo, and J. Cañavate, "Composites reinforced with reused tyres: Surface oxidant treatment to improve the interfacial compatibility," *Compos. Part A Appl. Sci. Manuf.*, vol. 38, no. 1, pp. 44–50, 2007, doi: 10.1016/j.compositesa.2006.01.022.

[22] M. Marín-Genescà, J. García-Amorós, R. Mujal-Rosas, X. S. Berna, and L. M. Vidal, "Comparison of mechanical and electrical characteristics of various polymers blended with ground tire rubber (GTR) and applications," *Appl. Sci.*, vol. 9, no. 8, 2019, doi: 10.3390/app9081564.

[23] B. Cotterell and J. Reddel, "The essential work of plane stress ductile fracture," *Int. J. Fract.*, vol. 13, pp. 267–277, 1977.

[24] P. S. Garcia, F. D. B. de Sousa, J. A. de Lima, S. A. Cruz, and C. H. Scuracchio, "Devulcanization of ground tire rubber: Physical and chemical changes after different microwave exposure times," *Express Polym. Lett.*, vol. 9, no. 11, pp. 1015–1026, 2015, doi: 10.3144/expresspolymlett.2015.91.

[25] K. Formela, J. Korol, and M. R. Saeb, "Interfacially modified LDPE/GTR composites with non-polar elastomers: From microstructure to macro-behavior," *Polym. Test.*, vol. 42, pp. 89–98, 2015, doi: 10.1016/j.polymertesting.2015.01.003.

[26] H. S. Liu, J. L. Mead, and R. G. Stacer, "Thermoplastic elastomers and rubber-toughened plastics from recycled rubber and plastics," *Rubber Chem. Technol.*, vol. 75, no. 1, pp. 49–64, 2002.

[27] S. M. Egodage, J. F. Harper, and S. Walpalage, "Effect of maleimide curing on mechanical properties of ground tyre rubber/waste polypropylene blends," *Plast. Rubber Compos.*, vol. 41, no. 8, pp. 332–340, 2012, doi: 10.1179/1743289811Y.0000000042.

[28] P. Fan and C. A. Lu, "A Study on Functionalization of Waste Tire Rubber Powder Through Ozonization," *J. Polym. Environ.*, vol. 19, pp. 943–949, 2011.

[29] J. Maris, S. Bourdon, J. M. Brossard, L. Cauret, L. Fontaine, and V. Montembault, "Mechanical recycling: Compatibilization of mixed thermoplastic wastes," *Polym. Degrad. Stab.*, vol. 147, pp. 245–266, 2018, doi: 10.1016/j.polymdegradstab.2017.11.001.

[30] X. . Zhang, C. . Lu, and M. Liang, "Preparation of rubber composites from ground tire rubber reinforced with waste-tire fiber through mechanical milling.," *J. Appl. Polym. Sci.*, vol. 103, pp. 4087–4094, 2006.

[31] J. Cañavate, "Influence of microwave treatment conditions of GTR on physico-mechanical and structural properties of NBR/NR/GTR composites.," *Afinidad*, vol. 76, pp. 171–179, 2019.

[32] G. Martínez-Barrera, J. J. del Coz-Díaz, F. P. Álvarez-Rabanal, F. López Gayarre, M. Martínez-López, and J. Cruz-Olivares, "Waste tire rubber particles modified by gamma radiation and their use as modifiers of concrete," *Case Stud. Constr. Mater.*, vol. 12, 2020, doi: 10.1016/j.cscm.2019.e00321.

[33] W. Feng and A. I. Isayev, "Recycling of tire-curing bladder by ultrasonic devulcanization," *Polym. Eng. Sci.*, vol. 46, no. 1, pp. 8–18, 2006, doi: 10.1002/pen.20449.

[34] A. M. Shanmugharaj, J. K. Kim, and S. H. Ryu, "UV surface modification of waste tire powder: Characterization and its influence on the properties of polypropylene/waste powder composites," *Polym. Test.*, vol. 24, no. 6, pp. 739–745, 2005, doi: 10.1016/j.polymertesting.2005.04.006.

[35] X. Zhang, X. Zhu, M. Liang, and C. Lu, "Improvement of the properties of ground tire rubber (GTR)-filled nitrile rubber vulcanizates through plasma surface modification of GTR powder.," *J. Appl. Polym. Sci.*, vol. 114, pp. 1118–1125, 2009.

[36] M. D. Romero-Sánchez, M. M. Pastor-Blas, and J. M. Martín-Martínez, "Treatment of a styrene-butadiene-styrene rubber with corona discharge to improve the adhesion to polyurethane adhesive," *Int. J. Adhes. Adhes.*, vol. 23, no. 1, pp. 49–57, 2003, doi: 10.1016/S0143-7496(02)00081-7.

[37] A. Fazli and D. Rodrigue, "Recycling Waste Tires into Ground Tire Rubber (GTR)/Rubber Compounds: A Review," *J. Compos. Sci.*, vol. 4, no. 3, p. 103, 2020, doi: 10.3390/jcs4030103.

[38] Y. H. Wang, Y. K. Chen, and D. Rodrigue, "Production of thermoplastic elastomers based on recycled pe and ground tire rubber: Morphology, mechanical properties and effect of compatibilizer addition," *Int. Polym. Process.*, vol. 33, no. 4, pp. 525–534, 2018, doi: 10.3139/217.3544.

[39] Y. A. Aggour, A. S. Al-Shihri, and M. R. Bazzt, "Surface Modification of Waste Tire by Grafting with Styrene and Maleic Anhydride," *Open J. Polym. Chem.*, vol. 02, no. 02, pp. 70–76, 2012,

doi: 10.4236/ojpchem.2012.22009.

[40] M. S. Satyanarayana, A. K. Bhowmick, and K. Dinesh Kumar, "Preferentially fixing nanoclays in the phases of incompatible carboxylated nitrile rubber (XNBR)-natural rubber (NR) blend using thermodynamic approach and its effect on physico mechanical properties," *Polymer (Guildf).*, vol. 99, pp. 21–43, 2016, doi: 10.1016/j.polymer.2016.06.063.

[41] A. K. Naskar, S. K. De, and A. K. Bhowmick, "Thermoplastic elastomeric composition based on maleic anhydride-grafted ground rubber tire," *J. Appl. Polym. Sci.*, vol. 84, no. 2, pp. 370–378, 2002, doi: 10.1002/app.10348.

[42] S. Coiai, E. Passaglia, F. Ciardelli, D. Tirelli, F. Peruzzotti, and E. Resmini, "Modification of cross-linked rubber particles by free radical polymerization," *Macromol. Symp.*, vol. 234, pp. 193–202, 2006, doi: 10.1002/masy.200650225.

[43] J. Zhang, H. Chen, C. Ke, Y. Zhou, H. Lu, and D. Wang, "Graft polymerization of styrene onto waste rubber powder and surface characterization of graft copolymer," *Polym. Bull.*, vol. 68, no. 3, p. 789, 2012, doi: 10.100786-9.

[44] A. Tolstov *et al.*, "Reactive compatibilization of polyethylene/ground tire rubber inhomogeneous blends via interactions of pre-functionalized polymers in interface," *Macromol. Symp.*, vol. 254, pp. 226–232, 2007, doi: 10.1002/masy.200750834.

[45] S. H. Lee, A. M. Shanmugharaj, V. Sridhar, Z. X. Zhang, and J. K. Kim, "Preparation and characterization of polypropylene and waste tire powder modified by allylamine blends," *Polym. Adv. Technol.*, vol. 20, no. 7, pp. 620–625, 2009, doi: 10.1002/pat.1307.

[46] A. M. Shanmugharaj, J. K. Kim, and S. H. Ryu, "Modification of rubber surface by UV surface grafting," *Appl. Surf. Sci.*, vol. 252, no. 16, pp. 5714–5722, 2006, doi: 10.1016/j.apsusc.2005.07.069.

[47] M. Du, B. Guo, and D. Jia, "Effects of thermal and UV-induced grafting of bismaleimide on mechanical performance of reclaimed rubber/natural rubber blends," *J. Polym. Res.*, vol. 12, no. 6, pp. 473–482, 2005, doi: 10.1007/s10965-005-3046-0.

[48] E. M. Abdel-Bary, A. M. Dessouki, E. M. El-Nesr, and M. M. Hassan, "Radiation-induced graft copolymerization of some vinyl monomers onto waste rubber powder," *Polym. - Plast. Technol. Eng.*, vol. 36, no. 2, pp. 241–256, 1997, doi:

10.1080/03602559708000617.

[49] H. Yamazaki, T. Nagasawa, W. Choi, and T. Endo, "Transformation of vulcanized natural rubber into lower molecular weight polymers and their application to grafted copolymer synthesis with some vinyl monomers," *J. Appl. Polym. Sci.*, vol. 101, no. 6, pp. 4003–4010, 2006, doi: 10.1002/app.22964.

[50] N. Shahidi, F. Teymour, and H. Arastoopour, "Amphiphilic particulate phase semi-interpenetrating polymer networks based on recycled rubber matrix," *Polymer (Guildf).*, vol. 45, no. 15, pp. 5183–5190, 2004, doi: 10.1016/j.polymer.2004.04.061.

[51] N. Shahidi, F. Teymour, and H. Arastoopour, "Dispersing behavior of recycled rubber particles," *Macromol. Symp.*, vol. 206, pp. 471–480, 2004, doi: 10.1002/masy.200450236.

[52] E. H. Hernández *et al.*, "Sulfuric acid treatment of ground tire rubber and its effect on the mechanical and thermal properties of polypropylene composites," *J. Appl. Polym. Sci.*, vol. 134, no. 21, pp. 1–7, 2017, doi: 10.1002/app.44864.

[53] T. Luo and A. I. Isayev, "Rubber/plastic blends based on devulcanized ground tire rubber," *J. Elastomers Plast.*, vol. 30, no. 2, pp. 133–160, 1998, doi: 10.1177/009524439803000204.

[54] J. K. Kim, S. H. Lee, and M. Balasubramanian, "A comparative study of effect of compatibilization agent on untreated and ultrasonically treated waste ground rubber tire and polyolefin blends," *Polimeros*, vol. 16, no. 4, pp. 263–268, 2006, doi: 10.1590/S0104-14282006000400004.

[55] B. Diao, A. Isayev, and V. Levin, "Basic Study of Continuous Ultrasonic Devulcanization of Unfilled Silicone Rubber," *Rubber Chem. Technol.*, vol. 72, no. 1, p. 152, 1999, doi: 10.5254/1.3538784.

[56] M. Van Duin, "Recent developments for EPDM-based thermoplastic vulcanisates," *Macromol. Symp.*, vol. 233, pp. 11–16, 2006, doi: 10.1002/masy.200690006.

[57] P. Song, S. Li, and S. Wang, "Interfacial interaction between degraded ground tire rubber and polyethylene," *Polym. Degrad. Stab.*, vol. 143, pp. 85–94, 2017, doi: 10.1016/j.polymdegradstab.2017.06.020.

[58] F. D. B. de Sousa *et al.*, "Blends Ground Tire Rubber devulcanized by microwaves/HDPE - Part B: Influence of clay addition," *Polim. E Tecnol.*, vol. 25, no. 4, pp. 382–391, 2015, doi: 10.1590/0104-1428.1747.

[59] C. R. Kumar, I. Fuhrmann, and J. Karger-Kocsis, "LDPE-based thermoplastic elastomers containing ground tire rubber with and without dynamic curing," *Polym. Degrad. Stab.*, vol. 76, no. 1, pp. 137–144, 2002, doi: 10.1016/S0141-3910(02)00007-1.

[60] P. Punnarak, S. Tantayanon, and V. Tangpasuthadol, "Dynamic vulcanization of reclaimed tire rubber and high density polyethylene blends," *Polym. Degrad. Stab.*, vol. 91, no. 12, pp. 3456–3462, 2006, doi: 10.1016/j.polymdegradstab.2006.01.012.

[61] A. K. Naskar, A. K. Bhowmick, and S. K. De, "Thermoplastic elastomeric composition based on ground rubber tire," *Polym. Eng. Sci.*, vol. 41, no. 6, pp. 1087–1098, 2001, doi: 10.1002/pen.10809.

[62] B. Maridass and B. R. Gupta, "Process Optimization of Devulcanization of Waste Rubber Powder From Syringe Stoppers by Twin Screw Extruder Using Response Surface Methodology," *Polym. Compos.*, vol. 29, no. 12, pp. 1350–1357, 2008, doi: 10.1002/pc.20379.

[63] B. Maridass and B. Gupta, "Recycling of waste tire rubber powder - Devulcanization in a counter rotating twin screw extruder," *Kautsch Gummi Kunstst*, vol. 56, no. 5, p. 232, 2003.

[64] B. Adhikari and S. Maiti, "Reclaimation and recycling of waste tyres.pdf," vol. 25, no. March, pp. 909–948, 2000.

[65] W. Cao, "Study on properties of recycled tire rubber modified asphalt mixtures using dry process," *Constr. Build. Mater.*, vol. 21, no. 5, pp. 1011–1015, 2007, doi: 10.1016/j.conbuildmat.2006.02.004.

[66] L. F. M. Leite and B. G. Soares, "Interaction of asphalt with ground tire rubber," *Pet. Sci. Technol.*, vol. 17, no. 9, pp. 1071–1088, 1999, doi: 10.1080/10916469908949766.

[67] P. Frantzis, "Development of crumb rubber reinforced bituminous binder under laboratory conditions," *J. Mater. Sci.*, vol. 38, no. 7, pp. 1397–1401, 2003, doi: 10.1023/A:1022991608114.

[68] F. J. Navarro, P. Partal, F. J. Martınez-Boza, and C. Gallegos, "Influence of Processing Conditions on the Rheological Behavior of Crumb Tire Rubber-Modified Bitumen," *J. Appl. Polym. Sci.*, vol. 104, pp. 1683–1691, 2007, doi: 10.1002/app.25800.

[69] O. Grigoryeva, Fainlei, O. Starostenko, I. Danilenko, N. Kozak, and G. Dudarenko, "Ground Tire Rubber (GTR) Reclamation: Virgin Rubber / Reclaimed GTR (RE)Vulcanizates," *Rubber*

Chem. Technol., vol. 77, no. 380, pp. 131–146, 2004, doi: 10.5254/1.3547806.

[70] U. S. Ishiaku, C. S. Chong, and H. Ismail, "Determination of optimum De-Link R concentration in a recycled rubber compound," *Polym. Test.*, vol. 18, no. 8, pp. 621–633, 1999, doi: 10.1016/S0142-9418(98)00060-9.

[71] A. A. Yehia, M. N. Ismail, Y. A. Hefny, E. M. Abdel-Bary, and M. A. Mull, "Mechano-chemical reclamation of waste rubber powder and its effect on the performance of NR and SBR vulcanizates," *J. Elastomers Plast.*, vol. 36, no. 2, pp. 109–123, 2004, doi: 10.1177/0095244304039896.

[72] B. Maridass and B. R. Gupta, "Performance optimization of a counter rotating twin screw extruder for recycling natural rubber vulcanizates using response surface methodology," *Polym. Test.*, vol. 23, no. 4, pp. 377–385, 2004, doi: 10.1016/j.polymertesting.2003.10.005.

[73] B. Maridass and B. R. Gupta, "Effect of extruder parameters on mechanical properties of revulcanized ground rubber tire powder," *Polimery*, vol. 52, no. 6, pp. 456–460, 2007.

[74] F. Cavalieri, F. Padella, and F. Cataldo, "Mechanochemical surface activation of ground tire rubber by solid-state devulcanization and grafting," *J. Appl. Polym. Sci.*, vol. 90, no. 6, pp. 1631–1638, 2003, doi: 10.1002/app.12829.

[75] K. Fukumori, M. Matsushita, M. Mouri, H. Okamoto, S. N, and S. Takeuchi K, "Dynamic devulcanization and dynamic vulcanization for recycling of crosslinked rubber," *KautschGummiKunstst*, vol. 59, no. 7–8, p. 405, 2006.

[76] S. Saiwari;, W. K. Dierkes, and J. W. M. Noordermeer, "COMPARATIVE INVESTIGATION OF THE DEVULCANIZATION PARAMETERS OF TIRE RUBBERS," *Rubber Chem. Technol.*, vol. 87, no. 1, pp. 31–42, 2014.

[77] M. Sabzekar, M. P. Chenar, S. M. Mortazavi, M. Kariminejad, S. Asadi, and G. Zohuri, "Influence of process variables on chemical devulcanization of sulfur-cured natural rubber," *Polym. Degrad. Stab.*, vol. 118, pp. 88–95, 2015, doi: 10.1016/j.polymdegradstab.2015.04.013.

[78] Ł. Zedler, M. Przybysz-Romatowska, J. Haponiuk, S. Wang, and K. Formela, "Modification of Ground Tire Rubber—Promising Approach for Development of Green Composites," *J. Compos. Sci.*, vol. 4, no. 1, p. 2, 2019, doi: 10.3390/jcs4010002.

[79] M. Myhre and D. A. MacKillop, "Rubber Recycling," *Rubber*

Chem. Technol., vol. 75, no. 3, pp. 429–474, 2002.

[80] T. A., S. D., and A. I. Isayev, "Devulcanization of Waste Tire Rubber by Powerful Ultrasound.," *Rubber Chem. Technol.*, vol. 69, no. 1, pp. 92–103, 1996.

[81] J. Yun;, J. S. Oh;, and A. I. Isayev, "Ultrasonic Devulcanization Reactors for Recycling of GRT: Comparative Study," *Rubber Chem. Technol.*, vol. 74, no. 2, pp. 317–330, 2001.

[82] B. Diao, A. I. Isayev, V. Y. Levin, and S. H. Kim, "Surface behavior of blends of SBR with ultrasonically devulcanized silicone rubber," *J. Appl. Polym. Sci.*, vol. 69, no. 13, pp. 2691–2696, 1998, doi: 10.1002/(sici)1097-4628(19980926)69:13<2691::aid-app19>3.3.co;2-5.

[83] O. Holst, B. Stenberg, and M. Christiansson, "Biotechnological possibilities for waste tyre-rubber treatment," *Biodegradation*, vol. 9, no. 3–4, p. 301, 1998.

[84] V. Pistor, C. H. Scuracchio, P. J. Oliveira, R. Fiorio, and A. J. Zattera, "Devulcanization of ethylene-propylene-diene polymer residues by microwave-Influence of the presence of paraffinic oil.," *Polym. Eng. Sci.*, vol. 51, no. 4, pp. 697–703, 2010, doi: doi:10.1002/pen.21875.

[85] C. H. Scuracchio, D. A. Waki, and R. E. S. Bretas, "Caracterização térmica e reológica de borracha de pneu desvulcanizada por microondas," *Polímeros*, vol. 16, no. 1, pp. 46–52, 2006, doi: 10.1590/s0104-14282006000100011.

[86] A. Bani, G. Polacco, and G. Gallone, "Microwave-Induced Devulcanization for Poly(ethylene– propylene–diene) Recycling," *J. Appl. Polym. Sci.*, vol. 120, pp. 2904–2911, 2011.

[87] A. I. Isayev, J. Chen, and A. Tukachinsky, "Novel Ultrasonic Technology for Devulcanization of Waste Rubbers," *Rubber Chem. Technol.*, vol. 68, no. 2, pp. 267–280, 1995, doi: https://doi.org/10.5254/1.3538741.

[88] A. Zanchet, L. N. Carli, M. Giovanela, R. N. Brandalise, and J. S. Crespo, "Use of styrene butadiene rubber industrial waste devulcanized by microwave in rubber composites for automotive application," *Mater. Des.*, vol. 39, pp. 437–443, 2012, doi: 10.1016/j.matdes.2012.03.014.

List of abbreviations

ABS	poly acryl butyl styrene
CB	carbon black
DCP	dicumyl peroxide
EPDM	ethylene-propylene diene monomer rubber
EPR	ethylene-propylene rubber
EVA	ethylene vinyl alcohol
EWF	Essential Work of Fracture
GMA	glycidyl methacrylate
GTR	ground tire rubber
HDPE	high-density polyethylene
HEBM	high energy ball milling
LDPE	low density polyethylene
LLDPE	linear low density polyethylene
MA	maleic anhydride
MAA	acrylic and methacrylic acids
NR	natural rubber
PA	polyamide
PP	polypropylene
PS	polystyrene
PVC	poly vinyl chloride
RTR	reclaimed tire rubber
SBR	styrene butadiene rubber
TDV	thermoplastic dynamic vulcanizate
TPE	thermoplastic elastomer
TPV	thermoplastic vulcanizated
w_e	specific essential work of fracture
w_f	specific work of fracture
w_p	specific non-essential work of fracture

10

Recycling of Waste Ground Tyre Rubber Particles in Thermoplastic Polymers: Effect of Macro- and Nano-Fillers on the Thermal and Mechanical Properties of Composites

Olawale Monsur Sanusi[a], Lucia Asaro[b], Pascal Médéric[c] and Nourredine Aït Hocine[a,*]

[a]*INSA CVL, Univ. Tours, Univ. Orléans, LaMé, 3 rue de la Chocolaterie, CS 23410, 41034 Blois, Cedex, France.*
[b]*Institute of Materials Science and Technology (INTEMA), University of Mar del Plata and National Research Council (CONICET), Av. Colón 10850, 7600 Mar del Plata, Argentina.*
[c]*IRDL, CNRS-UMR 6027, Univ. de Bretagne Occidentale, 6 avenue Victor Le Gorgeu, CS 93837, 29238 Brest Cedex 3, France.*
*nourredine.aithocine@insa-cvl.fr

10.1 Introduction

Commercially, tyre rubber is manufactured at a low cost through an efficient but irreversible process called vulcanization. Vulcanization can be defined as the creation of three-dimensional crosslinked network between rubber macromolecules via irreversible reactions, in the presence of curing agents (sulphur, peroxides), activators (stearic acid, zinc oxide), accelerators (thiazoles, sulphonamide), retarders (*N*-cyclohexylthio phthalimide) and other additives [1][2]. The sulphur vulcanization creates strong chemical bonds and the sulphur content strongly dictates the resulting structural and physical properties of the rubber vulcanizates [3]. Thus creating a deliberate complex chemical structure of rubbers required to achieve tyres with superior mechanical strength and high flexibility while putting resistance to harsh environmental conditions into consideration [4]. Hence, rubber products like tyres are insoluble and infusible, which makes them difficult to be reprocessed, unlike thermoplastics [5]. Meanwhile, global industrialization and population growth drive the rapid expansion in automotive sector leading to the increase in annual production of tyres, about 1.5 billion units globally [6][7]. This, consequently, generates a large volume of waste tyres after their lifespan, since they do not degrade naturally over centuries of years [8]. The traditional dumping or incineration of waste tyres without any measure poses great hazards to man and its environs [9]. Unlike thermoplastic waste, it is extremely difficult to completely recover elastomeric waste monomers [10], making reclaimed rubber less viable commercially. This results in the disposal of large quantities of such materials, which alters the conservation of the environment. Nevertheless, a laudable alternative means of using worn-out tyres is simply the incorporation of the GTR in thermoplastic matrices to form blends [7], Figure 10.1. This method receives premium scientific research interest as it seems to be the most straightforward rubber recycling means, it is low-cost, easy and environmentally friendly. Moreover, it consumes an appreciable quantity of the GTR [11]. More importantly, the renewability and recyclability of thermoplastic matrices support the ecosystem sustainability [12][13][14].

Although, there have been reviews on the recycling of tyres focusing on different areas of treatments and utilizations [4][14][15], nevertheless, this review uniquely classifies the utilization of waste tyres principally in thermoplastic matrices as well as revealed the contributions of additional phases (micro and nanofillers) incorporated to create high-performance composites.

10.2 Major concerns on waste tyre accumulation and methods of recycling

10.2.1 Major issues with waste tyres accumulation

More than 70% of total natural and synthetic rubber materials are utilized for tyre manufacturing [11][17]. When a tyre is no longer efficient or satisfactorily safe to be used, it becomes a waste tyre, that is equally referred to as end-of-life-tyre [18]. Every year, a huge volume of tyres is scrapped worldwide [19], that is over 17 million tons of worn-out tyres [20]. In fact, more than 180 million and 150 million tyres were scrapped in European Union countries and North America, respectively [21]. Unfortunately, far less than 1% of the entire waste tyres are naturally degraded over hundreds of years as a result of the complex cross-linked structure of rubber [8][10]. There are two critical problems associated with waste tyre rubbers. Number one is the economic loss in terms of the wastage of rubber when this valuable material is refused as waste, accompanied by an additional cost due to transportation of this waste to the disposal sites. Meanwhile, the unavailability of space has made tyre disposal by landfilling to be unrealistic with no potential in the nearest future [22]. Number two concern is the environmental pollution caused by the abandoned waste tyres, which is disastrous to aquatic and terrestrial organisms [14]. For instance, mass dumping of tyres breeds potential dangers to the man and his environs. These landfills harbor dangerous reptiles, accumulates disease causative organisms, or creates fire hazard zones since the abandoned elastomers are combustible materials. Consequently, the EU imposed a ban on landfilling of scrapped tyres [23]. Therefore, the attempts to recycle and re-utilize the waste tyre rubbers become necessary. Equally, tyres recycling represents a source of raw material and offers a possibility of earnings for people that collect the waste tyres, thus support the economic development [24].

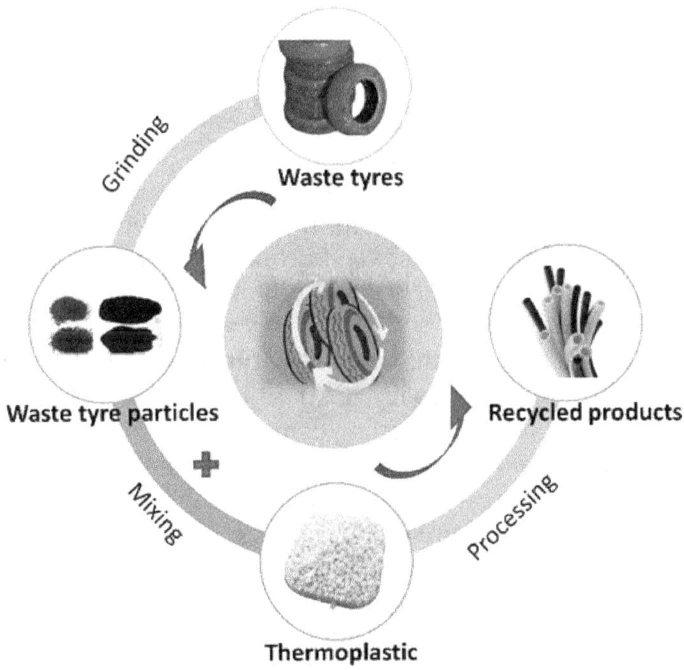

Figure 10.1 *Process scheme of recycling ground waste tyre particles in thermoplastics [16].*

10.2.2 Methods of waste tyres recycling

The complex composition and structure of tyres pose major restrictions to their recovery and recycling processes. Besides the 70-80% of sulphur-cured rubber component of tyres, fibre made up of polyester, nylon or cellulose, and steel cords are essential constituents used to reinforce tyres. Thus, separation of the fibres and steel components from the rubber is the first difficult step during recycling of tyres [25]. Then, the next stage of recycling of waste tyres is grinding to form GTR with different sizes and shapes [26].

Since manufactured tyres are infusible, insoluble and technically complex to be reprocessed [27], large amount of GTR is recycled as a source of energy. Incineration of waste tyres is seen as an approach of eliminating a substantial amount of this post-consumer product to generate or recover energy, but this is with significant negative environmental impacts such as air pollution and global warming [28].

Furthermore, several efforts are made by the government and scientists to deal with the problem of the ever-growing waste tyres [29][30]. The risks inherent in the scrap tyres necessitated the legislation of 1991 by United States Congress (Resource Conservation Recovery Act) and Intermodal Surface Transport Efficiency Act to mandate the use of a specified percentage of recycled tyres in any project funded by the federal government. In deviance to implementation, the funds would be withheld [31]. European Union (EU) issued the Landfill Directive "1999/31/EC" that prohibits the disposal of whole tyres and shredded tyres (50 – 300 mm size) in 2003 and 2006, respectively [11]. Thus, a significant use of tyres emerges in civil engineering for building and road construction, such as in insulation, sound deadening, soil stabilization, dam and road construction, artificial reefs, steel making and so on [32][33]. The use of tyres in this field is sometimes referred to as "downcycling" since the recycled materials have entire different applications from the original one, i.e. tyre [11]. However, except for road paving, most of the downcycling applications are saturated and outdated. Hence, there is a necessity for identifying innovative and value-added applications for waste tyres.

Recycling of tyres can be classified into physical reclaiming and chemical reclaiming processes. The physical reclaiming method involves the application of external energy on used tyres to breakdown the 3D cross-linked network of rubber. The applied energy transforms the molecular structure of rubber into fragments of smaller molecular weight availing it the miscibility with virgin matrix or blend during compounding. Applying sufficient energy, which is enough to cleave the rubber crosslink bonds, results in high quality reclaimed rubber. The physical reclaiming approach encompasses mechanical, cryo-mechanical grinding, thermo-mechanical, ultrasonic and microwave techniques [34]. Reclaiming by chemical approach involves the utilization of devulcanizing agents, solvents and elevated temperature [35]. In chemical devulcanization, the main goal is the penetration of devulcanizing reagent into the rubber matrix for cleavage reaction [36]. Some of the chemical medium used in rubber devulcanization include ionic liquid [37], deep eutectic solvent [38] and periodic acid [39]. Environmental concerns over the safety of the used chemicals initiated an innovative use of super-critical carbon dioxide as an alternative reaction medium [40]. Furthermore, thermochemical process has an extended applications in pyrolysis, whereby used tyres are converted to oil, gas or char that are either deployed as a chemical feedstock or energy sources [20][41]. Aside the complexity associated with control of the chemical reactions that limits its deployment in industry, this approach is expensive and does not completely eradicate toxic reagents [39]. For now, both chemical and physical approaches can lead to secondary pollution and involve urge energy consumption [42]. Complex instrumentation or the use of reactor plants and chemicals constitute environmental and cost concerns [43]. Moreover, the use of microorganisms in desulphurization of vulcanized rubber (microbial desulphurization) is another alternative but it only breaks the sulphur crosslinks on the surface of powdered waste tyre rubbers. In addition, it is a slow process [42][44].

During the last decade, one of the simple routes of tyre recycling that reduces energy consumption as well as addressing environmental challenges is incorporating GTR in thermoplastic polymers [15][45].

10.3 Recycling of waste tyre rubber in thermoplastic polymers

Mixing GTR with thermoplastics matrices or thermoplastic blends to elaborate thermoplastic elastomers (TPE) is the most promising practical application of GTR, as the thermoplastic phase in the composite makes the reprocessing or recyclability of the resulting compound feasible [46][47]. Even though the utilization of GTR in thermoplastic consumes a reasonable quantity of GTR (sometimes >50%

[46]) that addresses environmental concerns, the mechanical and thermal properties of the resulted TPE needs to be improved for competitiveness. Indeed, the composition of TPE, that is, the recycled tyre in the new thermoplastic-based matrix, affinity between the different phases and the GTR particle size are critical in order to achieve high-performance materials.

It is thus clear that recycling of GTR in different thermoplastic matrices/blends is a simple and inexpensive means of removing worn-out tyres from the environment with value addition, which supports ecological reasoning. More importantly, the resulting thermoplastic elastomer materials offer better material utilization as subsequent scraps and/or rejects can be easily recycled [48]. Nevertheless, several methods must be used to improve the interfacial adhesion and the degree of GTR dispersion: addition of compatibilizers/coupling agents, nanofillers, energy treatment, devulcanization and revulcanization [46][49].

10.3.1 Influence of GTR dispersion in polymer

Thermoplastics can be toughned by ensuring homogenous dispersion of GTR in the matrix [49]. The concentration of GTR in the matrix and its particle size influence the state of GTR dispersion and, thus, the final blend properties [50]. GTR of about 1 μm disperses more easily in matrix [15], but such particle size is not easy to produce even with cryogenic technique [51]. Equally, once the mass fraction of GTR in thermoplastic nears 10 – 20 wt%, the composite mechanical performance deteriorates [52]. Moreover, investigations revealed that for rubber to be effectively dispersed in thermoplastic, the surface energies of the two components should be identical [53]. Otherwise, pristine GTR will aggregate in matrix leading to deterioration of mechanical properties. Hence, delibrate efforts are adopted to improve the GTR dispersion – such as reduction of the GTR particle size [52], surface modification using compatibilizers/vulcanizing agents [54], encapsulation of GTR into neat rubber phase [55], or application of energy treatments [56], amongst others.

10.3.2 Influence of high-energy treatment

Energy treatments, like gamma-radiation [57], microwave, corona, plasma and electron beam radiation [58][59], increase concentration of oxygen on the GTR surface bringing better adhesion with host matrix [60]. Ramarad et al. [59] mixed GTR with ethylene vinyl acetate (EVA) and thereafter treated the composite with different doses of electron beam irradiation. The EVA/GTR material has lower processing torque compared to neat EVA, suggesting easier processability of the product due to the GTR rubbery feature. Although GTR entities are dispersed in EVA, the irradiation further improved its dispersion into smaller domain sizes within EVA matrix. They stated that the irradiation-induced formation of crosslink led to an increase of storage and loss moduli of EVA/GTR material. Also, Mohamad et al. [58] prepared TPE by mixing different ratio of natural rubber (NR), high density polyethylene (HDPE) and GTR to form NR/HDPE/GTR composite. They studied the influence of irradiation on the properties of different TPE blends and reported that irradiation enhanced the interparticle bond between the rubbery phases and HDPE matrix (Figure 10.2). Increasing the dosage leads to higher adhesion force and uniform surface interaction of the constituents that translated to improve mechanical strength, according to blend formulations given in Figure 10.3. Similarly, Hassan [57] reported that gamma irradiation improved the adhesion between PE and GTR leading to improvement in hardness, thermal and mechanical performance of the composite. De Sousa et al. [61] reported a less coarse morphology in HDPE/GTR composite prepared with GTR exposed to a longer period of microwave period than lower period, leading to better mechanical properties.

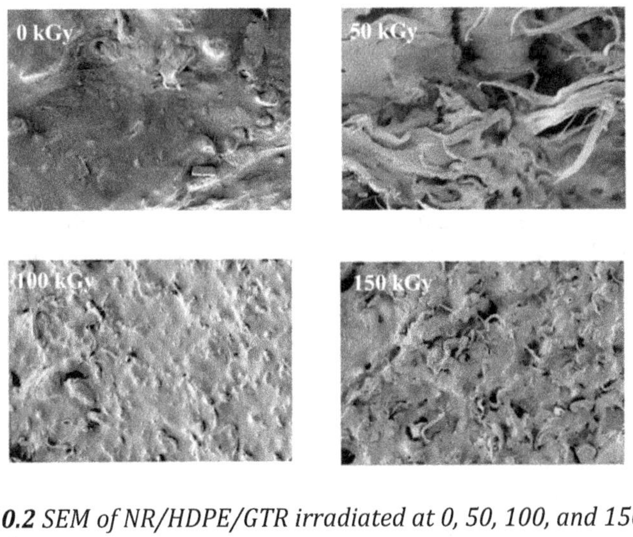

Figure 10.2 *SEM of NR/HDPE/GTR irradiated at 0, 50, 100, and 150 kGy absorption doses.*

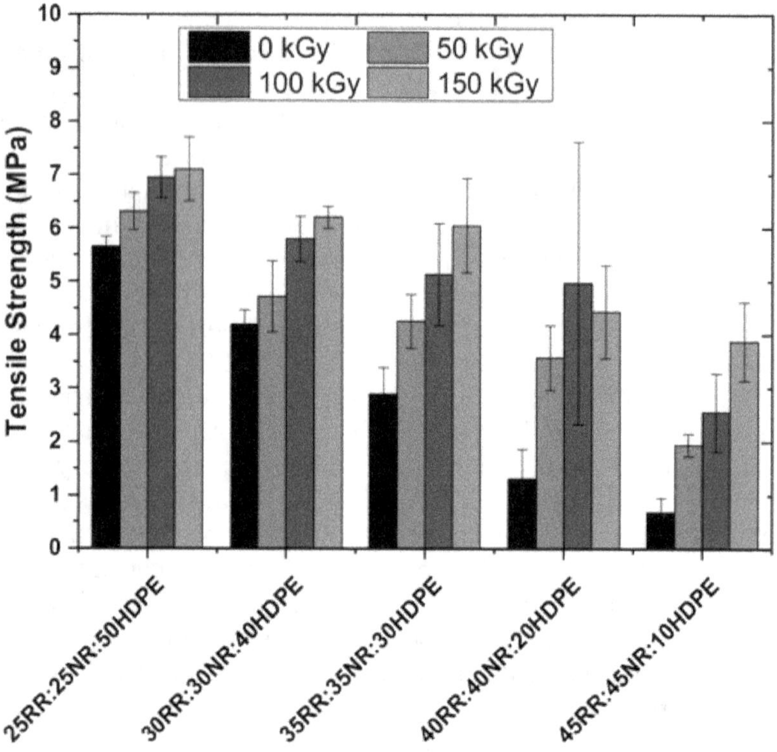

Figure 10.3 *Tensile strength of NR/HDPE/GTR irradiated at 0, 50, 100, and 150 kGy doses.*

10.3.3 Influence of additives

An additional phase may be introduced in the GTR/matrix composite to improve the processability and end-use properties, by creating, for instance, a thermoplastic blend as matrix, which increases the affinity between phases, playing the role of coupling agent. Datta and Włoch [50] utilized recycled polyethylene (rPE), GTR and thermoplastic poly(ester urethane) (TPU) to prepare rPE/GTR/TPU system by melt mixing. Increasing the concentration of neat TPU, while the rPE/GTR composition ratio was kept constant, improved the processability, due to the high melt flow index of TPU compared to rPE. Unfortunately, while hardness property was unaffected, both tensile strength and elastic modulus decreased with an increase of the TPU concentration. In the same way, rebounding resilience, that is the ability of a material to absorb high impact energy, was also decreased. They related the behavior to stronger interactions between GTR and rPE than between GTR and TPU. However, the decrease of TPU, while increasing rPE content, resulted in more permanent plastic deformation, but it improved abrasion resistance due to the best rPE/GTR interfacial interaction. This study shows the difficulty in obtaining an efficient balance between processability and mechanical performance.

Montagna et al. [62] mixed 10 wt% of GTR in polypropylene (PP), high impact polystyrene (HIPS), and their blend PP/HIPS via injection moulding, under an injection pressure of 60 bar and a temperature of 220 °C. It was reported that the decrease of GTR particle size leads to improved performances of such materials. For PP/GTR blend, the morphology study revealed voids created as a result of the pullout of GTR particles, and the void size increased with GTR particle size, confirming poor interfacial adhesion between PP and GTR. Consequently, both tensile and flexural moduli of PP/GTR blend are reduced compared to neat PP. However, good interfacial interaction was observed between HIPS and GTR, which was associated to the presence of polybutadiene in HIPS, resulting in increased tensile and flexural moduli. On other hand, PP/GTR absorbed the highest impact energy of 23 kJ/m^2 compared to HIPS/GTR (6.3 kJ/m^2) or ternary composite PP/HIPS/GTR (13.1 kJ/m^2). Low impact energy observed for HIPS/GTR was related to high rigidity of the material.

The efficient role played by polybutadiene was also highlighted by Wang et al. [63]. High impact polystyrene (HIPS), ethylene-vinyl acetate copolymer (EVA) and GTR were mixed and the effect of styrene-butadiene-styrene copolymer (SBS) as a compatibilizer was studied on the prepared HIPS/EVA/GTR blend. The addition of SBS increased the tensile strength by 145% and elongation at break by 1280%. Sample HIPS/EVA/GTR revealed highly stiff behavior, while those of HIPS/EVA/GTR/SBS indicate soft and tough profound behavior due to the presence of SBS inclusion. They explained that the GTR toughened the HIPS plastic phase while the SBS improved the interfacial adhesion of GTR and HIPS. Thus, the presence of a high concentration of GTR (55 - 70 wt%) in HIPS/EVA/GTR/SBS increased elongation at break and decreased initial modulus.

Wang et al. [64] reported a clear phase separation in the morphology study of uncompatibilized PE/GTR blend, which resulted in poor mechanical properties. So, two different compatibilizers were selected and their individual effect were studied on the properties of the blend: Ethylene-octene copolymer (Engage 8180) and Trans-polyoctenamer (Vestenamer 8012). Addition of Engage 8180 created homogenous morphological structure whereby the embedded GTR particles were difficult to be identified in the matrix. This signifies very good PE-GTR interaction translated to increased properties. Although Vestenamer 8012 also created a more homogenous structure, GTR particles were distinguished in the blend, leading weaker mechanical properties. The "ethylene" component in Engage 8180 is compatible with PE, while the other component "octene" interact with GTR styrene butadiene rubber (SBR). Hence, the good interaction of PE and GTR in the presence Engage 8180 that facilitate better stress transfer. However, in the case of Vestenamer 8012, trans-polyoctenamer has minute compatibility with PE. Thus, it is important to select the right compatibilizer for any give blend formulation.

Different coupling agents were also used, and their efficiency is attested. Kakroodi and Rodrigue [46] prepared maleated polyethylene (MAPE)/GTR thermoplastic elastomer using high concentration of GTR (50 - 90 wt%). The blend MAPE/GTR displayed more than 200% elongation at break, which was described as competitive material. Normally, a TPE is expected to display at least 100% elongation for it to be classified as a high-performance material. For comparison, GTR was incorporated in neat HDPE, but only 64% elongation at break was achieved in the HDPE/GTR composite. This was attributed to poor homogeneity of the morphology of the blend, which was improved with the incorporation of 5 wt% MAPE coupling agent to form the ternary HDPE/MAPE/GTR system. Similarly, the presence of MAPE improved the elastic properties.

Hamim et al. [65] examined the effect of polyethylene grafted maleic anhydride (PE-g-MAH) on the mechanical and swelling behavior of blend HDPE/ethylene-vinyl acetate (EVA) copolymer/GTR. The tensile strength of HDPE/EVA/GTR reduced with increasing GTR concentration because of incompatibility of HDPE/EVA and GTR, leading to fracture initiated from poor HDPE/EVA – GTR interface. A similar trend of property reduction with increasing GTR content was observed for HDPE/EVA/GTR/PE-g-MAH. But, at the same GTR fraction, the enhanced adhesion created by PE-g-MAH leads to strongly higher tensile strength of HDPE/EVA/GTR/PE-g-MAH material compared to HDPE/EVA/GTR one. On the other hand, the mass swell of both HDPE/EVA/GTR and HDPE/EVA/GTR/PE-g-MAH increased with GTR due to increase in elastomer content that absorbed the swelling liquid (toluene). However, for a given concentration of elastomeric phase, the mass of swollen HDPE/EVA/GTR was greater than that of HDPE/EVA/GTR/PE-g-MAH. The chemical bond created by the reaction of the PE-g-MAH acrylic anhydride group and EVA functional group established a good interaction with HDPE, thus, limited the swelling property in the grafted sample.

Da Costa et al. [49] formulated a ternary PP/EPDM/GTR system. Strong deterioration in mechanical strength, elongation and toughness with increasing GTR content was reported, owing to the poor interfacial adhesion between PP matrix and GTR. Indeed, the GTR without strong adhesion with the matrix could not play the role of transferring energy in material, hence, creating microvoids at blend interface, causing crack propagation under tensile loading. However, whenever the EPDM content is superior or equal to that of GTR in PP/EPDM/GTR, the elongation at break increased because EPDM plays a compatibilizing role between PP and GTR phases. Thus, EPDM served the role of compatibilizer, ensuring homogeneous morphology of the composite. However, since EPDM and GTR are softer materials compared to PP, an increase in their concentration resulted in a reduction of PP/EPDM/GTR stiffness.

10.3.4 Influence of cross-linking

Hejna et al. [66] studied the impact of three organic radical initiators (benzoyl peroxide (BP), dicumyl peroxide (DCP) and di-tert-butyl peroxide (DP)) on the cross-link density of HDPE/SBS/GTR blend. Irrespective of the peroxide kind, the measured sol fraction from the swelling test indicated increase in the cross-link densities. They identified 80% increase in the mechanical strength of the blend sample cross-linked with DCP, compared to 50% and 0% in samples with DB and BP, respectively. It was reported that the use of DCP in HDPE/SBS/GTR blend caused the least change in melt flow index, which created excellent crosslinking at the GTR-SBS interface.

Cañavate et al. [67] formulated a thermoplastic vulcanizate from HDPE, EPDM and GTR by a combination of melt-mixing and dynamic vulcanization in the presence of peroxides. HDPE/GTR samples, without EPDM, displayed weak performance due to incompatibility of HDPE and GTR that resulted in poor adhesion. On adding EPDM to the HDPE/GTR composite, toughness and elongation at break increased, but tensile strength and modulus reduced. It was showed that peroxide only vulcanized EPDM, but not GTR. As peroxide content increased, HDPE/EPDM/GTR mechanical properties increased due to greater crosslinking effect of peroxide on EPDM, creating a network of co-continuous phase that encapsulated the GTR particles. Authors reported a reduction in swelling degree with increasing peroxide content, which confirmed the increased crosslinking reaction. When peroxide concentration is sufficiently high, the elongation at break drops, which could be explained by the possible peroxide reaction with the thermoplastic HDPE. These results show the importance of the peroxide vulcanization.

Lu et al. [68] prepared thermoplastic vulcanizate (TPV) from PP/EPDM, with different particle sizes and fractions of GTR, by using a two-roll mill. They observed that tear strength, tensile strength and elongation at break improved with GTR particle size reduction, while the same properties reduced with increase in GTR content. The incorporation of polypropylene grafted maleic anhydride (PP-g-MAH) compatibilizer increased significantly PP/EPDM-GTR interfacial adhesion, hence the enhancement of above-mentioned properties. Moreover, the rheological study indicated a lower apparent viscosity of compatibilized TPV compared to uncompatibilized TPV, which depicted that PP-g-MAH improved the processing performance.

Zhang et al. [69] mixed crosslinked polyethylene (XLPE) and GTR in a high-shear mechanical mill to de-crosslink the constituents in solid-state before melt-mixing them. After 20 cycles of milling, the strength of XLPE/GTR increased from 5.5 to 6.0 MPa, with an elongation at break of 185%. Moreover, after dynamic vulcanization in the presence of sulphur, the strength and elongation at break increased to 9.1 MPa and 201%, respectively.

Li et al. [70] cryogenically mixed GTR with HDPE, and reported significant deterioration of impact strength and elongation at break of the obtained HDPE/GTR material, compared to HDPE matrix. Stress maxima occurring at the HDPE-GTR interface, due to incompatibility of the components, produced cracks and propagated, which resulted in the material break. By against, the impact strength and elongation at break of HDPE/GTR increased by 160% and 150%, respectively, when EPDM, peroxide and silicone oil additives were added. Meanwhile, HDPE/GTR and HDPE/EPDM binary materials present a reduced storage modulus, signifying the influence of the dynamic vulcanizates.

10.3.5 Influence of added fillers

To extend the applications of TPE/TPV while considering the clean environment without risking the targeted properties, incorporation of different fillers are currently practiced in composite development [47][71][72][73]. Incorporated fillers sometimes perform a dual role of compatibilizer and reinforcement. Fillers can reduce the size of GTR particles and facilitate their homogeneous distribution in matrix, creating a synergy among the composite constituents that consequently enhance the physical properties of final product [74].

Micrometric fillers

To obtain durable and more eco-friendly materials, natural cellulose fibres, such as hemp, flax and wood flours, are now incorporated in thermoplastics/GTR to create high-property composites [75][76].

Pang and Ismail [48] studied the influence of kenaf particles (KNF) on PP/GTR mixed. The authors reported that the thermal stability of the composites increased with the KNF concentration. Başboğa et al. [77] developed a wood-plastic composite using HDPE, GTR and red pine wood flours (RPF). Contrary to the general knowledge about lignocellulosic material in increasing the brittleness of composites, the incorporation of GTR augument the hardness and impact properties of lignocellulosic-based material by making it softer. Specifically, increasing the fraction of RPF raised the mechanical performance of the final HDPE/GTR/RPF composites, including the flexural strength with its modulus, and tensile strength/modulus. Kakroodi et al. [78] incorporated lignocellulosic flax fibre in HDPE/GTR in the presence of coupling agent (SEBS) to improve matrix-rubber-fibre surface adhesion. Substantial physical properties improvement was reported. Kakroodi et al. [79] prepared a ternary composite of maleic anhydride grafted polyethylene/GTR/hemp. The addition of hemp fibres increased the mechanical and thermal properties of the system. Rodríguez-Anda et al. [80] prepared composite based on HDPE/GTR and sugarcane bagasse (SB), and found that the mechanical properties were increased with material having 50/20/30 wt% (HDPE/GTR/SB) concentrations.

Other fillers like natural mineral (e.g., feldspar), industrial wastes (fly ash from coal thermal plants, brewers' spent grain) and glass fibre [81][82][83] are among other phases incorporated in thermoplastic/GTR blends to achieve high-performance composites. However, careful material design, selection and treatment process must be ensured to avoid properties degradation. For instance, Sridhar et al. [84] incorporated GTR in PP thermoplastic to form a blend and studied the effect of fly ash on performances of the resulting composite. The fly ash showed more affinity towards the rubbery GTR phase than PP, which created widespread debonding of fly ash particles. The weak interaction between the PP and fly ash particles leads to a reduction of strength by ~ 3% for 10 wt% added fly ash. The authors assumed that the trend could be reversed by treatment of fly ash with coupling agents. Formela et al. [85] prepared a composite foam constituted of a polyurethane (PU) filled with GTR and brewers' spent grain (BSG). Incorporation of GTR into PU/BSG composite raised its strength higher by 50% and accelerated its foaming reaction. In the same way, dynamic mechanical properties and thermal stability were increased with increasing GTR in ternary composite PU/GTR/BSG composite. The authors suggest that BSG fillers impose restrictions on the polymer chains movement thus increased the stiffness of the composite. Zainal and Ismail [86] used short glass fibre (SGF) in PP/GTR to develop a composite. With the use of silane-coupling agent (γ-aminopropyltrimethoxysilane, γ-APS), the thermal, tensile and swelling resistance properties of PP/GTR/SGF composite were significantly increased due to improved PP/GTR-SGF interaction.

Moghaddamzadeh and Rodrigue [87] utilized two waste components derived from worn-out tyres, i.e. GTR and recycled tyre fibre (RTF), and incorporated them in linear low-density polyethylene (LLDPE). High-impact energy LLDPE/GTR/RTF composite was achieved in the presence of compatibilizer MA-g-SEBS that facilitated GTR distribution in the LLDPE matrix.

Table 10.1 lists some of the thermoplastic/GTR blends incorporated with different micro-sized fillers.

Table 10.1 *Thermoplastic/GTR blends incorporated with different types of micrometric fillers*

Blend	GTR particle size	GTR concentration range	Filler type	Filler concentration range	Additional treatment/ agent	Processing method	Properties measured	Refs.
PP/GTR	-	65 phr	Fly ash	10-50 phr	SEBS-g-MAH	Extrusion	Morphology, Tensile, DMA, Thermal, Rheology, Magnetic property	[84]
PP/GTR	0.2-150 µm	30 phr	Kenaf	5-20 phr	-	Melt mixing	Morphology, Tensile, Thermal, Water uptake test	[48]
HDPE/GTR	250 µm	7.5-30 wt%	Flax	7.5-30 wt%	SEBS-g-MAH	Melt mixing	Morphology, Tensile, Flexural, Density	[78]
MAPE/GTR	50-850 µm	5-26 wt%	Hemp	10-60 wt%	-	Extrusion	Morphology, Tensile, Impact test Water uptake and ageing, Hardness, Thermal, Density	[79]
PU/GTR	1 mm	5-20 wt%	Brewers' spent grain	5-20 wt%	Flame retardant, Foaming agents	Mechanical stirring	Foaming parameter, Morphology, Tensile,	[85]

							Structure, DMA, Thermal stability	
HDPE/GTR	0.5-2 mm	10-30 wt%	Sugarcane bagasse	20-50 wt%	MAPE, sulphur	Thermo-compression	Flexural, Shear resistance, Swelling test, Morphology, Machining test	[80]
LLDPE/GTR	$90^{\pm 10}$ µm	5-25 wt%	*RTF	5-25 wt%	SEBS-g-MA	Injection molding	Density, Thermal, Morphology	[87]
PP/GTR	250-500 µm		Short glass fibre 5-20php	5-20 php	γ-APS	Melt mixing	Tensile, Thermal, Swelling resistance, Morphology	[86]
HDPE/GTR	≤4 mm	7.5, 15, 22.5 wt%.	Red pine wood flours (RPF) (0.074-0.841 mm)	15, 30 wt%	PE-g-MA, wax	Screw extruder Injection molding	Density, Swelling, Flexural, tensile, Impact test, Thermal, Morphology	[77]

Rigid nanofillers

The addition of nanofillers in polymeric blends leads to high-performance material, as there is the advantage of combining blends and nanocomposites properties [13][88][89]. It seems that the nanoparticles act as compatibilizer that reduces interfacial tension and promote more particle breakage during processing [61]. Thus, the particles coalescence is reduced, and increase in morphology refinement is achieved. Besides, dynamic vulcanization is adopted to significantly improve the interaction of matrix, GTR and nanoparticles [90]. Meanwhile, when nanoparticles such as carbon black (CB) are added in the matrix, carbon crystallites form, leading to particles aggregation, which causes reduction in the specific surface area of the nanofillers [91]. Therefore, appropriate preparation techniques, like surface treatment of the particles, should be achieved to facilitate agglomerate-free dispersion of nanoparticles in matrix.

Cazan et al. [28] mixed polyethylene terephthalate (PET), HDPE and GTR to prepare a ternary PET/HDPE/GTR composite. The incorporation of different oxides (TiO_2, CaO, ZnO and SiO_2) influenced the interface performance of the blend towards augmenting mechanical properties at different temperatures. Remarkably, SiO_2 gives the superior properties at negative temperature, while CaO supports high-temperature condition. The behaviors of the composites were correlated to specific ionic character of each oxide.

Sripornsawat et al. [47] mixed thermo-chemical devulcanized GTR and copolyester (COPE), and examined the effect of carbon black (CB) on the microstructure and mechanical behavior of COPE/GTR composite. They found that an increase in CB content facilitated the transformation towards the co-continuous structure of COPE/GTR and favored a uniform dispersion of smaller vulcanized GTR domains in the COPE matrix (Figure 10.4). The incorporated CB particles, which further facilitate the rubber-filler interaction, significantly increased the viscosity of the COPE/GTR blend. The resulting shear effects during mixing were enhanced, leading to better CB dispersion and, hence, strengthening of the composite. Thus, the composite COPE/GTR/CB exhibits increased stress relaxation, tensile strength and hardness compared to COPE/GTR, however, with little reduction in elongation at break. Authors also found that CB inclusion improved the damping property of composite.

Figure 10.4 *SEM showing the average size of GTR domain in COPE/GTR/CB at different concentration of CB (a) 0, (b) 5, (c) 10, (d) 15, (e) 20 wt%.*

Vahidifar et al. [74] incorporated multi-walled carbon nanotubes (MWCNT) to prepare PE/GTR/MWCNT nanocomposite, in the presence of a PE-g-MAH as a compatibilizer. The rubber phase was dynamically vulcanized using curing agents including sulphur. Though the GTR were well-dispersed within the PE/GTR composite, the MWCNT further reduced the GTR sizes and improved the homogeneity of PE/GTR/MWCNT morphology. MWCNT also served as a crystallization-nucleating agent in the PE matrix and restricted the mobility of the PE/GTR blend chains, thus, increasing the tensile and viscoelastic properties of the nanocomposite. Moreover, percolation threshold was identified at a low concentration of 1 wt% MWCNT. In fact, at this concentration, the nanofillers formed a network structure along blend to achieve high electrical conductivity property.

Kazemi et al. [92] developed MA-g-PE/GTR/MWCNT nanocomposite for electrical application. They observed that the characteristic peaks of maleic anhydride (1714 and 1791 cm^{-1}) in neat MA-g-PE are significantly reduced in comparison to MA-g-PE/GTR (MA-g-PE/PVR) blend. They concluded that the maleic anhydride groups grafted on PE formed a chemical bond with the C=C bonds of GTR particles, which forms a strong interfacial adhesion. On incorporation of MWCNT, the MA-g-PE-GTR miscibility/compatibility was improved. This resulted in good mechanical performance of the MA-g-

PE/GTR/MWCNT system, including crack propagation inhibition. Moreover, in terms of MWCNT percolation threshold value, the binary MA-g-PE/MWCNT requires a minimum of 1.0 wt% concentration of MWCNT. On the other hand, the ternary MA-g-PE/GTR/MWCNT nanocomposite requires only 0.5 wt% MWCNT and thereafter the conductivity growths rapidly. This was related to the localization of MWCNT in MA-g-PE phase of ternary system, as MWCNT cannot penetrate the complex 3D structure of GTR.

For nanoplatelets, Deriszadeh et al. [71] reinforced the PET/GTR blend with graphene oxide (GO) particles in presence of maleic-anhydride grafted styrene-ethylene-butylene styrene (SEBS-g-MAH) compatibilizer, aiming at developing epoxy/metal joining adhesive. The high stiffness of GO resulted in PET/GTR/GO nanocomposite exhibiting more than 33, 248 and 551% increases in tensile modulus, strength and toughness, respectively. De Sousa et al. [61] compared HDPE/GTR and HDPE/GTR/nanoclay composites. Contrary to mechanical properties deterioration attributed to poor adhesion in HDPE/GTR blend, the addition of nanoclay improved the stiffness, rheological and thermal properties in the ternary.

Bee et al. [93] analyzed the effect of montmorillonite (MMT) and electron beam irradiation on the properties of EVA/GTR blend. They found that the nanoclay, at low mass fractions, in EVA/GTR exhibited an intercalated structure that favored its dispersion and resulted in high mechanical strength, whereas an increase in the MMT concentration led to clay agglomeration that resulted in strength reduction. On the further treatment of EVA/GTR/MMT nanocomposite with irradiation, the interaction effect of the blend with MMT was improved. Hassan (2015) [94] studied the combined effect of MMT clay and gamma irradiation on the thermal and mechanical properties of polyamide (PA) and GTR blend. It was reported that the gamma radiation remarkably effected increase in the tensile strength and thermal stability of the PA/GTR/MMT nanocomposite compared to untreated nanocomposite. The gamma radiation treatment facilitated the release of free radicals on GTR surfaces that attacked the thermoplastic PA to form covalent bonds within the two phases. In addition, the radiation induced the cross-linking that created interatomic interaction by hydrogen bond, while the MMT was homogeneously dispersed, leading to improved properties.

Table 10.2 summarizes various thermoplastic/GTR nanocomposites incorporated with different shapes of rigid nanofillers.

Table 10.2 *Thermoplastic/GTR nanocomposites filled with different types of rigid nanofillers*

Blend	GTR particle size	GTR concentration range	Filler type	Filler concentration range	Additional treatment/ agent	Processing method	Properties measured	Ref.
COPE/GTR	40 mesh	50 wt%	CB	0-40 phr	Dynamic vulcanization	Melt mixing	Mechanical, DMA, Rheology, Morphology	[47]
EVA/GTR	-	20 phr	MMT	2-10 phr	Electron beam irradiation	Extrusion and compression molding	Structure, Gel content test, Tensile, Thermal, Morphology	[93]
HDPE/GTR	-	80 wt%	MMT	5 wt%	Dynamic vulcanization	Melt mixing	Tensile, Rheology, Morphology, DMA	[61]
LLDPE/GTR	11-14 µm	30-50 wt%	MWCNT	1-3 wt%	PE-g-MA Dynamic vulcanization	3-step melt mixing	Morphology, Tensile, Hardness, Thermal, Rheology	[74]
MAPE/GTR	<50 µm	50 wt%	MWCNT		PE-g-MA	Solution, melt casting	Structure, Morphology, Calometric, Tensile, Electrical properties	[92]
PET/HDPE/GTR	1 mm	60 wt%	TiO_2, CaO, ZnO,	0.25-2 wt%	-	Melt mixing	Tensile, Thermal, Contact angle,	[28]

			SiO_2				Structure, Morphology	
PET/HDPE/GTR	1 mm	78-80 wt%	CaO	0.5-2 wt%	-	Melt mixing	Physicochemical structure, Contact angle, Morphology, Mechanical	[75]
PET/GTR	-	20-80 wt%	GO	0.1-0.5 phr	SEBS-g-MAH	Extrusion	Single-lap shear test, Tensile test	[71]

10.4 Concluding Remarks

The utilization of waste tyre rubber (GTR) particles in polymers to fabricate thermoplastic elastomers proves to be an efficient, promising, and practical means of recycling large amount of the worn-out tyres. The development of recyclable composites is further possible with the incorporation of renewable reinforcements. Although the incorporation of GTR in thermoplastic consumes reasonable quantity that addresses environmental concerns, the properties of the final TPE composites need to be improved for competitive advantage, especially if higher percent of GTR (>50%) is to be used. Few methods have been emphasized to improve the interfacial adhesion: addition of compatibilizers/coupling agents, application of dynamic vulcanization techniques, high-energy irradiation treatments and so on. These approaches, individually or in combination with others, improve the compatibility of crosslinked rubber (GTR) molecules making them entangle with thermoplastic molecules and be dispersed homogeneously to form high-performance composites. More particularly, adding nanoparticles in thermoplastic/GTR blends improves the GTR dispersion in thermoplastic phases as well as facilitates the reduction of GTR size in the TPE, which results in better stress transfer in nanocomposites. Recyclability and renewability of thermoplastic/GTR can also be achieved by using natural cellulose fibre as a reinforcing phase without altering the mechanical properties of composites.

Exploring the method of hybridization/combination of two different dimensional rigid nanofillers in thermoplastic/GTR blends can be a great stride in composite development. As simultaneous incorporation of two rigid nanofillers are currently trending in polymer science. Researchers have shown that if two nanofillers have strong affinity, they can facilitate the dispersion of one another, compensate inherent weakness and ultimately combine their individual unique properties to improve the characteristics of the hosting matrix/blend. Moreover, numerical and analytical studies of thermoplastic/GTR nanocomposite behaviors would expand and reveal more phenomenon that is difficult to be captured from experimental investigations.

Besides the common applications of thermoplastic/GTR in automobile, construction and sports, high-impact application in defense technology may be of great benefit. Precisely, armour/ballistic engineering may find thermoplastic/GTR nanocomposites a potential alternative based on the reported impact resistance, thermal and mechanical performances. Thermoplastic/GTR nanocomposites may also be relevant for sound and acoustic applications. Finally, aggressive sensitization and research supports for more innovative research development on the waste tyres recycling should be advanced.

References

[1] P. Junkong, R. Morimoto, K. Miyaji, A. Tohsan, Y. Sakaki, and Y. Ikeda, "Effect of fatty acids on the accelerated sulfur vulcanization of rubber by active zinc/carboxylate complexes," *RSC Adv.*, vol. 10, no. 8, pp. 4772–4785, 2020, doi: 10.1039/c9ra10358a.

[2] C. B. B. Luna, E. M. Araújo, D. D. Siqueira, D. D. de S. Morais, E. A. dos S. Filho, and M. V. L. Fook, "Incorporation of a recycled rubber compound from the shoe industry in polystyrene: Effect of SBS compatibilizer content," *J. Elastomers Plast.*, vol. 52, no. 1, pp. 3–28, 2020, doi: 10.1177/0095244318819213.

[3] H. Yangthong, S. Pichaiyut, S. Wisunthorn, C. Kummerlöwe, N. Vennemann, and C. Nakason, "Role of geopolymer as a cure activator in sulfur vulcanization of epoxidized natural rubber," *J. Appl. Polym. Sci.*, vol. 137, no. 17, p. 48624, 2020, doi: 10.1002/app.48624.

[4] M. Sienkiewicz, H. Janik, K. Borzędowska-Labuda, and J. Kucińska-Lipka, "Environmentally friendly polymer-rubber composites obtained from waste tyres: A review," *J. Clean. Prod.*, vol. 147, pp. 560–571, 2017, doi: 10.1016/j.jclepro.2017.01.121.

[5] M. Sienkiewicz, J. Kucinska-Lipka, H. Janik, and A. Balas, "Progress in used tyres management in the European Union: A review," *Waste Manag.*, vol. 32, no. 10, pp. 1742–1751, 2012, doi: 10.1016/j.wasman.2012.05.010.

[6] D. Czajczyńska, R. Krzyżyńska, H. Jouhara, and N. Spencer, "Use of pyrolytic gas from waste tire as a fuel: A review," *Energy*, vol. 134, pp. 1121–1131, 2017, doi: 10.1016/j.energy.2017.05.042.

[7] L. Simon-Stőger and C. Varga, "PE-contaminated industrial waste ground tire rubber: How to transform a handicapped

resource to a valuable one," *Waste Manag.*, vol. 119, pp. 111–121, 2021, doi: 10.1016/j.wasman.2020.09.037.

[8] H. L. Liu, X. P. Wang, and D. M. Jia, "Recycling of waste rubber powder by mechano-chemical modification," *J. Clean. Prod.*, vol. 245, p. 118716, 2020, doi: 10.1016/j.jclepro.2019.118716.

[9] J. Ghosh *et al.*, "Cradle-to-cradle approach to waste tyres and development of silica based green tyre composites," *Resour. Conserv. Recycl.*, vol. 154, no. October 2019, 2020, doi: 10.1016/j.resconrec.2019.104629.

[10] B. Adhikari, D. De, and S. Maiti, "Reclamation and recycling of waste rubber," *Prog. Polym. Sci.*, vol. 25, no. 7, pp. 909–948, 2000, doi: 10.1016/S0079-6700(00)00020-4.

[11] J. Karger-Kocsis, L. Mészáros, and T. Bárány, "Ground tyre rubber (GTR) in thermoplastics, thermosets, and rubbers," *J. Mater. Sci.*, vol. 48, no. 1, pp. 1–38, 2013, doi: 10.1007/s10853-012-6564-2.

[12] O. M. Sanusi, A. Benelfellah, and N. Aït Hocine, "Clays and carbon nanotubes as hybrid nanofillers in thermoplastic-based nanocomposites – A review," *Appl. Clay Sci.*, vol. 185, p. 105408, 2020, doi: 10.1016/j.clay.2019.105408.

[13] O. M. Sanusi, A. Benelfellah, D. N. Bikiaris, and N. Aït Hocine, "Effect of rigid nanoparticles and preparation techniques on the performances of poly(lactic acid) nanocomposites: A review," *Polym. Adv. Technol.*, vol. 32, no. 2, pp. 444–460, 2021, doi: 10.1002/pat.5104.

[14] M. Nuzaimah, S. M. Sapuan, R. Nadlene, and M. Jawaid, "Recycling of waste rubber as fillers: A review," *IOP Conf. Ser. Mater. Sci. Eng.*, vol. 368, p. 012016, 2018, doi: 10.1088/1757-899X/368/1/012016.

[15] A. Fazli and D. Rodrigue, "Waste rubber recycling: A review on the evolution and properties of thermoplastic elastomers," *Materials (Basel).*, vol. 13, no. 3, p. 782, 2020, doi: 10.3390/ma13030782.

[16] F. N. Archibong, O. M. Sanusi, P. Médéric, and N. Aït Hocine, "An overview on the recycling of waste ground tyre rubbers in thermoplastic matrices: Effect of added fillers," *Resour. Conserv. Recycl.*, vol. 175, p. 105894, 2021, doi: 10.1016/j.resconrec.2021.105894.

[17] M. Myhre and D. A. MacKillop, "Rubber recycling," *Rubber Chem. Technol.*, vol. 75, no. 3, pp. 429–474, 2002, doi:

10.5254/1.3547678.

[18] I. Mangili *et al.*, "Mechanical and rheological properties of natural rubber compounds containing devulcanized ground tire rubber from several methods," *Polym. Degrad. Stab.*, vol. 121, pp. 369–377, 2015, doi: 10.1016/j.polymdegradstab.2015.10.004.

[19] V. L. Shulman, "Tire Recycling," in *Waste*, Second., T. M. Letcher and D. A. Vallero, Eds. Belgium: The European Tyre Recycling Association (ETRA), Brussels, 2019, pp. 489–515. doi: 10.1016/b978-0-12-815060-3.00026-8.

[20] K. Formela, M. Sulkowski, M. R. Saeb, X. Colom, and J. T. Haponiuk, "Assessment of microstructure, physical and thermal properties of bitumen modified with LDPE/GTR/elastomer ternary blends," *Constr. Build. Mater.*, vol. 106, pp. 160–167, 2016, doi: 10.1016/j.conbuildmat.2015.12.108.

[21] A. Ayanoglu and R. Yumrutaş, "Rotary kiln and batch pyrolysis of waste tire to produce gasoline and diesel like fuels," *Energy Convers. Manag.*, vol. 111, pp. 261–270, 2016, doi: 10.1016/j.enconman.2015.12.070.

[22] B. S. Thomas, R. C. Gupta, and V. J. Panicker, "Recycling of waste tire rubber as aggregate in concrete: Durability-related performance," *J. Clean. Prod.*, vol. 112, pp. 504–513, 2016, doi: 10.1016/j.jclepro.2015.08.046.

[23] P. Grammelis, N. Margaritis, P. Dallas, D. Rakopoulos, and G. Mavrias, "A Review on Management of End of Life Tires (ELTs) and Alternative Uses of Textile Fibers," *Energies*, vol. 14, no. 3, p. 571, 2021, doi: 10.3390/en14030571.

[24] F. D. B. De Sousa, "Vulcanization of natural rubber: Past, present and future perspectives," in *Natural Rubber: Properties, Behavior and Applications*, J. L. Hamilton, Ed. Nova Science Publishers, New York, 2016, pp. 47–88.

[25] S. Ramarad, M. Khalid, C. T. Ratnam, A. L. Chuah, and W. Rashmi, "Waste tire rubber in polymer blends: A review on the evolution, properties and future," *Progress in Materials Science*, vol. 72. pp. 100–140, 2015. doi: 10.1016/j.pmatsci.2015.02.004.

[26] H. J. Manuel, "Standards for rubber granulates and powders," *KGK-Kautschuk und Gummi Kunststoffe*, vol. 54, no. 3, pp. 101–105, 2001.

[27] S. L. Zhang, Z. X. Xin, Z. X. Zhang, and J. K. Kim, "Characterization of the properties of thermoplastic elastomers containing

waste rubber tire powder," *Waste Manag.*, vol. 29, no. 5, pp. 1480–1485, 2009, doi: 10.1016/j.wasman.2008.10.004.

[28] C. Cazan, M. Cosnita, and L. Isac, "The influence of temperature on the performance of rubber - PET-HDPE waste -based composites with different inorganic fillers," *J. Clean. Prod.*, vol. 208, pp. 1030–1040, 2019, doi: 10.1016/j.jclepro.2018.10.045.

[29] M. Liang, C. Sun, Z. Yao, H. Jiang, J. Zhang, and S. Ren, "Utilization of wax residue as compatibilizer for asphalt with ground tire rubber/recycled polyethylene blends," *Constr. Build. Mater.*, vol. 230, p. 116966, 2020, doi: 10.1016/j.conbuildmat.2019.116966.

[30] J. Maris, S. Bourdon, J. M. Brossard, L. Cauret, L. Fontaine, and V. Montembault, "Mechanical recycling: Compatibilization of mixed thermoplastic wastes," *Polym. Degrad. Stab.*, vol. 147, pp. 245–266, 2018, doi: 10.1016/j.polymdegradstab.2017.11.001.

[31] X. Shu and B. Huang, "Recycling of waste tire rubber in asphalt and portland cement concrete: An overview," *Constr. Build. Mater.*, vol. 67, no. PART B, pp. 217–224, 2014, doi: 10.1016/j.conbuildmat.2013.11.027.

[32] S. Wang, Q. Wang, X. Wu, and Y. Zhang, "Asphalt modified by thermoplastic elastomer based on recycled rubber," *Constr. Build. Mater.*, vol. 93, pp. 678–684, 2015, doi: 10.1016/j.conbuildmat.2015.06.047.

[33] N. F. Medina, R. Garcia, I. Hajirasouliha, K. Pilakoutas, M. Guadagnini, and S. Raffoul, "Composites with recycled rubber aggregates: Properties and opportunities in construction," *Constr. Build. Mater.*, vol. 188, pp. 884–897, 2018, doi: 10.1016/j.conbuildmat.2018.08.069.

[34] F. D. B. De Sousa, C. H. Scuracchio, G. H. Hu, and S. Hoppe, "Devulcanization of waste tire rubber by microwaves," *Polym. Degrad. Stab.*, vol. 138, pp. 169–181, 2017, doi: 10.1016/j.polymdegradstab.2017.03.008.

[35] J. I. Gumede, B. G. Hlangothi, C. D. Woolard, and S. P. Hlangothi, "Organic chemical devulcanization of rubber vulcanizates in supercritical carbon dioxide and associated less eco-unfriendly approaches: A review," *Waste Manag. Res.*, pp. 1–14, 2021, doi: 10.1177/0734242X211008515.

[36] Z. Liu *et al.*, "Devulcanization of waste tire rubber in supercritical carbon dioxide," *Gaofenzi Cailiao Kexue Yu Gongcheng/Polymeric Mater. Sci. Eng.*, vol. 31, no. 12, pp. 92–97,

2015.

[37] S. Seghar *et al.*, "Devulcanization of styrene butadiene rubber by microwave energy: Effect of the presence of ionic liquid," *Express Polym. Lett.*, vol. 9, no. 12, pp. 1076–1086, 2015, doi: 10.3144/expresspolymlett.2015.97.

[38] R. Saputra, R. Walvekar, M. Khalid, C. T. Ratnam, N. M. Mubarak, and S. Dharaskar, "Devulcanisation of ground rubber tyre by novel ternary deep eutectic solvents," *J. Mol. Liq.*, vol. 306, p. 112913, 2020, doi: 10.1016/j.molliq.2020.112913.

[39] F. Sadaka, I. Campistron, A. Laguerre, and J. F. Pilard, "Controlled chemical degradation of natural rubber using periodic acid: Application for recycling waste tyre rubber," *Polym. Degrad. Stab.*, vol. 97, no. 5, pp. 816–828, 2012, doi: 10.1016/j.polymdegradstab.2012.01.019.

[40] I. Mangili, E. Collina, M. Anzano, D. Pitea, and M. Lasagni, "Characterization and supercritical CO2 devulcanization of cryo-ground tire rubber: Influence of devulcanization process on reclaimed material," *Polym. Degrad. Stab.*, vol. 102, no. 1, pp. 15–24, 2014, doi: 10.1016/j.polymdegradstab.2014.02.017.

[41] Z. Guangjian and W. Jincheng, "Study on application behavior of pyrolysis char from waste tires in silicone rubber composites," *E-Polymers*, vol. 16, no. 3, pp. 255–264, 2016, doi: 10.1515/epoly-2015-0285.

[42] Y. Li, S. Zhao, and Y. Wang, "Microbial desulfurization of ground tire rubber by Thiobacillus ferrooxidans," *Polym. Degrad. Stab.*, vol. 96, no. 9, pp. 1662–1668, 2011, doi: 10.1016/j.polymdegradstab.2011.06.011.

[43] G. G. Choi, S. J. Oh, and J. S. Kim, "Clean pyrolysis oil from a continuous two-stage pyrolysis of scrap tires using in-situ and ex-situ desulfurization," *Energy*, vol. 141, pp. 2234–2241, 2017, doi: 10.1016/j.energy.2017.12.015.

[44] C. Yao, S. Zhao, Y. Wang, B. Wang, M. Wei, and M. Hu, "Microbial desulfurization of waste latex rubber with Alicyclobacillus sp.," *Polym. Degrad. Stab.*, vol. 98, no. 9, pp. 1724–1730, 2013, doi: 10.1016/j.polymdegradstab.2013.06.002.

[45] S. Satapathy, A. Nag, and G. B. Nando, "Thermoplastic elastomers from waste polyethylene and reclaim rubber blends and their composites with fly ash," *Process Saf. Environ. Prot.*, vol. 88, no. 2, pp. 131–141, 2010, doi: 10.1016/j.psep.2009.12.001.

[46] A. R. Kakroodi and D. Rodrigue, "Highly filled thermoplastic elastomers from ground tire rubber, maleated polyethylene and high density polyethylene," *Plast. Rubber Compos.*, vol. 42, no. 3, pp. 115–122, 2013, doi: 10.1179/1743289812Y.0000000042.

[47] B. Sripornsawat, S. Saiwari, and C. Nakason, "Thermoplastic vulcanizates based on waste truck tire rubber and copolyester blends reinforced with carbon black," *Waste Manag.*, vol. 79, pp. 638–646, 2018, doi: 10.1016/j.wasman.2018.08.038.

[48] A. L. Pang and H. Ismail, "Influence of kenaf form and loading on the properties of kenaf-filled polypropylene/waste tire dust composites: A comparison study," *J. Appl. Polym. Sci.*, vol. 131, no. 19, pp. 2–7, 2014, doi: 10.1002/app.40877.

[49] H. M. Da Costa, V. D. Ramos, W. S. Da Silva, and A. S. Sirqueira, "Analysis and optimization of polypropylene (PP)/ethylene-propylene-diene monomer (EPDM)/scrap rubber tire (SRT) mixtures using RSM methodology," *Polym. Test.*, vol. 29, no. 5, pp. 572–578, 2010, doi: 10.1016/j.polymertesting.2010.04.003.

[50] J. Datta and M. Włoch, "Morphology and properties of recycled polyethylene/ground tyre rubber/thermoplastic poly(ester-urethane) blends," *Macromol. Res.*, vol. 23, no. 12, pp. 1117–1125, 2015, doi: 10.1007/s13233-015-3155-5.

[51] E. Lievana and J. Karger-Kocsis, "Use of ground tyre rubber (GTR) in thermoplastic polyolefin elastomer compositions," *Prog. Rubber, Plast. Recycl. Technol.*, vol. 20, no. 1, pp. 1–10, 2004, doi: 10.1177/147776060402000101.

[52] N. Candau, O. Oguz, N. León Albiter, G. Förster, and M. L. Maspoch, "Poly (Lactic Acid)/Ground Tire Rubber Blends Using Peroxide Vulcanization," *Polymers (Basel).*, vol. 13, no. 9, p. 1496, 2021, doi: 10.3390/polym13091496.

[53] I. Z. Halász, D. Kocsis, D. Á. Simon, A. Kohári, and T. Bárány, "Development of Polypropylene-based Thermoplastic Elastomers with Crumb Rubber by Dynamic Vulcanization: A Potential Route for Rubber Recycling," *Period. Polytech. Chem. Eng.*, vol. 64, no. 2, pp. 248–254, 2020, doi: 10.3311/PPch.13962.

[54] S. Li, H. Tian, H. Wu, N. Ning, M. Tian, and L. Zhang, "Coupling effect of molecular weight and crosslinking kinetics on the formation of rubber nanoparticles and their agglomerates in EPDM/PP TPVs during dynamic vulcanization," *Soft Matter*, vol. 16, pp. 2185–2198, 2020, doi: 10.1039/c9sm02059d.

[55] P. Lima, J. Oliveira, and V. Costa, "Partial replacement of EPDM by GTR in thermoplastic elastomers based on PP/EPDM: Effects on morphology and mechanical properties," *J. Appl. Polym. Sci.*, vol. 131, no. 8, pp. 1–10, 2014, doi: 10.1002/app.40160.

[56] P. S. Garcia, J. A. de Lima, C. H. Scuracchio, and S. A. Cruz, "The effect of adding devulcanized rubber on the thermomechanical properties of recycled polypropylene," *J. Appl. Polym. Sci.*, vol. 138, no. 29, pp. 1–11, 2021, doi: 10.1002/app.50703.

[57] M. M. Hassan, R. O. Aly, J. A. Hasanen, and E. S. F. El Sayed, "The effect of gamma irradiation on mechanical, thermal and morphological properties of glass fiber reinforced polyethylene waste/reclaim rubber composites," *J. Ind. Eng. Chem.*, vol. 20, no. 3, pp. 947–952, 2014, doi: 10.1016/j.jiec.2013.06.027.

[58] S. F. Mohamad *et al.*, "Insight on the properties of thermoplastic elastomer-based natural rubber and recycled rubber post-treated with electron beam irradiation," *Mater. Res. Express*, vol. 8, p. 025302, 2021, doi: 10.1088/2053-1591/abe064.

[59] S. Ramarad, C. T. Ratnam, M. Khalid, A. L. Chuah, and S. Hanson, "Improved crystallinity and dynamic mechanical properties of reclaimed waste tire rubber/EVA blends under the influence of electron beam irradiation," *Radiat. Phys. Chem.*, vol. 130, pp. 362–370, 2017, doi: 10.1016/j.radphyschem.2016.09.023.

[60] R. Sonnier, E. Leroy, L. Clerc, A. Bergeret, and J. M. Lopez-Cuesta, "Polyethylene/ground tyre rubber blends: Influence of particle morphology and oxidation on mechanical properties," *Polym. Test.*, vol. 26, no. 2, pp. 274–281, 2007, doi: 10.1016/j.polymertesting.2006.10.011.

[61] F. D. B. De Sousa *et al.*, "Blends of ground tire rubber devulcanized by microwaves/HDPE-Part B: Influence of clay addition," *Polimeros*, vol. 25, no. 4, pp. 382–391, 2015, doi: 10.1590/0104-1428.1955.

[62] L. S. Montagna, L. S. Bento, M. R. Da Silva Silveira, and R. M. C. Santana, "Evaluation of the effect of the incorporation of rubber tire waste particles on the properties of PP, HIPS and PP/HIPS matrices," *Polimeros*, vol. 23, no. 2, pp. 169–174, 2013, doi: 10.4322/polimeros.2013.077.

[63] Z. Wang, Y. Zhang, F. Du, and X. Wang, "Thermoplastic elastomer based on high impact polystyrene/ethylene-vinyl acetate copolymer/waste ground rubber tire powder composites

compatibilized by styrene-butadiene-styrene block copolymer," *Mater. Chem. Phys.*, vol. 136, no. 2–3, pp. 1124–1129, 2012, doi: 10.1016/j.matchemphys.2012.08.063.

[64] Y. H. Wang, Y. K. Chen, and D. Rodrigue, "Production of thermoplastic elastomers based on recycled pe and ground tire rubber: Morphology, mechanical properties and effect of compatibilizer addition," *Int. Polym. Process.*, vol. 33, no. 4, pp. 525–534, 2018, doi: 10.3139/217.3544.

[65] F. A. R. Hamim, I. H. A. Aziz, and S. A. Ghani, "Effect of pegmah on tensile properties and swelling behaviour of recycled high density polyethylene/ethylene vinyl acetate/waste tyre dust (r-HDPE/EVA/WTD) composites," *Appl. Mech. Mater.*, vol. 554, pp. 137–140, 2014, doi: 10.4028/www.scientific.net/AMM.554.137.

[66] A. Hejna, M. Klein, M. R. Saeb, and K. Formela, "Towards understanding the role of peroxide initiators on compatibilization efficiency of thermoplastic elastomers highly filled with reclaimed GTR," *Polym. Test.*, vol. 73, no. September 2018, pp. 143–151, 2019, doi: 10.1016/j.polymertesting.2018.11.005.

[67] J. Cañavate, P. Casas, X. Colom, and F. Nogués, "Formulations for thermoplastic vulcanizates based on high density polyethylene, ethylene-propylene-diene monomer, and ground tyre rubber," *J. Compos. Mater.*, vol. 45, no. 11, pp. 1189–1200, 2011, doi: 10.1177/0021998310369596.

[68] X. Lu, W. Wang, and L. Yu, "Waste ground rubber tire powder/thermoplastic vulcanizate blends: Preparation, characterization, and compatibility," *J. Appl. Polym. Sci.*, vol. 131, no. 3, p. 39868, 2014, doi: 10.1002/app.39868.

[69] X. Zhang, C. Lu, and M. Liang, "Preparation of thermoplastic vulcanizates based on waste crosslinked polyethylene and ground tire rubber through dynamic vulcanization," *J. Appl. Polym. Sci.*, vol. 122, pp. 2110–2120, 2011, doi: 10.1002/app.34293.

[70] Y. Li, Y. Zhang, and Y. Zahang, "Mechanical properties of high-density polyethylene/scrap rubber powder composites modified with ethylene-propylene-diene terpolymer, dicumyl peroxide, and silicone oil," *J. Appl. Polym. Sci.*, vol. 88, no. 8, pp. 2020–2027, 2003, doi: 10.1002/app.11907.

[71] A. Deriszadeh, F. Shahraki, D. M. Kalhori, and O. M. Jazani, "Epoxy/GO nanocomposite adhesives reinforced with waste blends of PET/GTR: Evaluation of mechanical properties,"

Prog. Org. Coatings, vol. 136, p. 105292, 2019, doi: 10.1016/j.porgcoat.2019.105292.

[72] M. Favakeh, S. Bazgir, and M. Karbasi, "Dynamically vulcanized thermoplastic elastomer nanocomposites based on linear low-density polyethylene/styrene-butadiene rubber/nanoclay/bitumen: morphology and rheological behavior," *Iran. Polym. J.*, vol. 29, no. 3, pp. 209–217, 2020, doi: 10.1007/s13726-020-00786-9.

[73] O. M. Sanusi, A. Benelfellah, Z. Terzopoulou, D. N. Bikiaris, and N. Aït Hocine, "Properties of polylactide reinforced with montmorillonite/multiwalled carbon nanotube hybrid," *Mater. Today Proc.*, vol. 47, pp. 3247–3250, 2021, doi: 10.1016/j.matpr.2021.06.448.

[74] A. Vahidifar, E. Esmizadeh, M. Elahi, M. H. R. Ghoreishy, G. Naderi, and D. Rodrigue, "Thermoplastic vulcanizate nanocomposites based on polyethylene/reclaimed rubber: A correlation between carbon nanotube dispersion state and electrical percolation threshold," *J. Appl. Polym. Sci.*, vol. 136, no. 30, p. 47795, 2019, doi: 10.1002/app.47795.

[75] M. Cosnita, C. Cazan, and A. Duta, "The influence of inorganic additive on the water stability and mechanical properties of recycled rubber, polyethylene terephthalate, high density polyethylene and wood composites," *J. Clean. Prod.*, vol. 165, pp. 630–636, 2017, doi: 10.1016/j.jclepro.2017.07.103.

[76] A. K. Mohanty, S. Vivekanandhan, J. M. Pin, and M. Misra, "Composites from renewable and sustainable resources: Challenges and innovations," *Science (80-.).*, vol. 362, no. 6414, pp. 536–542, 2018, doi: 10.1126/science.aat9072.

[77] İ. H. Başboğa, İ. Atar, K. Karakuş, and F. Mengeloğlu, "Determination of some technological properties of injection molded pulverized-HDPE based composites reinforced with micronized waste tire powder and red pine wood wastes," *J. Polym. Environ.*, vol. 28, pp. 1776–1794, 2020, doi: 10.1007/s10924-020-01726-7.

[78] A. R. Kakroodi, J. Bainier, and D. Rodrigue, "Mechanical and morphological properties of flax fiber reinforced high density polyethylene/recycled rubber composites," *Int. Polym. Process.*, vol. 27, no. 2, pp. 196–204, 2012, doi: 10.3139/217.2473.

[79] A. R. Kakroodi, Y. Kazemi, and D. Rodrigue, "Mechanical, rheological, morphological and water absorption properties of maleated polyethylene/hemp composites: Effect of ground tire

rubber addition," *Compos. Part B Eng.*, vol. 51, pp. 337–344, 2013, doi: 10.1016/j.compositesb.2013.03.032.

[80] R. Rodríguez-Anda, S. García-Enriquez, M. G. Lomelí-Ramírez, F. J. Fuentes-Talavera, J. A. Silva-Guzmán, and J. G. Torres-Rendón, "Composite based on recycled high-density polyethylene-sugarcane bagasse-tire rubber waste: formulation, mechanical and workability evaluation," *Interciencia J.*, vol. 45, no. 9, pp. 80–103, 2020.

[81] M. Allouch, M. Kamoun, J. Mars, M. Wali, and F. Dammak, "Experimental investigation on the mechanical behavior of recycled rubber reinforced polymer composites filled with aluminum powder," *Constr. Build. Mater.*, vol. 259, p. 119845, 2020, doi: 10.1016/j.conbuildmat.2020.119845.

[82] N. Bukit, E. M. Ginting, I. S. Pardede, E. Frida, and B. F. Bukit, "Mechanical properties of composite thermoplastic hdpe / natural rubber and palm oil boiler ash as a filler," *J. Phys. Conf. Ser.*, vol. 1120, p. 012003, 2018, doi: 10.1088/1742-6596/1120/1/012003.

[83] O. M. Sanusi, O. D. Komolafe, T. O. Ogundana, M. O. Olaleke, and Y. Y. Sanni, "Development of Wood-ash/Resin Polymer Matrix Composite for Body Armour Application," *FUOYE J. Eng. Technol.*, vol. 1, no. 1, pp. 10–14, 2016, doi: 10.46792/fuoyejet.v1i1.4.

[84] V. Sridhar *et al.*, "Fly ash reinforced thermoplastic vulcanizates obtained from waste tire powder," *Waste Manag.*, vol. 29, no. 3, pp. 1058–1066, 2009, doi: 10.1016/j.wasman.2008.08.013.

[85] K. Formela *et al.*, "Structural, thermal and physico-mechanical properties of polyurethane/brewers' spent grain composite foams modified with ground tire rubber," *Ind. Crops Prod.*, vol. 108, pp. 844–852, 2017, doi: 10.1016/j.indcrop.2017.07.047.

[86] Z. Zainal and H. Ismail, "The effects of short glass fibre (SGF) loading and a silane coupling agent on properties of polypropylene/waste tyre dust/short glass fibre (PP/WTD/SGF) composites," *Polym. Plast. Technol. Eng.*, vol. 50, pp. 297–305, 2011, doi: 10.1080/03602559.2010.531868.

[87] S. Moghaddamzadeh and D. Rodrigue, "The effect of polyester recycled tire fibers mixed with ground tire rubber on polyethylene composites. Part I : Morphological analysis," *Prog. Rubber, Plast. Recycl. Technol.*, vol. 34, no. 4, pp. 200–220, 2018, doi: 10.1177/1477760618798267.

[88] O. M. Sanusi *et al.*, "Influence of montmorillonite/carbon nanotube hybrid nanofillers on the properties of poly(lactic acid)," *Appl. Clay Sci.*, vol. 201, p. 105925, 2021, doi: 10.1016/j.clay.2020.105925.

[89] O. M. Sanusi, A. Benelfellah, L. Papadopoulos, Z. Terzopoulou, D. N. Bikiaris, and N. Aït Hocine, "Properties of poly(lactic acid)/montmorillonite/carbon nanotubes nanocomposites: determination of percolation threshold," *J. Mater. Sci.*, vol. 56, pp. 16887–16901, 2021, doi: 10.1007/s10853-021-06378-z.

[90] S. M. R. Paran, G. Naderi, M. H. R. Ghoreishy, and A. Heydari, "Enhancement of mechanical, thermal and morphological properties of compatibilized graphene reinforced dynamically vulcanized thermoplastic elastomer vulcanizates based on polyethylene and reclaimed rubber," *Compos. Sci. Technol.*, vol. 161, pp. 57–65, 2018, doi: 10.1016/j.compscitech.2018.04.006.

[91] J. C. Huang, "Carbon black filled conducting polymers and polymer blends," *Adv. Polym. Technol.*, vol. 21, no. 4, pp. 299–313, 2002, doi: 10.1002/adv.10025.

[92] Y. Kazemi, A. Ramezani Kakroodi, A. Ameli, T. Filleter, and C. B. Park, "Highly stretchable conductive thermoplastic vulcanizate/carbon nanotube nanocomposites with segregated structure, low percolation threshold and improved cyclic electromechanical performance," *J. Mater. Chem. C*, vol. 6, no. 2, pp. 350–359, 2018, doi: 10.1039/c7tc04501h.

[93] S. T. Bee, L. T. Sin, T. T. Hoe, C. T. Ratnam, S. L. Bee, and A. R. Rahmat, "Study of montmorillonite nanoparticles and electron beam irradiation interaction of ethylene vinyl acetate (EVA)/de-vulcanized waste rubber thermoplastic composites," *Nucl. Instruments Methods Phys. Res. Sect. B Beam Interact. with Mater. Atoms*, vol. 423, no. December 2017, pp. 97–110, 2018, doi: 10.1016/j.nimb.2018.03.013.

[94] M. M. Hassan, "Synergistic effect of montmorillonite-clay and gamma irradiation on the characterizations of waste polyamide copolymer and reclaimed rubber powder nanocomposites," *Compos. Part B Eng.*, vol. 79, pp. 28–34, 2015, doi: 10.1016/j.compositesb.2015.01.046.

11

The Novelty of Rubber in Concrete Mix

Camille A. Issa

Lebanese American University, Byblos, Lebanon
cissa@lau.edu.lb

11.1 Introduction

Disposal of waste rubber tires has developed into a main environmental topic all over the globe. Annually millions of tires are abandoned, thrown away or buried all over the globe, symbolizing a very grave threat to the ecosystem. It was anticipated that almost 1000 million tires end their service life annually and out of that, more than 50% are abandoned to landfills or trash without any treatment. By the year 2030, there would be 5000 million tires to be abandoned on a regular basis. Tire burning, which was the simplest and cheapest method of disposal, causes severe fire hazards. Temperature in that region rises and the poisonous smoke with uninhibited emissions of possibly harmful compounds is very hazardous to humans, animals, and plants. The residue powder left after burning pollutes the soil.

As a solution to this predicament, the need to incorporate scrap tire rubber as recycled materials in the construction industry are turning into an essential endeavor and this is classified as sustainable materials. The utilization of waste rubber in concrete mixtures leads to landfill avoidance and decreases the exhaustion of naturally derived materials. Waste rubber can be utilized as fine aggregate, coarse aggregate, or both.

This chapter summarizes the status of the body of knowledge realized so far on the utilization of waste rubber as partially or fully replacing natural fine aggregates in conventional concrete mixtures. This chapter summarizes the effects of waste sand rubber on the workability, setting time, bleeding, density, strength, impact energy, impact load, toughness, ductility, shrinkage, abrasion resistance, freeze/thaw resistance, fire resistance, thermal insulation, carbonation resistance, corrosion resistance, water absorption, porosity, chloride ion penetration, resistance to aggressive environmental, energy absorption, sound absorption, electrical resistance, and cracking resistance.

As known, waste generation in the EU was estimated to stand at over 1.43 billion tons per year and was increasing at rates comparable to those of economic growth [70]. Consequently, waste decline and recycling are very important aspects in a waste management context since they assist in preserving natural assets and diminish the necessity for precious landfill space. Waste rubber is one of the significant wastes which has been a key concern in the planet. Data explored from the literature has shown that in 2005, over 10 billion tires are discarded worldwide yearly [5]. According to Colom et al. [24], it was estimated that around 1 billion tires are retracted from consumption yearly. It was estimated that 1000 million tires reach the end of their useful life yearly. By the year 2030, the number can reach up to 1200 million tires representing almost 5000 million tires (including stockpiled) to be abandoned on a routine basis [86]. In the United States, for example, there were 2–3 billion tires discarded in landfills annually [53] and 275 million scrap tires stockpiled across the Nation, with an increase of 290 million tires produced yearly [14]. It was estimated that one car tire per person was discarded each year in the developed world and hence 1 billion waste tires were disposed globally each year [70]. It was projected that roughly 4 billion of waste tires were in landfills and stockpiles globally [20].

The US Environmental Protection Agency (EPA) reported that 290 million scrap tires were produced in 2003. Of the 290 million, 45 million of these scrap tires were utilized to produce automotive and truck tire re-treads [3]. In Europe every year, 355 million tires are manufactured in 90 plants, embodying 24% of world production [96]. In addition, the EU has millions of used tires that have been unlawfully discarded or stockpiled. The ineffective disposal of tires may, in some cases, pose a potential threat to human health due to fire risk, being haven for rodents, or other pests such as mosquitoes and the surge in environmental hazards. Most countries all over the globe have depend on land filling to dispose of used tires but the limited space and their potential for reuse has led to many countries enforcing a ban on this procedure. The present estimate for these remarkable stockpiles all over the EU stands at 5.5 million tons and the projected yearly cost for the administration of end-of-life tires

(ELT) is estimated at € 600 million [96].

In UK, roughly 37 million tires were used annually in 2002. This number continues to increase [73]. In Thailand, the record of the year 2000 alone revealed a utilization of approximately 250,000 metric tons of rubber products. About 38% of this were vehicle tires. These numbers keep on growing every year with the numbers of vehicles, as do the future problems relating to waste tires [107]. In Taiwan, over 100,000 tons of waste tires are generated yearly, and this number is increasing [127]. In Australia, the trend for amassed waste tires was growing at a rate of 2% and it was anticipated more than 20 million tires wereamassed in landfills by year 2010 [75]. In France, over 10 million scrap-tires per year were generated [104]. In Singapore, rubber tires are disposed of by sending them to incineration plants. Based on waste statistics and recycling rate for 2008 [130], from a total scrap tire waste output of 25,100 tons, 3000 tons were disposed of, whereas 22,100 tons were recycled such that the recycling rate of this waste was approximately 88%. In Spain, 42% of worn tires generated have been destined to energy recovery, mainly in cement kilns, 10% have been reused and 48% have been destined to material recovery in 2011[28]. In 2010 the EU27 plus Turkey produced about 4.5 million tons of tires, estimated at 17 million tons. From this quantity, it is judged that more than 3.2 million tons of waste tires are abandoned annually and for this reason the disposal of waste tires is considered as an increasing environmental and economic problem [128].

Most of waste tires are landfill disposed, which is the most common method (Figure 11.1). This method will be drastically reduced soon because of the recent introduction of European Union decrees that include significant restraints on this procedure in favor of alternatives leaning towards material and energy recovery. In addition, the disposal of used tires in landfills, stockpiles or illegal dumping grounds increases the risk of accidental fires with uncontrolled emission of potentially harmful compounds [133]. To adequately dispose of these millions of waste tires, the utilization of inventive procedures to recycle them is essential. However, a significant percentage of the waste tires are used in civil engineering applications such as road and rail foundations and embankments (0.24 million tons) retreaded (0.26million tons) or exported (0.33 million tons) each year [31]. Recycled waste tires are used for energetic purposes in cement kilns [104], incinerated to produce electricity [83], used as an additive to PC mortar/concrete [4,15,74], as a light weight filler [21], as crash barriers, bumpers and artificial reefs, etc. [54,103].

Figure 11.1 *Typical waste tire disposal scenery.*

The global utilization of natural sand as a fine aggregate in concrete mixture is steep and several developing countries have faced some strain in the source of natural sand to meet the growing needs of infrastructural development in recent years. In several countries there is a scarcity of natural fine aggregate which is suitable for construction. Overall, during the last 15 years, it has become evident that the accessibility of suitable natural sand is diminishing [98-99]. The scarcity of sources of natural sand has led for the utilization of by-products as fine aggregate. Reuse of waste rubber as a partial or full replacement of fine aggregate in construction projects not only decreases the necessity for extraction of natural raw materials, but also preserves landfill space and the environment. The published literature has useful review papers related to properties of concrete containing scrap-tire rubber [104], mechanical properties of rubber [95], fresh/hardened properties of rubberized and self-compacting rubberized concrete [48], pyrolysis of waste tires [129], recycled tire rubber modified bitumen for road asphalt mixtures [94] and recycling of waste tire rubber in asphalt and Portland cement concrete [103].

This chapter aims at summarizing the novelty of the studies conducted on the effect of partial or full replacement of fine aggregates with waste rubber on the properties of customary concrete. More specifically, the effects of waste rubber sand on the workability, setting time, bleeding, density, strength, impact energy, impact load, toughness, ductility, shrinkage, abrasion resistance, freeze/thaw resistance, fire resistance, thermal insulation, carbonation resistance, corrosion resistance, water absorption, porosity, chloride ion penetration, resistance to aggressive environmental, energy absorption, sound absorption, electrical resistance and cracking resistance of rubberized concrete are reviewed.

11.2 Workability, Setting Time, Segregation, and Bleeding

From the review of the literature, it could be deduced that numerous studies reported that the addition of waste rubber sand in the concrete mixture reduced the workability (Figure 11.2). This is resulted from the higher water absorption of rubber sand versus natural sand. The decline in the workability mainly depends on the rubber content and particle size. On the contrary, few studies reported that the inclusion of rubber into the concrete mixture has increased the workability (Figure 11.2). It is worth stating that a few other studies believed that some rubber sand contents reduced workability, whilst other contents increased it [13,60,92]. The addition of rubber sand in the concrete mixture would normally result in a decline in the workability and this is normally expected from using this type of recycled material. Table 11.1 summarizes the effect of rubber sand as replacement percentage of mass on the workability, setting time, segregation, and bleeding of concrete mixtures.

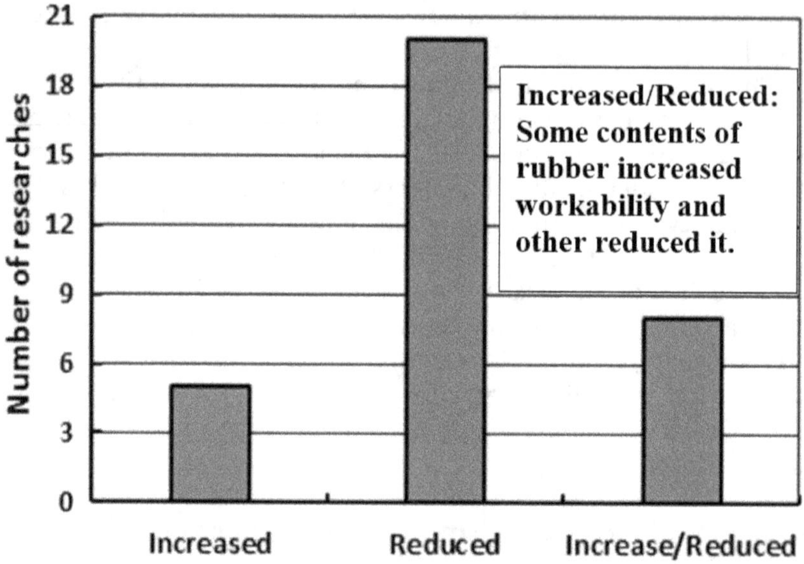

Figure 11.2 *Number of research studies versus the effect of waste rubber sand on the workability of concrete mixtures [99].*

Table 11.1 *Effects of rubber replacement of sand on the workability, setting time, segregation, and bleeding of concrete mixtures*

References	Rubber content (%)	Size (mm)	Effects
Wang et al. [127]	10, 20, 30, and 40	4.75	- Increased bleeding
Mohammadi et al. [75]	20 and 40	–	- Reduced workability
Albano et al. [6]	5 and 10	0.29 and 0.59	- Reduced workability - No segregation
Holmes et al. [51]	7.5	0.425–4.75	- Reduced workability
Guo et al. [47]	4, 8, 12, and 16	0.85–1.4	- Reduced workability
Youssf et al. [132]	5, 10, and 20	1.18 and 2.36	- Reduced workability
Raj et al. [97]	5, 10, 15, and 20	4.75	- Reduced workability
Ganesan et al. [35]	15 and 20	4.75	- Reduced workability
Gesoglu and Guneyisi [37]	5, 15, and 25	Sand Gradation	- Reduced workability
Ozbay et al. [85]	5, 15, and 25	0–3	- Reduced workability
Guneyisi [44]	5, 15, and 25	Sand Gradation	- Reduced workability - Amended fresh properties - Initial and final setting time
Rahman et al. [96]	28	1–4	- Reduced workability

Topcu and Demir [114]	10, 20, and 30	1–4	- Reduced workability
Karahan et al. [59]	10, 20, and 30	0.15–4.75	- Reduced workability
Grdic et al. [42]	10, 20, and 30	0.5–4	- Reduced workability
Batayneh et al. [14]	20, 40, 60, 80, and 100	0.15–4.75	- Reduced workability
Taha et al. [110]	25–100	1–5	- Reduced workability
Khatib and Bayomy [61]	5–100	Close to sand	- Reduced workability
Pelisser et al. [91]	10	<4.8	- Increased workability
Bravo and de Brito [19]	5, 10, and 15	Like sand gradation	- 5% and 15% reduced workability - 10% increased workability
Onuaguluchi and Panesar [84]	5, 10, and 15	~86% smaller than 2.3 mm	- Increased workability
Antil et al. [8]	5, 10, 15, and 20	0.075–4.75	- 5% and 10% increased workability - 15% and 20% reduced workability
Parveen et al. [89]	5, 10, 15 and 20	0.075–4.75	- 5% and 10% increased workability - 15% and 20% reduced workability
Balaha et al. [13]	5, 10, 15, and 20	<4	- Increased workability
Azmi et al. [12]	10, 15, 20, and 30	2-2.36	- Increased workability
Wang et al. [127]	10, 20, 30, and 40	4.75	- Increased initial setting time - 10% and 20% increased workability - 30% and 40% reduced workability
Topcu and Sarıdemir [115]	15, 30 and 45	0–1and 1–4	- Increased workability
Turgut and Yesilata [121]	10–70	0.075–4.75	- Up to 10% increased workability - 50–70% reduced workability
Khaloo et al. [60]	25, 50, 75, and 100	4.75	- 25%, 50%, and 75% increased workability - 100% reduced workability

11.3 Density

11.3.1 Fresh density

Table 11.2 summarizes the reduction percentage in the fresh unit weight when natural sand is partially replaced with rubber sand in concrete mixtures.

Table 11.2 *Effect of rubber sand on the reduction of the fresh unit weight of concrete*

References	Rubber content (%)	Size (mm)	% Reduction
Skripkiunas et al. [105]	3.2	0–1	0.66
Albano et al. [6]	5 and 10	0.59	20.33 and 29.58
Albano et al. [6]	5 and 10	0.29	22.51 and 38.2
Pedro et al. [90]	5, 10, and 15	0–2	4.17, 7.21 and 10
Gesoglu et al. [39]	10 and 20	4	1.41 and 6.85
Thomas et al. [111]	2.5–20	Mesh 30, 0.8–2mm, 2–4 mm	From 1 to 13.22
Ozbay et al. [85]	5, 15, and 25	0–3	0.91, 3.32 and 5
Grdic et al. [42]	10, 20, and 30	0.5–4	3.8, 9.3 and 13.3
Fadiel et al. [32]	10, 20, 30, and 40	0–0.6	10, 20.94, 27.1 and 35.37
Fadiel et al. [32]	10, 20, 30, and 40	0.84–2	6.1, 15.88, 21.59 and 29.46
Mohammadi et al. [75]	10, 20, 30, and 40	–	2.47, 4.81, 7.42 and 11.81
Topcu and Sarıdemir [115,116]	45	0–1	1.13
Topcu and Sarıdemir [115,116]	45	1–4	16.92
Taha et al. [110]	25, 50, 75, and 100	1–5	11.57, 14.35, 17.13 and 21.48
Khaloo et al. [60]	25, 50, 75, and 100	≤ 4.75	15.46, 28.66, 34 and 34.79
Batayneh et al. [14]	20, 40, 60, 80, and 100	0.15–4.75	7.59, 13.78, 17.17, 23.69 and 27.44

11.3.2 Hardened density

Table 11.3 summarizes the reduction percentage in the dry unit weight when natural sand is partially replaced with rubber sand in concrete mixtures and some water to cement ratios(w/c).

Table 11.3 *Effect of rubber content on the dry unit weight of mortars and concrete*

References	Rubber content (%)	Size (mm)	Type	% Reduction
Skripkiunas et al. [105]	3.2	0-1	Concrete	0.85
Pelisser et al. [91]	10	<4.8	Concrete	13
Pedro et al. [90]	5, 10 and 15	0–2	Mortar	4.32, 7.45 and 9.9

Onuaguluchi and Panesar [84]	5, 10 and 15	~86% smaller than 2.3	Concrete	1.75, 2.58 and 7
Corinaldesi et al. [25]	10 and 30(SBR)	0–12	Mortar	3.84 and 16.35
Corinaldesi et al. [25]	10 and 30 (SR)	0–8	Mortar	2.86 and 13.54
Sukontasukkul [106]	10, 20, and 30	Passing sieve # 6	Concrete	14.23, 16.6 and 19.76
Sukontasukkul [106]	10, 20, and 30	Passing sieve # 26	Concrete	17.39, 22.13 and 28.06
Sukontasukkul and Tiamlom [108]	10, 20, and 30	Passing sieve # 6	Concrete	3.66, 9.76 and 12.19
Sukontasukkul and Tiamlom [108]	10, 20 and 30	Passing sieve # 26	Concrete	9.76, 13.42 and 16.46
Grdic et al. [42]	10, 20 and 30	0.5–4	Concrete	4.64, 9.49 and 13.2
Hilal [49]	20 and 30	0.7–5	Concrete	6.4 and 10.4
Fadiel et al. [32]	10, 20, 30 and 40	0–0.6	Mortar	8, 20.66, 28 and 39
Fadiel et al. [32]	10, 20, 30 and 40	0.84–2	Mortar	3.25, 13.94, 21.21 and 30.1
Gisbert et al. [41]	10, 20, 30 and 40	80% returned 0.25	Mortar	9.62, 12.14, 14 and 17.14
Gisbert et al. [41]	10, 20, 30 and 40	80% returned 2	Mortar	17.35, 11.74, 21.48 and 28.64
Turki et al. [122]	10, 20, 30, 40, and 50	1–4	Concrete	2.76, 6.4, 13.43, 16.78 and 22.3
Uygunoglu and Topcu]124]	10, 20, 30, 40 and 50	1–4	Mortar	3.9, 4.44, 8.81, 11.35 and 16 (w/c = 0.4)
Ling [66]	5, 10, 15, 20, 25, 30, 40, and 50	1–5	Concrete	0.88, 1.95, 2.59, 3.71, 4.59, 6.54, 7.28 and 7.65 (w/c = 0.45)
Ling [66]	5, 10, 15, 20, 25, 30, 40, and 50	1–5	Concrete	1.27, 2.08, 1.72, 3.21, 3.57, 5.34, 5.83 and 8.41(w/c = 0.55)
Turki et al. [123]	10, 30, and 50	1–4	Mortar	3.64, 16.3 and 20.85
Eiras et al. [28]	40, 50, and 60	~0.08–1.3	Mortar	18.69, 25.23 and 27.1
Turgut and Yesilata [121]	10–70	4.7-0.075	Concrete Block	2.76–29.49
El-Gammal et al. [30]	50 and 100	~0–5	Concrete	9.42 and 13.45
Taha et al. [110]	25, 50, 75, and 100	1–5	Concrete	9.35%, 13.36%, 16.44% and 19.16%

From Sections 3.1 and 3.2, it could be deduced that the inclusion of rubber sand in the concrete mixtures resulted in the reduction of both fresh and hardened density concrete density. This reduction in the concrete density is related to the lower density of rubber compared to natural sand [6]. This effect is more noticeable for smaller rubber particles, due to the greater porosity of the composite obtained, suggesting that a greater quantity of spaces filled with water within the interface rubber concrete. Albano et al. [6], Gesoglu and Guneyisi [37], Turki et al. [122], and Thomas et al. [111] related the reduction in the unit weight of the rubberized mixture to the lower specific gravity of rubber. Raj et al. [97] related the reduction in the hardened density of rubberized concrete to the relatively lower density of rubber compared to natural sand. Taha et al. [110] attributed the reduction in the fresh and hardened unit weight of rubberized concrete to two factors: first, the ability of rubber particles to entrap air in its rough surface texture; and second, to the low specific gravity of the rubber particles compared to natural aggregates. The reduction in the density with the inclusion of rubber sand is one advantage of using this recycled material. Thus, the use of rubber sand in concrete mixture in some engineering applications would result into light-weight concrete.

11.4 Mechanical strength

The predominant reported effects of the addition of sand rubber as natural sand replacement were a reduction in the concrete compressive strength, static modulus of elasticity and dynamic modulus of elasticity. This reduction increased with the increase of rubber sand content. The reduction in the mechanical strength may be related to the bond defects between rubber sand and the matrix according to Turatsinze et al. [118]. Corinaldesi et al. [25] related the low strength of rubberized mortar to the low quality of the interfacial transition zone between rubber particles and cement paste, while Turki et al. [122] related this phenomenon to the void space between rubber aggregate and cement matrix. Albano et al. [6] related the low strength to the increased porosity or weakness points in rubberized concrete matrix. Raj et al. [97] related the low strength of rubberized concrete to the weak interface or the transition zone of the rubberized mortar and the conventional coarse aggregates. These weak interfaces acted as the originators of microcracks which eventually grew to macro size, leading to failure under compression. Thomas et al. [111] related the low strength of rubberized concrete to the smooth surfaces of the rubber particles that led to a weak bond with the cement paste. Taha et al. [110] related the low strength of rubberized concrete to three main reasons:

- The deformability of the rubber particles compared with the surrounding cement paste, that results in initiating cracks around the rubber particles in a pattern identical to that occurring with air voids in normal concrete.
- The weak bond between rubber particles and the cement paste.
- The possible reduction of the mixture density which depends greatly on the density, size, and hardness of the aggregates.

Several researchers [1,2,22,29,58,61,62,85,110] related the low strength of rubberized mixture to the weak bond between rubber particles and cement paste, and increased matrix porosity. The reduction in the strength by using rubber sand is one of the main shortcomings of using this recycled material which limits its wide use by engineers. Some studies [37,45] proposed to replace portion of cement with silica fume to alleviate the degradation in strength caused by rubber sand. Others [18] proposed to blend the rubber particles with latex prior so that they are added to the concrete/mortar seeking to improve the strength. Few studies have proposed to improve the rubber-cementitious matrix bond [64], notably treating the rubber particles with NaOH aqueous solution [101]. However, the results that they obtained showed that the strength benefit due to the rubber treatment was minimal.

11.5 Impactenergy and impact loading

From the review of the literature [7,9,39,69,90,100,109,110,125], it could be deduced that the addition of rubber sand in the concrete mixture, up to 50%, improved impact energy. Rubber sand, up to 20% enhanced impact loading behavior. The reason behind this improvement is that under the impact compression, the macroscopic cracks of rubber concrete specimens first appeared on the local weak surface of the edge and propagated with the increase of impact pressure and impact times [36,54]. Also, under the dynamic splitting, when the central crack initiation was not satisfied, the stress was concentrated in the damage zone caused by impact the splitting direction extended to the center of the specimen along the direction of external force due to the action of stress concentration. Rubber concrete not only has outstanding performance in impact resistance and energy dissipation thus having a wide application prospect in many fields of civil engineering, but such also as concrete used in bridge/highway guardrails and anti-seismic materials.

11.6 Toughness

The reported research results [13,46,52,60,68,79,107,110], indicate that the addition of rubber particles led to a substantial reduction in fracture toughness of about 50% compared to the control mix. This reduction may be related to increasing porosity of the concrete mix with tire rubber content which weakens the matrix. The weak interfacial bond between tire rubber particles and surrounding cement paste allowed a crack to easily develop around the tire rubber particles. The increase of the fracture toughness of concrete with the inclusion of rubber particles can be explained by the ability of the rubber to add a few toughening mechanisms to the conventional concrete including crack bridging by rubber particles and rubber particles' bending, compressing, and twisting. The tire rubber particles absorb part of the energy the matrix is subjected to, and therefore the composite material can absorb more energy before fracturing compared to the bare concrete matrix [110]. Table 11.4 summarizes studies about the effect of rubber sand on the toughness of mortars and concretes.

Table 11.4 *Effect of Rubber Sand on the Toughness of Concrete*

References	Rubber content (%)	Size (mm)	Increased toughness
Najim and Hall [79]	5, 10, and 15	2–6	Yes
Liu et al. [68]	5, 10, and 15	≤5	Yes
Balaha et al. [13]	20	<4	Yes
Taha et al. [110]	25, 50, 75, and 100	1–5	Yes
Khaloo et al. [60]	25	≤4.75	Yes
Khaloo et al. [60]	75 and 100	≤4.75	Yes
Guo et al. [46]	4 and 8	0.85–1.40	No
Guo et al. [46]	16	0.85–1.40	No
Huang et al. [52]	10, 20, 30, and 40	Average 0.135	No

11.7 Ductility and strain capacity

From the review of the literature [35,42,46,49,50,52,55-57,60,65,76,79,93,97,112,125], it could be deduced that the addition of rubber sand in the concrete mixture increased its ductility.

The failure of rubberized concrete did not occur instantly and did not cause any disconnection in the specimens in comparison with the control specimens [60]. As reported by Raj et al. [97] the addition of rubber sand in concretes diminished the brittleness index value and enhanced the ductility of concrete, thus enabling a transforming concrete from a brittle material to a more ductile material. This is resulted from the superior energy absorption capacity of rubber, which led to plastic deformation during the loading. Li et al. [64] related the enhancement in the ductility of rubberized concrete to the incorporation of small ductile particles into concrete. At the same line, the inclusion of rubber aggregate in the matrix increased its strain capacity as reported by Turatsinze et al. [119], Nguyen et al. [81] and Huang et al. [52].

11.8 Shrinkage

Table 11.5 summarizes a review of the literature which clearly indicates that the addition of rubber sand as mass percentage replacement of sand in concrete mixtures resulted in the increase of the shrinkage. The higher shrinkage of rubberized mortar/concrete is partly due to the lower compressive strength and elastic modulus. At the same token, Turatsinze et al. [119], reported that the benefit of the higher straining capacity of rubberized cement-based composites due to their higher elasticity and lower compressive strength could be offset by their higher shrinkage. The shrinkage of rubberized mortar/concrete seemed to be depending on the particle size and the content of rubber sand in the mixture. The shrinkage increased with increasing rubber sand content in which rubber sand is more compressible and highly flexible than natural sand. As the replacement rate increased, the lack of fine aggregate caused a greater decrease of internal restraints and led to higher shrinkage [108]. In terms of rubber sand particle size, the smaller size appeared to shrink much more than the larger size [108]. Sukontasukkul and Tiamlom [108] reported that large shrinkage might come from two combined effects: (1) the lower internal restraint due to lack of sand and (2) the increase of more flexible material due to the presence of rubber crumb. Uygunoglu and Topcu [124] reported that the increase in the total shrinkage with the addition of rubber sand is a consequence of the increase of open porosity.

Table 11.5 *Effect of rubber sand on the shrinkage of concrete mixtures*

References	Rubber Content (%)	Size (mm)	Increased shrinkage
Turatsinze et al. [119]	20 and 30	≤4	Yes
Bravo and de Brito [19]	5, 10 and 15	Same sand size	Yes
Pedro et al. [90]	15	≤2	Yes
Yung et al. [133]	5, 10 and 20	0.3, 0.6 and 0.6 + 0.3	Yes
Sukontasukkul and Tiamlom [108]	10, 20 and 30	Sieve # 6 and # 26	Yes
Turatsinze et al. [118,120]	20 and 30	≤4	Yes
Huang et al. [52]	10, 20, 30 and 40	≤0.15	Yes
Uygunoglu and Topcu [124]	10, 20, 30, 40 and 50	1- 4	Yes
Uygunoglu and Topcu [124]	10, 20 and 30	(w/c = 0.4) 1–4	Yes
Uygunoglu and Topcu [124]	40 and 50	(w/c = 0.51) 1–4	No
Aules [10]	Up to 30	≤4.75	No
Chunlin et al. [23]	10	1–5	No

11.9 Abrasion Resistance

From the literature review [33,42,85,107,111,126], it could be deduced that the abrasion resistance of concrete with the addition of rubber sand is not conclusive and requires additional attention. However, it seems that the addition of rubber sand increases the abrasion resistance if appropriate rubber sand content and suitable water cement ratio (w/c) were used.

11.10 Freeze/thaw and ageing resistance

From the review of the published literature [4,59,87,88,90,114,121], it could be deduced that the resistance to freezing and thawing cycles increased with the increase of rubber sand content. The sand rubber resulted in more air entrapment than regular concrete and thus provided the rubberized concrete with the increased capacity to resist freezing–thawing cycles.

11.11 Fire resistance and Thermal Insulation

From the review of the literature [27,28,32,47,48,55,56,80,72,77,87,88,92,93,106,114], it could be deduced that the addition of rubber sand in the concrete or mortar mixture decreased its thermal conductivity. This means that the inclusion of rubber sand in the mixture increased its thermal insulation. The thermal insulation increased with increasing rubber sand content. Theoretically, the thermal conductivity is inversely proportional to the density of the material. The reduction in the thermal conductivity of the rubberized matrix could be partly attributed to increasing air entrapment caused by non-wetting rubber particles during mixing and partly to the lower thermal conductivity of the crumb rubber particles. As the addition of crumb rubber increased, there was a greater moisture-dependent effect on the saturated state thermal conductivity due to the increase apparent porosity caused by air entrapment [48]. Increasing thermal insulation of mortar/concrete with the addition of rubber sand is one advantage of using these recycled rubbers. where thermal insulation is required. On the other hand, the addition of rubber sand in the mixture decreased its fire resistance but decreased the risk of spalling caused by exposure to elevated temperatures.

11.12 Carbonation Resistance

Bravo and de Brito [19] reported a decline in the carbonation resistance of concrete specimens containing waste tire rubber aggregates, with the same gradation as natural sand, at replacement levels of 5%, 10% and 15%, by volume. The carbonation depth marginally increased with increasing rubber sand content. The increase in the carbonation depth may be due to the greater void volume between rubber aggregates and the cement paste.

It is obvious that there is a distinct lack in researching of the effect of rubber sand on the carbonation resistance of concrete. Actually, the carbonation resistance of concrete containing rubber sand still needs more investigations. This can be a major topic for future investigations. However, according to the available study, it can be concluded that the inclusion of rubber sand in the matrix decreased its carbonation resistance. This is one of the disadvantages of using the recycled rubbers.

11.13 Corrosion resistance

Karahan et al. [59] reported that reinforcing bar mass loss of concrete without rubber and concrete containing 10% crumb rubber (size 0.15–4.75 mm), by volume, as natural sand replacement, was almost the same. However, when crumb rubber content reached 30%, the reinforcing bar mass loss was almost two times greater than the control concrete. Yung et al. [133] reported that partially replacing natural sand in concretes with 5% rubber (size 0.6 or 0.3 mm), by volume, led to anti-sulphate corrosion resistance.

It can be deduced that there is a distinct lack in researching on the effect of rubber sand on the corrosion resistance. The corrosion resistance of bar imbedded in concrete containing rubber sand still needs more studies. This can be a major topic for future investigations.

11.14 Water Absorption, Porosity, and Chloride Ion Penetration

Table 11.6 summarizes the effects of the inclusion of rubber sand in the concrete mixture, in terms of water absorption, permeability, chloride ion penetration, and porosity. Definitely there is a rise in its percentage of water absorption and porosity, as reported by several studies, but it mainly depended on the rubber particle size and its content in the matrix. In general, the cause of high absorption could be the result of the formation of porosity during mixing process. As known, rubber particles are non-polar by nature (water insolvable), so, during mixing, they would trap air bubbles at the particle surfaces. This phenomenon causes the interface between cement paste and rubber to be porous and highly absorptive. On the other hand, small particle size of rubber (size < 0.5 mm) reduced the absorption. This is because the smaller particles of rubber acts as fillers to fill up capillary pores in the matrix [108]. Karahan et al. [59] reported that the reason behind the increased porosity could be related to the higher amounts of air that were trapped during mixing [17]. This may have occurred due to the tendency of rubber particles to entrap air in their rough surfaces because of their non-polar nature [16,117] and/or because the hydrophobic nature of rubber increases air content when rubber particles are added. Uygunoglu and Topcu [124] related the reduction in the absorption to the entrapment of air by the rubber particles at the particle-paste and particle-particle interfaces. Gesoglu and Guneyisi [38], related the increase in the absorption with increasing rubber content to the increase in porosity of rubber in mixtures. This is probably due to some deviations of rubber particles from sand grain size distribution and/or a significant higher air amount trapped during mixing procedure of rubberized mixtures. The increase in the water absorption of the mortar/concrete with the inclusion of rubber sand is one disadvantage of using this recycled material. On the other hand, Ling [67] reported that 10% of rubber was distorted and filled the voids between the natural aggregates under a compression force of a plant-made machine. This filling mechanism was found to reduce the porosity by filling up the free pore volume in the crete mixture. The chloride ion penetration depth increased with increasing rubber sand content. However, the magnitudes of the chloride penetration depth can be reduced by replacing 10% cement with silica fume [37] or by replacing part of cement with fly ash [38,39].

Table 11.6 *Effect of rubber sand on the water absorption, porosity, and chloride ion penetration of mortars and concretes*

References	Rubber content (%)	Size (mm)	Effect
Segre et al. [102]	10	0.2	- Reduced water absorption
Marques et al. [71]	12	Passing sieve 0.8	- Reduced water absorption
Pedro et al. [90]	5, 10 and 15	0–2	- 5% reduced water absorption - 10% increased water absorption - 15% better permeability
Ganesan et al. [33]	15	<4.75	- Reduced water permeability, water absorption and chloride ion penetration
Gesoglu et al. [39]	10 and 20	2 and 4	- Reduced permeability coefficient
Ling [67]	10, 20 and 30	3–31 and 1–5	- 10% reduced porosity - 20% and 30% increased porosity
Sukontasukkul and Tiamlom [108]	10, 20 and 30	Passing sieve No. 26	- Reduced water absorption
Sukontasukkul and Tiamlom [108]	10, 20 and 30	Passing sieve No. 6	- Increased water absorption
Hilal [49]	10 and 20	0.7–5	- Increased water absorption
Azevedo et al. [11]	5, 10 and 15	1–2.4	- Increased capillary water absorption
Bravo and de Brito [19]	5, 10 and 15	Like sand gradation	- Increased water absorption
Onuaguluchi and Panesar [84]	5, 10 and 15	~86% smaller than 2.3 mm	- Increased water absorption and porosity
Thomas et al. [111]	2.5–20%	Mesh 30, 0.8–2 mm, 2–4 mm	- Increased water absorption
Bignozzi and Sandrolini [17]	22.2 and 33.3	55% 0.5-2and 45% 0.5-0.7	- Increased water absorption
Karahan et al. [59]	10, 20 and 30	0.15–4.75	- Increased porosity
Fadiel et al. [32]	10, 20, 30 and 4 0	0–0.6	- 10%, 20% decreased water absorption - 30%, 40% increased water absorption
Fadiel et al. [32]	10, 20, 30 and 4 0	0.84–2	- 10%, 20%, 30% decreased water absorption - 40% increased water absorption
Mohammed et al. [77]	10, 25 and 50	0.6	- Increased water absorption
Eiras et al. [28]	40, 50 and 60	~0.08–1.13	- Increased water absorption

Turgut and Yesilata [121]	10–70	0.75–4.75	- Increased water absorption and porosity
Turki et al. [123]	10, 30 and 50	1-4	- Increased porosity
Uygunoglu and Topcu [124]	10–50	1-4	- Increased water absorption and porosity
Gesoglu and Guneyisi [38]	5, 15 and 25	<4	- Increased chloride ion permeability
Al-Akhras and Samadi [4]	5 and 10	0.15	- Reduced chloride ion penetration
Oikonomou and Mavridou [83]	2.5, 5, 7.5,10, 12.5 and 15	0.75-1.18	- Reduced chloride ion penetration depth
Onuaguluchi and Panesar [84]	5, 10 and 15	~86% smaller than 2.3 mm	- Reduced RCPT
Gesoglu and Guneyisi [37]	5, 15 and 25	like sand gradation	- Increased chloride ion penetration depth

11.15 Resistance to aggressive environmental conditions

From the review of the literature [11,33,102,114], it could be deduced that there is lack of published work on the effect of aggressive environmental conditions on the properties of concrete containing rubber sand. This aspect still needs more studies. It can be a major topic for future investigations. However, according to the available studies, it could be observed that the addition of rubber sand in the concrete mix enhanced its resistance against hydrochloric acid and seawater.

11.16 Energy absorption

Studies by [9,34,36,54,85] concluded that the inclusion of rubber sand in the matrix increased its energy absorption capacity. They reported that the energy dissipated by the rubber concrete specimens at maximum load increased drastically as rubber content increased. The increase of the energy absorption with the addition of rubber sand is one advantage of using this recycled material.

11.17 Sound absorption

From the studies by Sukontasukkul [106], Khaloo et al. [60], Mohammed et al. [77], Najim and Hall [78], Bowland et al. [18], and Gisbert et al. [41], it could be concluded that the addition of rubber sand in the concrete mixture increased its sound insulation. This results in a benefit of using rubberized concrete as sound barriers, turning the waste rubber into a valuable material.

11.18 Electrical resistance

From the studies by Yung et al. [133], Issa and Salem [56], and Mohammed et al. [77], it can be deduced that there is lack of research on the effect of rubber sand on the electrical resistance of concrete. The electrical resistance of concrete containing rubber sand still needs more investigations. But, according to the available studies, it can be concluded that the electrical resistance increased with the inclusion of rubber sand.

11.19 Cracking resistance

Studies by Jingfu and Yongqi [57], Huang et al. [52], Ganesan et al. [35], Nguyen et al. [82], and Khaloo et al. [60], reported that cracking width, length, and number in rubberized concretes was smaller than that of the plain concrete and the propagation of failure patterns was more gradual and uniform. In short, the literature results revealed that the addition of rubber sand exhibited a higher cracking resistance than the control specimens.

11.20 Conclusion

Using rubber as fine aggregate in concrete exhibited certain advantages and some disadvantages. The advantages of utilizing rubber sand are:

- decreasing density
- improving impact energy
- improving impact loading
- increasing toughness
- increasing ductility
- increasing freeze/thaw resistance

- increasing thermal insulation
- increasing sound insulation
- increasing damping capacity
- increasing strain capacity
- reducing micro-cracks after firing
- increasing abrasion resistance depending on the rubber sand content and w/c ratio
- increasing resistance against hydrochloric acid attack
- improving energy absorption
- increasing electrical resistance
- increasing cracking resistance.

The disadvantages of utilizing rubber sand are:

- decreasing workability
- increasing bleeding
- decreasing mechanical strength
- increasing drying shrinkage
- decreasing carbonation resistance
- decreasing corrosion resistance
- increasing water absorption
- increasing chloride ion penetration depth
- decreasing sulfuric acid resistance.

It is safe to conclude that rubber sand can be used for manufacturing:

- lightweight concrete
- non-structural works
- decreasing the risk of spalling due to exposure to elevated temperatures
- as safety barriers with improved resistance to traffic noise
- as sound barriers
- as a sound absorber
- as a shock absorber in highway construction
- as an earthquake shock-wave absorber in buildings
- as construction material for thermal insulation
- as sound insulation
- for acoustic anti-vibration properties
- as construction material where high resistance for impact load is required.

Moreover, rubberized concrete can be used:

- in foundation pads for machinery
- in the railway station where vibration damping is required
- as railway buffers
- as jersey barriers
- as bridge abutment fill
- in bunkers where resistance to impact or explosion is required
- for pipe bedding and trench filling
- in artificial roof construction
- for pile heads and as paving slabs.

References

1. Abdulla, Aziz Ibrahim, Ahmed, Salwa Hadi, 2011. Effect of rubber treated by acidic solution on some mechanical properties of rubberize cement mortar. Eng. Technol. J. 29 (13), 2793–2793.

2. Abdulla, Aziz Ibrahim, Aules, Wisam Amer, Ahmed, Salwa Hadi, 2010. Cement mortar properties contain crumb rubber treated with alkaline materials. Mod. Appl. Sci. 4 (12), 146–163.

3. Association of US Tire Manufacturers, 2018, Sustainability Report, 28 pages.

4. Al-Akhras, Nabil, Samadi, Mohammed M., 2004. Properties of tire rubber ash mortar. Cement Concr. Compos. 26, 821–826.

5. Alamo-Nole Luis, A., Oscar, Perales-Perez, Roman-Velazquez Felix, R., 2011. Sorption study of toluene and xylene in aqueous solutions by recycled tires crumb rubber. J. Hazard. Mater. 185, 107–111.

6. Albano, C., Camacho, N., Reyes, J., Feliu, J.L., Hernandez, M., 2005. Influence of scrap rubber to Portland I concrete composites: destructive and non-destructive testing. Compos. Struct. 71, 439–446.

7. Al-Tayeb, M.M., Abu Bakar, B.H., Akil, H.M., 2013. Performance of rubberized and hybrid rubberized concrete structures under static and impact load conditions. Exp. Mech. 53, 377–384.

8. Antil, Yogender, Verma, Vivek, Singh, Bhupinder, 2014. Rubberized concrete with crumb rubber. Int. J. Sci. Res. (IJSR) 3 (5), 1481–1483.

9. Atahan, Ali O., Yucel, Ayhan Oner, 2012. Crumb rubber in concrete: static and dynamic evaluation. Constr. Build. Mater. 36, 617–622.
10. Aules, Wesam Amer, 2011. Utilization of crumb rubber as partial replacement in sand for cement mortar. Eur. J. Sci. Res. 51 (2), 203– 210.
11. Azevedo, F., Pacheco-Torgal, F., Jesus, C., de Aguiar Barroso, J.-L., Cano es, A.F., 2012. Properties and durability of HPC with tire rubber wastes. Constr. Build. Mater. 34, 186–191.
12. Azmi, N.J., Mohammed, B.S., Al-Mattaeneh, H.M.A., 2008. Engineering properties of concrete containing recycled tire rubber. ICCBT, B 34, 373–382.
13. Balaha, M.M., Badawy, A.A.M., Hashish, M., 2007. Effect of using ground tire rubber as fine aggregate on the behavior of concrete mixes. Indian J. Eng. Mater. Sci. 14, 427–435.
14. Batayneh, Malek K., Marie, Iqbal, Asi, Ibrahim, 2008. Promoting the use of crumb rubber concrete in developing countries. Waste Manage. 28, 2171–2176.
15. Benazzouk, A., Douzane, O., Queneudec, M., 2004. Transport of fluids in cement-rubber composites. Cement Concr. Compos. 26, 21–29.
16. Benazzouk, A., Douzane, O., Langlet, T., Mezzeb, K., Roucoult, J.M., Queneudec, M., 2007. Physico-mechanical properties and water absorption of cement composite containing shredded rubber wastes. Cement Concr. Compos. 29, 732–740.
17. Bignozzi, M.C., Sandrolini, F., 2006. Tire rubber waste recycling in self- compacting concrete. Cem. Concr. Res. 36, 735–739.
18. Bowland, Adam G., Weyers, Richard E., Charney, Finley A., Dowling, Norman E., Murray, Thomas M., Andrei, Ramniceanu, 2012. Effect of vibration amplitude on concrete with damping additives. ACI Mater. J. 109–M35, 371–378.
19. Bravo, Miguel, de Brito, Jorge, 2012. Concrete made with used tire aggregate: durability-related performance. J. Clean. Prod. 25, 42–50.
20. World Business Council for Sustainable Development, Managing End-of- Life Tires. Available at<http://www.bir.org/assets/Documents/industry/ManagingEndOfLifeTires.pdf>(accessed15.3.2021).
21. Chen, Shy-Haur, Wang, Her-Yung, Jhou, Jhou-Wei, 2013. Investigating the properties of lightweight concrete containing high

contents of recycled green building materials. Constr. Build. Mater. 48, 98–103.

22. Chung, K.H., Hong, Y.K., 1999. Introductory behavior of rubber concrete. J. Appl. Polym. Sci. 72, 35–40.

23. Chunlin, Liu, Kunpeng, Zha, Depeng, Chen, 2011. Possibility of concrete prepared with steel slag as fine and coarse aggregates: a preliminary study. Proc. Eng. 24, 412–416.

24. Colom, X., Carrillo, F., Canavate, J., 2007. Composites reinforced with reused tires: surface oxidant treatment to improve the interfacial compatibility. Compos. A Appl. Sci. Manuf. 38 (1), 44–50.

25. Corinaldesi, Valeria, Mazzoli, Alida, Moriconi, Giacomo, 2011. Mechanical behavior and thermal conductivity of mortars containing waste rubber particles. Mater. Des. 32, 1646–1650.

26. Correia, Sivaldo L., Partala, Tania, Loch, Relipe C., Segadaes, A.M., 2010. Factorial design used to model the compressive strength of mortars containing recycled rubber. Compos. Struct. 92, 2047–2051.

27. Correia, J.R., Marques, A.M., Pereira, C.M.C., de Brito, J., 2012. Fire reaction properties of concrete made with recycled rubber aggregate. Fire Mater. 36 (2), 139–152.

28. Eiras, J.N., Segovia, F., Borrachero, M.V., Monzo, J., Bonilla, M., Paya, J., 2014. Physical and mechanical properties of foamed Portland cement composite containing crumb rubber from worn tires. Mater. Des. 59, 550–557.

29. Eldin, N.N., Senouci, A.B., 1993. Rubber tire particles as concrete aggregate. J. Mater. Civ. Eng. ASCE 5, 478–496.

30. El-Gammal, A., Abdel-Gawas, A.K., El-Sherbini, Y., Shalaby, A., 2010. Compressive strength of concrete utilizing waste tire rubber. J. Eng. Trends Eng. Appl. Sci. JETEAS 1 (1), 96–99.

31. ETRMA, 2011. End of Life Tires: A Valuable Resource with Growing Potential. European tire and rubber manufacturers association, Brussels, Belgium, 2011.

32. Fadiel, Ashraf, Al Rifaie, Fouad, Taher, Abu-Lebdeh, Fini, Ellie, 2014. Use of crumb rubber to improve thermal efficiency of cement-based materials. Am. J. Eng. Appl. Sci. 7 (1), 1–11.

33. Ganesan, N., Raj, B., Shashikala, A.P., 2012. Strength and durability of self-compacting rubberized concrete. Indian Concr. J., 15–24

34. Ganesan, N., Raj, B., Shashikala, A.P., 2013. Behavior of self-consolidating rubberized concrete beam-column joints. ACI Mater. J. 110 (6), 697–704.

35. Ganesan, N., Raj, J. Bharati, Shashikala, A.P., 2013. Flexural fatigue behavior of self-compacting rubberized concrete. Constr. Build. Mater. 44, 7–14.
36. Gerges, Najib N., Issa, Camille A., and Fawaz, Samer A., 2018. Rubber concrete: Mechanical and dynamical properties, Case Studies in Construction Materials, 9,
37. Gesoglu, Mehamet, Guneyisi, Erhan, 2007. Strength development and chloride penetration in rubberized concretes with and without silica fume. Mater. Struct. 40, 953–964.
38. Gesoglu, Mehmet, Guneyisi, Erhan, 2011. Permeability of self-compacting rubberized concrete. Constr. Build. Mater. 25, 3319–3326.
39. Gesoglu, Mehmet, Guneyisi, Erhan, Khoshnaw, Ganjeena, Ipek, Suleyman, 2014. Investigating properties of pervious concretes containing waste tire rubbers. Constr. Build. Mater. 63, 206–213.
40. Ghaly, Ashraf, Cahill, Janes D., 2005. Correlation of strength, rubber content, and water to cement ratio in rubberized concrete. Can. J. Eng. 32, 1075–1081.
41. Gisbert, A. Nadal, Borrell, J.M. Gadea, Garcıa, F. Parres, Sanchis, E. Julia, Amoros, J.E. Crespo, Alcaraz, J. Alcaraz, Segura, J., Vicente, F. Salas, 2014. Analysis behavior of static and dynamic properties of Ethylene–Propylene–Diene–Methylene crumb rubber mortar. Constr. Build. Mater. 50, 671–682.
42. Grdic, Zoran, Toplicic-Curcic, Godana, Ristic, Nenad, Grdic, Dusan, Mitkovic, Petar, 2014. Hydro-abrasive resistance and mechanical properties of rubberized concrete. GRAĐEINAR 66 (1), 11–20.
43. Grinys, Audrius, Sivilevicius, Henrikas, Dauksys, Mindaugas, 2012. Type Rubber additive effect on concrete mixture strength. J. Civil Eng. Manag. 18 (3), 393–401 (Taylor & Francis).
44. Guneyisi, Erhan, 2010. Fresh properties of self-compacting rubberized concrete incorporated with fly ash. Mater. Struct. 43, 1037– 1048.
45. Guneyisi, Erhan, Gesoglu, Mehamet, Ozturan, Turan, 2004. Properties of rubberized concretes containing silica fume. Cem. Concr. Res. 34, 2309–2317.
46. Guo, Y.C., Zhang, J.H., Chen, G., Chen, G.M., Xie, Z.H., 2014. Fracture behaviors of a new steel fiber reinforced recycled aggregate concrete with crumb rubber. Constr. Build. Mater. 53, 32–39.

47. Guo, Yong-chang, Zhang, Jian-hong, Chen, Guang-ming, Xie, Zhi-hong, 2014. Compressive behavior of concrete structures incorporating recycled concrete aggregates, rubber crumb and reinforced with steel fibre, subjected to elevated temperatures. J. Clean. Prod. 72, 103–203.

48. Hall, Matthew R., Najim, Khalid B., Hopfe, Christina J., 2012. Transient thermal behavior of crumb rubber-modified concrete and implications for thermal response and energy efficiency in buildings. Appl. Therm. Eng. 33–34, 77–85.

49. Hilal, Ameer Abdulrahman, 2011. Effect of crumb tires rubber on some properties of foamed concrete. Anbar J. Eng. Sci. AJES 4 (2), 1–17.

50. Ho, Anh Cuong, Turatsinze, Anaclet, Hameed, Rashid, Vu, Duc Chinh, 2012. Effects of rubber aggregates from grinded used tires on the concrete resistance to cracking. J. Clean. Prod. 23, 209–215.

51. Holmes, Niall, Dunne, Kevin, O'Donnell, John, 2014. Longitudinal shear resistance of composite slabs containing crumb rubber in concrete toppings. Constr. Build. Mater. 55, 365–378.

52. Huang, Xiaoyan, Ranade, Ravi, Ni, Wen, Li, Victor C., 2013. On the use of recycled tire rubber to develop low E-modulus ECC for durability concrete repairs. Constr. Build. Mater. 46, 134–141

53. Humphrey, D., 1995. Civil Engineering Applications of Chipped Tires. University of Maine, Department of Civil and Environmental Engineering, Maine, USA, 2–20.

54. Issa, Camille A., Gerges, Nagib N., and Fawaz, Samer, "Recycled Rubber Aggregates in Concrete Highway Barrier Construction", Proceedings of the American Advanced Materials Congress (AAMC), December 4-9, 2016, Miami, USA.

55. Issa, Camille A., "Utilization of Recycled Rubber in Concrete Mix Design", IABSE Symposium Report, IABSE Symposium, Venice 2010, pp. 41-48.

56. Issa, Camille A., Salem, George, 2013. Utilization of recycled crumb rubber as fine aggregates in concrete mix design. Constr. Build. Mater. 42, 48–52.

57. Jingfu, Kang, Yongqi, Jiang, 2008. Improvement of cracking-resistance and flexural behavior of cement-based materials by addition of rubber particles. J. Wuhan Univ. Technol. Mater. Sci. Ed., 379–383

58. Jingfu, Kang, Han, Chuncui, Zhang, Zhenli, 2009. Strength and shrinkage behaviors of roller-compacted concrete with rubber additives. Mater. Struct. 42, 1117–1124.
59. Karahan, Okan, Ozbay, Erdogan, Hossain Khandaker, M.A., Lachemi, Mohamed, Atios Cengiz, D., 2012. Fresh, mechanical, transport, and durability properties of self-consolidating rubberized concrete. ACI Mater. J. 109–M39, 413–420.
60. Khaloo, Ali R., Dehestani, M., Rahmatabadi, P., 2008. Mechanical properties of concrete containing a high volume of tire-rubber particles. Waste Manage. 28, 2472–2482.
61. Khatib, Zaher K., Bayomy, Fouad M., 1999. Rubberized Portland cement concrete. J. Mater. Civ. Eng., 206–213
62. Lee, H.S., Lee, H., Moon, J.S., Jung, H.W., 1998. Development of tire- added latex concrete. ACI Mater. J. 95 (4), 356–364.
63. Li, Z., Li, F., Li, J.S.L., 1998. Properties of concrete incorporating rubber tire particles. Mag. Concr. Res. 50 (4), 297–304.
64. Li, Guoqiang, Pang, Su-Seng, Samuel, I. Ibekwe, 2011. FRP tube encased rubberized concrete cylinders. Mater. Struct. 44, 233–243.
65. Lijuan, Li, Shenghua, Ruan, Lan, Zeng, 2014. Mechanical properties and constitutive equations of concrete containing a low volume of tire rubber particles. Constr. Build. Mater. 70, 291–308.
66. Ling, Tung-Chai, 2011. Prediction of density and compressive strength for rubberized concrete blocks. Constr. Build. Mater. 25, 4303–4306.
67. Ling, Tung-Chai, 2012. Effects of compaction method and rubber content on the properties of concrete paving blocks. Constr. Build. Mater. 28, 164–175.
68. Liu, Feng, Zheng, Wanhu, Li, Lijuan, Feng, Wenxian, Ning, Guofang, 2013. Mechanical and fatigue performance of rubber concrete. Constr. Build. Mater. 47, 711–719.
69. Maher, Al-Tayeb Mustafa, Abu Bakar, B.H., Hanafi, Ismail, Hazizan Md, Akil, 2013. Effect of partial replacement of sand by fine crumb rubber on impact load behavior of concrete beam: experiment and nonlinear dynamic analysis. Mater. Struct. 46, 1299–1307.
70. Martınez, Juan Daniel, Neus, Puy, Ramon, Murillo, Tomas, Garcıa, Victoria, Navarro Marıa, Maria, Mastral Ana, 2013. Waste tire pyrolysis – a review. Renew. Sustain. Energy Rev. 23, 179–213.

71. Marques, A.C., Akasaki, J.L., Trigo, A.P.M., Marques, M.L., 2008. Influence of the surface treatment of tire rubber residues added in mortars. IBRACON Struct. Mater. J. 1 (1), 113–120.
72. Marques, A.M., Correia, J.R., de Brito, J., 2013. Post-fire residual mechanical properties of concrete made with recycled rubber aggregate. Fire Saf. J. 58, 49–57.
73. Martin, W., 2001. Tire Crack-Down to Help the Environment. UK Government Environment Agency.
74. Merino, Mercedes del Rıo, Jaime Santa Cruz, Astorqui, Mariano Gonzalez, Cortina, 2007. Viability analysis and constructive applications of lightened mortar (rubber cement mortar). Constr. Build. Mater. 21, 1785–1791.
75. Mohammadi, Iman, Khabbaz, Hadi, Vessalas, Kirk, 2014. In-depth assessment of Crumb Rubber Concrete (CRC) prepared by water- soaking treatment method for rigid pavements. Constr. Build. Mater. 71, 456–471.
76. Mohammed, Bashar S., 2010. Structural behavior and m-k value of composite slab utilizing concrete containing crumb rubber. Constr. Build. Mater. 24, 1214–1221.
77. Mohammed, Bashar S., Hossain, Khandaker M. Anwar, Swee, Jackson Ting Eng, Wong, Grace, 2012. Properties of crumb rubber hollow concrete block. J. Clean. Prod. 23, 57–67.
78. Najim, K.B., Hall, M.R., 2010. A review of the fresh/hardened properties and applications for plain-(PRC) and self-compacting rubberized concrete (SCRC). Constr. Build. Mater. 24, 2043–2051.
79. Najim, Khalid B., Hall, Matthew R., 2012. Mechanical and dynamic properties of self-compacting crumb rubber modified concrete. Constr. Build. Mater. 27, 521–530.
80. Netinger, I., Kesegic, I., Guljas, I., 2011. The effect of high temperatures on the mechanical properties of concrete made with different types of aggregates. Fire Saf. J. 46 (7), 425–430.
81. Nguyen, T.-H., Toumi, A., Turatsinze, A., 2010. Mechanical properties of steel fibre reinforced and rubberized cement-based mortars. Mater. Des. 31, 641–647.
82. Nguyen, T.-H., Toumi, A., Turatisnze, A., Tazi, F., 2012. Restrained shrinkage cracking in steel fibre reinforced and rubberized cement- based mortars. Mater. Struct. 45, 899–904.
83. Oikonomou, N., Mavridou, S., 2009. Improvement of chloride ion penetration resistance in cement mortars modified with rubber from worn automobile tires. Cement Concr. Compos. 31, 403–407.

84. Onuaguluchi, Obinna, Panesar, Daman K., 2014. Hardened properties of concrete mixtures containing pre-coated crumb rubber and silica fume. J. Clean. Prod. 82, 125–131.

85. Ozbay, Erdogan, Lachemi, Mohamed, Umur Korkut, Sevim, 2011. Compressive strength, abrasion resistance and energy absorption capacity of rubberized concretes with and without slag. Mater. Struct. 44, 1297–1307.

86. Pacheco-Torgal, F., Yining, Ding, Said, Jalali, 2012. Properties and durability of concrete containing polymeric waste (tire rubber and polyethylene terephthalate bottles): an overview. Constr. Build. Mater. 30, 714–724.

87. Paine, K.A., Dhir, R.K., 2010. Research on new applications for granulated rubber in concrete. Constr. Mater. 163 (CM1), 7–17.

88. Paine K., Dhir R., Moroney R., Kopasakis K., Use of crumb rubber to achieve freeze thaw resisting concrete, International Conference on concrete for extreme conditions, Dundee, UK, 2012, pp. 486–498.

89. Parveen, Sachin, Dass, Ankit, Sharma, 2013. Rubberized concrete: needs of good environment (overview). Int. J. Emerg. Technol. Adv. Eng. 3 (3), 192–196.

90. Pedro, D., de Brito, J., Veiga, R., 2013. Mortars made with fine granulate from shredded tires. J. Mater. Civ. Eng. ASCE 25, 519–529.

91. Pelisser, Fernando, Zavarise, Nilomar, Longo, Tiago Arent, Bernardin, Adriano Michael, 2011. Concrete made with recycled tire rubber: effect of alkaline activation and silica fume addition. J. Clean. Prod. 19, 757– 763.

92. Pelisser, Fernando, Barcelos, Airton, Santos, Diego, Peterson, Michael, Bernardin, Adriano Michael, 2012. Lightweight concrete production with low Portland cement consumption. J. Clean. Prod. 23, 68–74.

93. Pierce, C.E., Blackwell, M.C., 2003. Potential of scrap tire rubber as lightweight aggregate in flowable fill. Waste Manage. 23, 197–208.

94. Presti, Davide Lo, 2013. Recycled tire rubber modified bitumen for road asphalt mixtures: a literature review. Constr. Build. Mater. 49, 863– 881.

95. Puşca, Al., Bobancu, S., Duta, A., 2010. Mechanical properties of rubber – An overview. Bull. Transilvania Univ. Brasov 3 (52), 107–114.

96. Rahman, M.M., Usama, M., Al-Ghalib Ali, A., 2012. Fundamental properties of rubber modified self-compacting concrete (RMSCC). Constr. Build. Mater. 36, 630–637.
97. Raj, Bharati, Ganesan, N., Shashikala, A.P., 2011. Engineering properties of self-compacting rubberized concrete. J. Reinf. Plast. Compos. 30 (23), 1923–1930.
98. Rashad Alaa, M., 2013. A preliminary study on the effect of fine aggregate replacement with metakaolin on strength and abrasion resistance of concrete. Constr. Build. Mater. 44, 487–495.
99. Rashad Alaa, M., 2016. A comprehensive overview about recycling rubber as fine aggregate replacement in traditional cementitious materials. International Journal of Sustainable Built Environment, 5(1), 46-82.
100. Sallam, H.E.M., Sherbibi, A.S., Seleem, M.H., Balaha, M.M., 2008. Impact resistance of rubberized concrete. Eng. Res. J. (ERJ) 31 (3), 265–271.
101. Segre, N., Joekes, I., 2000. Use of tire rubber as addition to cement paste. Cem. Concr. Res. 30 (9), 1421–1425.
102. Segre, N., Joekes, I., Galves, A.D., Rodrigues, J.A., 2004. Rubber-mortar composites: effect of composition on properties. J. Mater. Sci. 39, 3319–3327.
103. Shu, Xiang, Huang, Baoshan, 2014. Recycling of waste tire rubber in asphalt and Portland cement concrete: an overview. Constr. Build. Mater. 67, 217–224.
104. Siddique, Rafat, Naik Tarun, R., 2004. Properties of concrete containing scrap-tire rubber – as overview. Waste Manage. 24, 563–569.
105. Skripkiunas, Gintautas, Grinys, Audrius, Cernius, Benjaminas, 2007. Deformation properties of concrete with rubber waste additives. Mater. Sci. 13 (3), 219–223.
106. Sukontasukkul, Piti, 2009. Use of crumb rubber to improve thermal and sound properties of pre-cast concrete panel. Constr. Build. Mater. 23, 1084–1092.
107. Sukontasukkul, Piti, Chaikaew, Chalermphol, 2006. Properties of concrete pedestrian block mixed with crumb rubber. Constr. Build. Mater. 20, 450–457.
108. Sukontasukkul, Piti, Tiamlom, Koshi, 2012. Expansion under water and drying shrinkage of rubberized concrete mixed with crumb rubber with different size. Constr. Build. Mater. 29, 520–526.

109. Sukontasukkul, Piti, Jamanam, Sitisak, Sappakittipakorn, Manote, Banthia, Nemkumar, 2013. Preliminary study on bullet resistance of double-layer concrete panel made of rubberized and steel fiber reinforced concrete. Mater. Struct. http://dx.doi.org/10.1617/s11527- 013-0049-x.
110. Taha, M.M., El-Dieb, A.S., Abd El-Wahab, M.A., Abdel-Hameed, M.E., 2008. Mechanical, fracture, and microstructural investigations of rubber concrete. J. Mater. Civ. Eng. ASCE, 640–649.
111. Thomas, Blessen Skariah, Gupta, Ramesh Chandra, Kalla, Pawan, Cseteneyi, Laszlo, 2014. Strength, abrasion, and permeation characteristics of cement concrete containing discarded rubber fine aggregates. Constr. Build. Mater. 59, 205–212.
112. Topcu, I., 1995. The properties of rubberized concretes. Cem. Concr. Res.25 (2), 304–310.
113. Topcu, I.B., Avcular, N., 1997. Analysis of rubberized concrete as a composite material. Cem. Concr. Res. 27 (8), 1135–1139.
114. Topcu, Ilker Bekir, Demir, Abdullah, 2007. Durability of rubberized mortar and concrete. J. Mater. Civ. Eng. ASCE, 173–178.
115. Topcu, Ilker Bekir, Sandemir, Mustafa, 2008. Prediction of rubberized mortar properties using artificial neural network and fuzzy logic. J. Mater. Process. Technol. 199, 108–118.
116. Topcu, Ilker Bekir, Sandemir, Mustafa, 2008. Prediction of rubberized concrete properties using artificial neural network and fuzzy logic. Constr. Build. Mater. 22, 532–540.
117. Turatsinze, A., Garros, M., 2008. On the modulus of elasticity and strain capacity of self-compacting concrete incorporating rubber aggregates. Resour. Conserv. Recycl. 52, 1209–1215.
118. Turatsinze, A., Bonnet, S., Granju, J.-L., 2005. Mechanical characterization of cement-based mortar incorporating rubber aggregates from recycled worn tires. Build. Environ. 40, 221–226.
119. Turatsinze, A., Granju, J.-L., Bonnet, S., 2006. Positive synergy between steel-fibers and rubber aggregates: effect on the resistance of cement- based mortars to shrinkage cracking. Cem. Concr. Res. 36, 1692–1697.
120. Turatsinze, A., Bonnet, S., Granju, J.-L., 2007. Potential of rubber aggregates to modify properties of cement-based mortars: improvement in cracking shrinkage resistance. Constr. Build. Mater. 21, 176–181.
121. Turgut, Paki, Yesilata, Bulent, 2008. Physico-mechanical and thermal performance of newly developed rubber-added bricks. Energy Build. 40, 679–688.

122. Turki, M., Bretagne, E., Rouis, M.J., Queneudec, M., 2009. Microstructure, physical and mechanical properties of mortar-rubber aggregates mixtures. Constr. Build. Mater. 23, 2715–2722.

123. Turki, M., Ben, Naceur I., Makni, M., Rouis, J., Sai, K., 2009. Mechanical and damage behavior of mortar-rubber aggregates mixtures: experiments and simulations. Mater. Struct. 42, 1313–1324.

124. Uygunoglu, Tayfun, Topcu, Ilker Bekir, 2010. The role of scrap rubber particles on the drying shrinkage and mechanical properties of self- consolidating mortars. Constr. Build. Mater. 24, 1141–1150.

125. Vadivel, T. Senthil, Thenmozhi, R., Dodurani, M., 2014. Experimental behavior of waste tire rubber aggregate concrete under impact loading. IJST, Trans. Civ. Eng. 38, 251–259.

126. Valadares, Filipe, Bravo, Miguel, de Brito, Jorge, 2012. Concrete with used tire rubber aggregates: mechanical performance. ACI Mater. J. 109–M26, 283–292.

127. Wang, Her-Yung, Chen, Bo-Tsun, Wu, Yu-Wu, 2013. A study of the fresh properties of controlled low-strength rubber lightweight aggregate concrete (CLSRLC). Constr. Build. Mater. 41, 526–531.

128. Williams, P.T., Besler, S., Taylor, D.T., Bottrill, R.P., 1995. Pyrolysis of automotive tire waste. J. Inst. Energy 68, 11–21.

129. Williams Paul, T., 2013. Pyrolysis of waste tires: a review. Waste Manage. 33, 1714–1728.

130. Wong, Sook-Fun, Ting, Seng-Kiong, 2009. Use of recycled rubber tires in normal- and high strength concretes. ACI Mater. J. 106–M37, 325– 332.

131. Yi, Xianren, Fan, Xiaochun, 2009. Flexural-tensile properties of layered steel fiber reinforced rubber concrete. J. Wuhan Univ. Technol. Mater. Sci. Ed. 24 (5), 846–848.

132. Youssf, Osama, El Gaeady, Mohamed A., Mills, Julie E., Ma, Xing, 2014. An experimental investigation of crumb rubber concrete confined by fibre reinforced polymer tubes. Constr. Build. Mater. 53, 522–532.

133. Yung, Wang Her, Yung, Lin Chin, Hua, Lee Hsien, 2013. A study of the durability properties of waste tire rubber applied to self-compacting concrete. Constr. Build. Mater. 41, 665–672.

www.ingramcontent.com/pod-product-compliance
Lightning Source LLC
LaVergne TN
LVHW011530200926
848484LV00001B/211